BIOMATERIALS

BIOMATERIALS

Edited by

Joyce Y. Wong
Joseph D. Bronzino

CRC Press
Taylor & Francis Group
Boca Raton London New York

CRC Press is an imprint of the
Taylor & Francis Group, an **informa** business

This material was previously published in *Biomedical Engineering Fundamentals* © 2006 by Taylor & Francis Group, LLC.

CRC Press
Taylor & Francis Group
6000 Broken Sound Parkway NW, Suite 300
Boca Raton, FL 33487-2742

© 2007 by Taylor & Francis Group, LLC
CRC Press is an imprint of Taylor & Francis Group, an Informa business

No claim to original U.S. Government works
Printed in the United States of America on acid-free paper
10 9 8 7 6 5 4 3 2 1

International Standard Book Number-10: 0-8493-7888-5 (Hardcover)
International Standard Book Number-13: 978-0-8493-7888-1 (Hardcover)

Library of Congress Cataloging-in-Publication Data

Biomaterials / editors, Joyce Y. Wong and Joseph D. Bronzino.
    p. ; cm.
    "A CRC title."
    Includes bibliographical references and index.
    ISBN-13: 978-0-8493-7888-1 (alk. paper)
    ISBN-10: 0-8493-7888-5 (alk. paper)
    1. Biomedical materials. I. Wong, Joyce Y. II. Bronzino, Joseph D., 1937- III. Title.
    [DNLM: 1. Biocompatible Materials--chemistry. 2. Prostheses and Implants. 3. Prosthesis Implantation--methods. QT 37 B614035 2007]

R857.M3B5683 2007
610.28'4--dc22
                                                  2007007400

Visit the Taylor & Francis Web site at
http://www.taylorandfrancis.com

and the CRC Press Web site at
http://www.crcpress.com

Introduction and Preface

A Brief Note Regarding This Book

Due to unforeseen circumstances, the bulk of this book is largely identical to the previous edition with the notable exception of the addition of a chapter relating to micro- and nanotechnologies in advanced **biomaterials**. The content remains current, and the following introduction is taken from the previous edition with minor changes describing the content of this book. In addition, a number of relevant biomaterials journals have been added to the list.

Biomaterial is used to make devices to replace a part or a function of the body in a safe, reliable, economic, and physiologically acceptable manner [Hench and Erthridge, 1982]. A variety of devices and materials are used in the treatment of disease or injury. Commonplace examples include sutures, needles, catheters, plates, tooth fillings, etc. A biomaterial is a synthetic material used to replace part of a living system or to function in intimate contact with living tissue. The Clemson University Advisory Board for Biomaterials has formally defined a biomaterial to be "a systemically and pharmacologically inert substance designed for implantation within or incorporation with living systems." Black defined biomaterials as "a nonviable material used in a medical device, intended to interact with biological systems" [Black, 1992]. Others include "materials of synthetic as well as of natural origin in contact with tissue, blood, and biological fluids, and intended for use for prosthetic, diagnostic, therapeutic, and storage applications without adversely affecting the living organism and its components" [Bruck, 1980]. Still another definition of biomaterials is stated as "any substance (other than drugs) or combination of substances, synthetic or natural in origin, which can be used for any period of time, as a whole or as a part of a system which treats, augments, or replaces any tissue, organ, or function of the body" [Williams, 1987] adds to the many ways of looking the same but expressing in different ways. By contrast, a **biological material** is a material such as skin or artery, produced by a biological system. Artificial materials that simply are in contact with the skin, such as hearing aids and wearable artificial limbs are not included in our definition of biomaterials since the skin acts as a barrier with the external world.

According to these definitions, one must have a vast field of knowledge or collaborate with different specialties in order to develop and use biomaterials in medicine and dentistry as Table I.1 indicates. The uses of biomaterials, as indicated in Table I.2, include replacement of a body part which has lost function due to disease or trauma, to assist in healing, to improve function, and to correct abnormalities. The role of biomaterials has been influenced considerably by advances in many areas of biotechnology and science. For example, with the advent of antibiotics, infectious disease is less of a threat than in former times so that degenerative disease assumes a greater importance. Moreover, advances in surgical technique and instruments have permitted materials to be used in ways which were not possible previously. This book is intended to develop in the reader a familiarity with the uses of materials in medicine and dentistry and with some rational basis for these applications.

The performance of materials in the body can be classified in many ways. First, biomaterials may be considered from the point of view of the problem area which is to be solved, as in Table I.2. Second, we may consider the body on a tissue level, an organ level (Table I.3), or a system level (Table I.4). Third,

TABLE I.1   Fields of Knowledge to Develop Biomaterials

Discipline	Examples
Science and engineering	Materials sciences: structure–property relationship of synthetic and biological materials including metals, ceramics, polymers, composites, tissues (blood and connective tissues), etc.
Biology and physiology	Cell and molecular biology, anatomy, animal and human physiology, histopathology, experimental surgery, immunology, etc.
Clinical sciences	All the clinical specialties: dentistry, maxillofacial, neurosurgery, obstetrics and gynecology, ophthalmology, orthopedics, otolaryngology, plastic and reconstructive surgery, thoracic and cardiovascular surgery, veterinary medicine, and surgery, etc.

Source: Modified from von Recum, A.F. [1994] Boston, MA. Biomaterials Society.

TABLE I.2   Uses of Biomaterials

Problem area	Examples
Replacement of diseased or damaged part	Artificial hip joint, kidney dialysis machine
Assist in healing	Sutures, bone plates, and screws
Improve function	Cardiac pacemaker, intraocular lens
Correct functional abnormality	Cardiac pacemaker
Correct cosmetic problem	Augmentation mammoplasty, chin augmentation
Aid to diagnosis	Probes and catheters
Aid to treatment	Catheters, drains

TABLE I.3   Biomaterials in Organs

Organ	Examples
Heart	Cardiac pacemaker, artificial heart valve, total artificial heart
Lung	Oxygenator machine
Eye	Contact lens, intraocular lens
Ear	Artificial stapes, cochlea implant
Bone	Bone plate, intramedullary rod
Kidney	Kidney dialysis machine
Bladder	Catheter and stent

we may consider the classification of materials as polymers, metals, ceramics, and composites as is done in Table I.5. In that vein, the role of such materials as biomaterials is governed by the interaction between the material and the body; specifically, the effect of the body environment on the material and the effect of the material on the body [Williams and Roaf, 1973; Bruck, 1980; Hench and Erthridge, 1982; von Recum, 1986; Black, 1992; Park and Lakes, 1992; and Greco, 1994].

It should be evident from any of these perspectives that most current applications of biomaterials involve structural functions, even in those organs and systems which are not primarily structural in their nature, or very simple chemical or electrical functions. Complex chemical functions such as those of the liver and complex electrical or electrochemical functions such as those of the brain and sense organs cannot be carried out by biomaterials at this time.

Historical Background

The use of biomaterials did not become practical until the advent of an aseptic surgical technique developed by Dr. J. Lister in the 1860s. Earlier surgical procedures, whether they involved biomaterials or not, were generally unsuccessful as a result of infection. Problems of infection tend to be exacerbated in the

TABLE I.4 Biomaterials in Body Systems

System	Examples
Skeletal	Bone plate, total joint replacements
Muscular	Sutures, muscle stimulator
Circulatory	Artificial heart valves, blood vessels
Respiratory	Oxygenator machine
Integumentary	Sutures, burn dressings, artificial skin
Urinary	Catheters, stent, kidney dialysis machine
Nervous	Hydrocephalus drain, cardiac pacemaker, nerve stimulator
Endocrine	Microencapsulated pancreatic islet cells
Reproductive	Augmentation mammoplasty, other cosmetic replacements

TABLE I.5 Materials for Use in the Body

Materials	Advantages	Disadvantages	Examples
Polymers (nylon, silicone rubber, polyester, polytetrafluoroethylene, etc.)	Resilient Easy to fabricate	Not strong Deforms with time May degrade	Sutures, blood vessels, hip socket, ear, nose, other soft tissues, sutures
Metals (Ti and its alloys, Co–Cr alloys, stainless steels, Au, Ag, Pt, etc.)	Strong, tough, ductile	May corrode Dense Difficult to make	Joint replacements, bone plates and screws, dental root implants, pacer and suture wires
Ceramics (aluminum oxide, calcium phosphates including hydroxyapatite, carbon)	Very biocompatible, Inert Strong in compression	Brittle Not resilient Difficult to make	Dental; femoral head of hip replacement, coating of dental and orthopedic implants
Composites (carbon–carbon, wire or fiber reinforced bone cement)	Strong, tailor-made	Difficult to make	Joint implants, heart valves

presence of biomaterials, since the implant can provide a region inaccessible to the body's immunologically competent cells. The earliest successful implants, as well as a large fraction of modern ones, were in the skeletal system. Bone plates were introduced in the early 1900s to aid in the fixation of long-bone fractures. Many of these early plates broke as a result of unsophisticated mechanical design; they were too thin and had stress-concentrating corners. Also, materials such as vanadium steel, which was chosen for its good mechanical properties corroded rapidly in the body and caused adverse effects on the healing processes. Better designs and materials soon followed. Following the introduction of stainless steels and cobalt chromium alloys in the 1930s, greater success was achieved in fracture fixation, and the first joint-replacement surgeries were performed. As for polymers, it was found that warplane pilots in World War II who were injured by fragments of plastic PMMA (polymethyl methacrylate) aircraft canopy did not suffer adverse chronic reactions from the presence of the fragments in the body. PMMA became widely used after that time for corneal replacement and for replacements of sections of damaged skull bones. Following further advances in materials and in surgical technique, blood vessel replacements were tried in the 1950s and heart valve replacements and cemented joint replacements in the 1960s. Table I.6 lists notable developments relating to implants. Recent years have seen many further advances.

Performance of Biomaterials

The success of biomaterials in the body depends on factors such as the material properties, design, and **biocompatibility** of the material used, as well as other factors not under the control of the engineer, including the technique used by the surgeon, the health and condition of the patient, and the activities

TABLE I.6 Notable Developments Relating to Implants

Year	Investigators	Development
Late 18-19th century		Various metal devices to fix bone fractures; wires and pins from Fe, Au, Ag, and Pt
1860–1870	J. Lister	Aseptic surgical techniques
1886	H. Hansmann	Ni-plated steel bone fracture plate
1893–1912	W.A. Lane	Steel screws and plates (Lane fracture plate)
1912	W.D. Sherman	Vanadium steel plates, first developed for medical use; lesser stress concentration and corrosion (Sherman plate)
1924	A.A. Zierold	Introduced Stellites® (CoCrMo alloy)
1926	M.Z. Lange	Introduced 18-8sMo stainless steel, better than 18-8 stainless steel
1926	E.W. Hey-Groves	Used carpenter's screw for femoral neck fracture
1931	M.N. Smith-Petersen	First femoral neck fracture fixation device made of stainless steel
1936	C.S. Venable, W.G. Stuck	Introduced Vitallium® (19-9 stainless steel), later changed the material to CoCr alloys
1938	P. Wiles	First total hip replacement prosthesis
1939	J.C. Burch, H.M Carney	Introduced tantalum (Ta)
1946	J. and R. Judet	First biomechanically designed femoral head replacement prosthesis. First plastics (PMMA) used in joint replacements
1940s	M.J. Dorzee, A. Franceschetti	First used acrylics (PMMA) for corneal replacement
1947	J. Cotton	Introduced Ti and its alloys
1952	A.B. Voorhees, A. Jaretzta, A.B. Blackmore	First successful blood vessel replacement made of cloth for tissue ingrowth
1958	S. Furman, G. Robinson	First successful direct heart stimulation
1958	J. Charnley	First use of acrylic bone cement in total hip replacement on the advice of Dr. D. Smith
1960	A. Starr, M.L. Edwards	First commercial heart valves
1970s	W.J. Kolff	Total heart replacement

Source: Park, J.B. [1984] New York: Plenum Publishing Co.

of the patient. If we can assign a numerical value f to the probability of failure of an implant, then the reliability can be expressed as

$$r = 1 - f$$

If, as is usually the case, there are multiple modes of failure, the total reliability r_t is given by the product of the individual reliabilities $r_1 = (1 - f_1)$, etc.

$$r_t = r_1 \cdot r_2 \cdots r_n$$

Consequently, even if one failure mode such as implant fracture is perfectly controlled so that the corresponding reliability is unity, other failure modes such as infection could severely limit the utility represented by the total reliability of the implant. One mode of failure which can occur in a biomaterial, but not in engineering materials used in other contexts, is an attack by the body's immune system on the implant. Another such failure mode is an unwanted effect of the implant upon the body; for example, toxicity, inducing allergic reactions, or causing cancer. Consequently, biocompatibility is included as a material requirement in addition to those requirements associated directly with the function of the implant.

Biocompatibility involves the acceptance of an artificial implant by the surrounding tissues and by the body as a whole. Biocompatible materials do not irritate the surrounding structures, do not provoke an abnormal inflammatory response, do not incite allergic or immunologic reactions, and do not cause cancer. Other compatibility characteristics that may be important in the function of an implant device made of biomaterials include (1) adequate mechanical properties such as strength, stiffness, and fatigue properties; (2) appropriate optical properties if the material is to be used in the eye, skin, or tooth; and (3) appropriate density. Sterilizability, manufacturability, long-term storage, and appropriate engineering design are also to be considered.

The failure modes may differ in importance as time passes following the implant surgery. For example, consider the case of a total joint replacement in which infection is most likely soon after surgery, while loosening and implant fracture become progressively more important as time goes on. Failure modes also depend on the type of implant and its location and function in the body. For example, an artificial blood vessel is more likely to cause problems by inducing a clot or becoming clogged with thrombus than by breaking or tearing mechanically.

With these basic concepts in mind, the chapters in this book focus on biomaterials consisting of different materials such as metallic, ceramic, polymeric, and composite. The impact of recent advances in the area of nano- and microtechnology on biomaterial design is highlighted in this book.

Defining Terms

Biocompatibility: Acceptance of an artificial implant by the surrounding tissues and by the body as a whole.

Biological material: A material produced by a biological system.

Biomaterial: A synthetic material used to make devices to replace part of a living system or to function in intimate contact with living tissue.

References

Black, J. (1992) *Biological Performance of Materials*, 2nd ed. New York: M. Dekker, Inc.

Bruck, S.D. (1980) *Properties of Biomaterials in the Physiological Environment.* Boca Raton, FL: CRC Press.

Greco, R.S. (1994) *Implantation Biology.* Boca Raton, FL: CRC Press.

Hench, L.L. and Erthridge, E.C. (1982) *Biomaterials — An Interfacial Approach*, Vol. 4, A. Noordergraaf, Ed. New York: Academic Press.

Park, J.B. (1984) *Biomaterials Science and Engineering.* New York: Plenum Publishing Co.

Park, J.B. and Lakes, R.S. (1992) *Biomaterials: An Introduction*, 2nd ed. NY: Plenum Publishing Co.

von Recum, A.F. (1994) Biomaterials: educational goals. In: *Annual Biomaterials Society Meeting.* Boston, MA. Biomaterials Society.

von Recum, A.F. (1986) *Handbook of Biomaterials Evaluation.* New York: Macmillan Publishing Co., pp. 97–158 and 293–502.

Williams, D.F. (1987) Definition in biomaterials. In: *Progress in Biomedical Engineering.* Amsterdam: Elsevier, p. 67.

Williams, D.F. and Roaf, R. (1973) *Implants in Surgery.* London: W.B. Saunders.

Further Information

(Most important publications relating to the biomaterials area are given for further reference.)

Allgower, M., Matter, P., Perren, S.M., and Ruedi, T. 1973. *The Dynamic Compression Plate*, DCP, Springer-Verlag, New York.

Bechtol, C.O., Ferguson, A.B., and Laing, P.G. 1959. *Metals and Engineering in Bone and Joint Surgery*, Balliere, Tindall and Cox, London.

Black, J. 1992. *Biological Performance of Materials*, 2nd ed., Marcel Dekker, New York.

Bloch, B. and Hastings, G.W. 1972. *Plastic Materials in Surgery*, 2nd ed., C.C. Thomas, Springfield, IL.

Bokros, J.C., Arkins, R.J., Shim, H.S., Haubold, A.D., and Agarwal, N.K. 1976. Carbon in prosthestic devices. In: *Petroleum Derived Carbons*, M.L. Deviney and T.M. O'Grady, Eds. *Am. Chem. Soc. Symp.*, Series No. 21, American Chemical Society, Washington, D.C.

Boretos, J.W. 1973. *Concise Guide to Biomedical Polymers*, C.C. Thomas, Springfield, IL.

Boretos, J.W. and Eden, M. (Eds.) 1984. *Contemporary Biomaterials*, Noyes, Park Ridge, NJ.

Brown, P.W. and Constantz, B. 1994. *Hydroxyapatite and Related Materials*, CRC Press, Boca Raton, FL.

Bruck, S.D. 1974. *Blood Compatible Synthetic Polymers: An Introduction*, C.C. Thomas, Springfield, IL.

Bruck, S.D. 1980. *Properties of Biomaterials in the Physiological Environment*, CRC Press, Boca Raton, FL.

Chandran, K.B. 1992. *Cardiovascular Biomechanics*, New York University Press, New York.

Charnley, J. 1970. *Acrylic Cement in Orthopedic Surgery*, Livingstone, Edinborough and London.

Cooney, D.O. 1976. *Biomedical Engineering Principles*, Marcel Dekker, New York.

Cranin, A.N., Ed. 1970. *Oral Implantology*, C.C. Thomas, Springfield, IL.

Dardik, H., Ed. 1978. *Graft Materials in Vascular Surgery*, Year Book Medical Publishing, Chicago.

de Groot, K., Ed. 1983. *Bioceramics of Calcium Phosphate*, CRC Press, Boca Raton, FL.

Ducheyne, P., Van der Perre, G., and Aubert, A.E., Eds. 1984. *Biomaterials and Biomechanics*, Elsevier Science, Amsterdam.

Dumbleton, J.H. and Black, J. 1975. *An Introduction to Orthopedic Materials*, C.C. Thomas, Springfield, IL.

Edwards, W.S. 1965. *Plastic Arterial Grafts*, C.C. Thomas, Springfield, IL.

Edwards, W.S. 1965. *Plastic Arterial Grafts*, C.C. Thomas, Springfield, IL.

Eftekhar, N.S. 1978. *Principles of Total Hip Arthroplasty*, C.V. Mosby, St. Louis, MO.

Frost, H.M. 1973. *Orthopedic Biomechanics*, C.C. Thomas, Springfield, IL.

Fung, Y.C. 1993. *Biomechanics: Mechanical Properties of Living Tissues*, 2nd ed., Springer-Verlag, New York.

Ghista, D.N. and Roaf, R., Eds. 1978. *Orthopedic Mechanics: Procedures and Devices*, Academic Press, London.

Greco, R.S., Ed. 1994. *Implantation Biology*, CRC Press, Boca Raton, FL.

Guidelines for Blood–Material Interactions, Revised 1985. Report of the National Heart, Lung, and Blood Institute Working Group, Devices and Technology Branch, NHLBI, NIH Publication No. 80-2185.

Gyers, G.H. and Parsonet, V. 1969. *Engineering in the Heart and Blood Vessels*, J. Wiley & Sons, New York.

Hastings, G.W. and Williams, D.F., Eds. 1980. *Mechanical Properties of Biomaterials*, John Wiley & Sons, New York.

Hench, L.L. and Ethridge, E.C. 1982. *Biomaterials: An Interfacial Approach*, Academic Press, New York.

Hench, L.L. and Wilson, J., Eds. 1993. *An Introduction to Bioceramics*, World Scientific, Singapore.

Heppenstall, R.B., Ed. 1980. *Fracture Treatment and Healing*, W.B. Saunders, Philadelphia, PA.

Homsy, C.A. and Armeniades, C.D., Eds., 1972. Biomaterials for skeletal and cardiovascular applications, *J. Biomed. Mater. Symp.*, No. 3, John Wiley & Sons, New York.

Hulbert, S.F., Young, F.A., and Moyle, D.D., Eds. 1972. *J. Biomed. Mater. Res. Symp.*, No. 2.

Kawahara, H., Ed. 1989. *Oral Implantology and Biomaterials*, Elsevier Science, Armsterdam.

Kronenthal, R.L. and Oser, Z., Eds. 1975. *Polymers in Medicine and Surgery*, Plenum Press, New York.

Kuntscher, G. 1947. *The Practice of Intramedullary Nailing*, C.C. Thomas, Springfield, IL.

Lee, H. and Neville, K. 1971. *Handbook of Biomedical Plastics*, Pasadena Technology Press, Pasadena, CA.

Lee, S.M., Ed. *Advances in Biomaterials*, Technomic Pub. AG, Lancaster, PA, 1987.

Leinninger, R.I. 1972. Polymers as surgical implants, CRC *Crit. Rev. Bioeng.*, 2: 333–360.

Levine, S.N., Ed. 1968. Materials in Biomedical Engineering, *Ann. NY Acad. Sci.*, 146.

Levine, S.N., Ed. 1968. Polymers and Tissue Adhesives, *Ann. NY Acad. Sci.*, Part IV, 146.

Lynch, W. 1982. *Implants: Reconstructing Human Body*, Van Nostrand Reinhold, New York.

Martz, E.O., Goel, V.K., Pope, M.H., and Park, J.B., 1997 Materials and design of spinal implants — A review. *J. Biomed. Mat. Res. (App. Biomater.)*, 38: 267–288.

Mears, D.C. 1979. *Materials and Orthopedic Surgery*, Williams & Wilkins, Baltimore, MD.

Oonishi, H. Aoki, H. and Sawai, K., Eds. 1989. *Bioceramics*, Ishiyaku EuroAmerica, Tokyo.

Park, J.B. 1979. *Biomaterials: An Introduction*, Plenum Press, New York.

Park, J.B. 1984. *Biomaterials Science and Engineering*, Plenum Press, New York.

Park, J.B. and Lakes, R.S. 1992. *Biomaterials: An Introduction*, 2nd ed., Plenum Press, New York.

Park, K., Shalaby, W.S.W., and Park, H. 1993. *Biodegradable Hydrogels for Drug Delivery*, Technomic, Lancaster, PA.

Rubin, L.R., Ed. 1983. *Biomaterials in Reconstructive Surgery*, C.V. Mosby, St. Louis, MO.

Savastano, A.A., Ed. 1980. *Total Knee Replacement*, Appleton-Century-Crofts, New York.

Sawyer, P.N. and Kaplitt, M.H. 1978. *Vascular Grafts*, Appleton-Century-Crofts, New York.

Schaldach, M. and Hohmann, D., Eds. 1976. *Advances in Artificial Hip and Knee Joint Technology*, Springer-Verlag, Berlin.

Schnitman, P.A. and Schulman, L.B., Eds. 1980. Dental Implants: Benefits and Risk, A *NIH-Harvard Consensus Development Conference*, NIH Pub. No. 81-1531, U.S. Dept. Health and Human Services, Bethesda, MD.

Sharma, C.P. and Szycher, M., Eds. 1991. *Blood Compatible Materials and Devices*, Technomic, Lancaster, PA.

Stanley, J.C., Burkel, W.E., Lindenauer, S.M., Bartlett, R.H., and Turcotte, J.G., Eds. 1972. *Biologic and Synthetic Vascular Prostheses*, Grune & Stratton, New York.

Stark, L. and Agarwal, G., Eds. 1969. *Biomaterials*, Plenum Press, New York.

Swanson, S.A.V. and Freeman, M.A.R., Eds. 1977. *The Scientific Basis of Joint Replacement*, John Wiley & Sons, New York.

Syrett, B.C. and Acharya, A., Eds. 1979. *Corrosion and Degradation of Implant Materials*, ASTM STP 684, American Society for Testing and Materials, Philadelphia, PA.

Szycher, M. and Robinson, W.J., Eds. *Synthetic Biomedical Polymers, Concepts and Applications*, Technomic, Lancaster, PA.

Szycher, M., Ed. 1991. *High Performance Biomaterials*, Technomic, Lancaster, PA.

Taylor, A.R. 1970. *Endosseous Dental Implants*, Butterworths, London.

Uhthoff, H.K., Ed. 1980. *Current Concepts of Internal Fixation of Fractures*, Springer-Verlag, Berlin.

Venable, C.S. and Stuck, W.C. 1947 *The Internal Fixation of Fractures*, C.C. Thomas, Springfield, IL.

Webster, J.G., Ed. 1988. *Encyclopedia of Medical Devices and Instrumentation*, John Wiley & Sons, New York.

Williams, D.F. and Roaf, R. 1973. *Implants in Surgery*, W.B. Saunders, London.

Williams, D.F., Ed. 1976. *Compatibility of Implant Materials*, Sector Pub. Ltd., London, 1976.

Williams, D.F., Ed. 1981. *Fundamental Aspects of Biocompatibility*, vols I and II, CRC Press, Boca Raton, FL.

Williams, D.F., Ed. 1981. *Systemic Aspects of Blood Compatibility*, CRC Press, Boca Raton, FL.

Williams, D.F., Ed. 1982. *Biocompatibility in Clinical Practice*, vols I and II, CRC Press, Boca Raton, FL.

Wright, V., Ed. 1969. *Lubrication and Wear in Joints*, J.B. Lippincott, Philadelphia, PA.

Yamamuro, T., Hench, L.L., and Wilson J., Eds. 1990. *CRC Handbook of Bioactive Ceramics*, vols I and II, CRC Press, Boca Raton, FL.

Journals of Interest

Acta Biomaterialia
Annals of Biomedical Engineering
Bioconjugate Chemistry
Biomacromolecules
Biomaterials
Biomedical Materials and Engineering
CRC Critical Review in Bioengineering
Journal of Arthoplasty
Journal of Biomechanics
Journal of Biomedical Materials Research
Journal of Controlled Release
Journal of Applied Biomaterials
Journal of Medical Engineering and Technology

Journal of Orthopaedic Research
Journal of Biomaterials Science, Polymer Edition
Journal of Biomedical Engineering
Langmuir
Acta Orthopaedica Scandinavica
Clinical Orthopaedics and Related Research
Journal of Bone and Joint Surgery
International Orthopaedics
Medical Engineering and Physics
American Association of Artificial Internal Organs: Transactions
Tissue Engineering
Transactions of the Orthopaedic Research Society Meeting (annually held during February): Abstracts
Transactions of the Society for Biomaterials (annually held during April and May): Abstracts
Transactions of the American Society of Artificial Internal Organs (annually held in spring): Extended Abstracts
Society for Biomaterials: http://www.biomaterials.org/index.html

Editors

Dr. Joyce Y. Wong is a Clare Boothe Luce associate professor in biomedical engineering (BME) and associate chair of graduate studies in the department of biomedical engineering at Boston University. Dr. Wong's research focuses on the development of biomaterials to probe how structure, material properties and composition of the cell-biomaterial interface affect fundamental cellular processes. Her current research interests include tissue engineering of small diameter blood vessels for bypass and intravascular pharmacology (e.g. stents); development of targeted nano- and micro-particle contrast agents for multimodal (magnetic resonance, ultrasound, and optical) detection of atherosclerotic and vulnerable plaque; and engineering biomimetic systems to study restenosis and breast cancer. Awards she has received include a NSF Career Award and Dupont Young Professor Award. Dr. Wong is currently the associate director of the Center for Nanoscience and Nanobiotechnology at Boston University. She has served on NIH Study Section panels, and is on the editorial advisory board of the journal *Polymer Reviews* and co-editor of *Biointerphases*. She is also an active member of the American Chemical Society, Biomedical Engineering Society, Materials Research Society, AVS Science and Technology (executive committee member, Biomaterials Interfaces Division), Biophysical Society, Society for Biomaterials, and American Society of Cell Biology.

Joseph D. Bronzino received the B.S.E.E. degree from Worcester Polytechnic Institute, Worcester, MA, in 1959, the M.S.E.E. degree from the Naval Postgraduate School, Monterey, CA, in 1961, and the Ph.D. degree in electrical engineering from Worcester Polytechnic Institute in 1968. He is presently the Vernon Roosa Professor of Applied Science, an endowed chair at Trinity College, Hartford, CT and President of the Biomedical Engineering Alliance and Consortium (BEACON), which is a nonprofit organization consisting of academic and medical institutions as well as corporations dedicated to the development and commercialization of new medical technologies (for details visit www.beaconalliance.org).

He is the author of over 200 articles and 11 books including the following: *Technology for Patient Care* (C.V. Mosby, 1977), *Computer Applications for Patient Care* (Addison-Wesley, 1982), *Biomedical Engineering: Basic Concepts and Instrumentation* (PWS Publishing Co., 1986), *Expert Systems: Basic Concepts* (Research Foundation of State University of New York, 1989), *Medical Technology and Society: An Interdisciplinary Perspective* (MIT Press and McGraw-Hill, 1990), *Management of Medical Technology* (Butterworth/Heinemann, 1992), *The Biomedical Engineering Handbook* (CRC Press, 1st ed., 1995; 2nd ed., 2000; Taylor & Francis, 3rd ed., 2005), *Introduction to Biomedical Engineering* (Academic Press, 1st ed., 1999; 2nd ed., 2005).

Dr. Bronzino is a fellow of IEEE and the American Institute of Medical and Biological Engineering (AIMBE), an honorary member of the Italian Society of Experimental Biology, past chairman of the Biomedical Engineering Division of the American Society for Engineering Education (ASEE), a charter member and presently vice president of the Connecticut Academy of Science and Engineering (CASE), a charter member of the American College of Clinical Engineering (ACCE) and the Association for the Advancement of Medical Instrumentation (AAMI), past president of the IEEE-Engineering in Medicine and Biology Society (EMBS), past chairman of the IEEE Health Care Engineering Policy

Committee (HCEPC), past chairman of the IEEE Technical Policy Council in Washington, DC, and presently editor-in-chief of Elsevier's BME Book Series and Taylor & Francis' *Biomedical Engineering Handbook*.

Dr. Bronzino is also the recipient of the Millennium Award from IEEE/EMBS in 2000 and the Goddard Award from Worcester Polytechnic Institute for Professional Achievement in June 2004.

Contributors

W.C. Billotte
University of Dayton
Dayton, Ohio

K.J.L. Burg
Carolinas Medical Center
Charlotte, North Carolina

K.B. Chandran
Department of Biomedical
  Engineering
College of Engineering
University of Iowa
Iowa City, Iowa

Chih-Chang Chu
TXA Department
Cornell University
Ithaca, New York

Vijay K. Goel
Department of Biomedical
  Engineering
University of Iowa
Iowa City, Iowa

Jessica Kaufman
Department of Biomedical
  Engineering
Boston University
Boston, Massachusetts

J.C. Keller
University of Iowa
Iowa City, Iowa

Gilson Khang
Department of Polymer Science
  and Technology
Chonbuk National University
Seoul, South Korea

Young Kon Kim
Inje University
Kyungnam, North Korea

Catherine Klapperich
Departments of Manufacturing
  and Biomedical Engineering
Boston University
Boston, Massachusetts

Roderic S. Lakes
University of Wisconsin-Madison
Madison, Wisconsin

Hai Bang Lee
Biomaterials Laboratory
Korea Research Institute of
  Chemical Technology
Yusung Taejon, North Korea

Jin Ho Lee
Department of Polymer Science
  and Engineering
Hannam University
Taejon, North Korea

Shu-Tung Li
Collagen Matrix, Inc.
Franklin Lakes, New Jersey

Chien-Chi Lin
Department of Bioengineering
Clemson University
Clemson, South Carolina

Adolfo Llinás
Pontificia Universidad Javeriana
Bogota, Colombia

Andrew T. Metters
Department of Chemical and
  Biomolecular Engineering
Department of Bioengineering
Clemson University
Clemson, South Carolina

Joon B. Park
Department of Biomedical
  Engineering
University of Iowa
Iowa City, Iowa

Sang-Hyun Park
Orthopedic Research Center
Orthopedic Hospital
Los Angeles, California

S.W. Shalaby
Poly-Med, Inc.
Anderson, South Carolina

Joyce Y. Wong
Department of Biomedical
  Engineering
Boston University
Boston, Massachusetts

Contents

1

Metallic Biomaterials

Joon B. Park
University of Iowa

Young Kon Kim
Inje University

1.1 Introduction

Metals are used as biomaterials due to their excellent electrical and thermal conductivity and mechanical properties. Since some electrons are independent in metals, they can quickly transfer an electric charge and thermal energy. The mobile free electrons act as the binding force to hold the positive metal ions together. This attraction is strong, as evidenced by the closely packed atomic arrangement resulting in high specific gravity and high melting points of most metals. Since the metallic bond is essentially nondirectional, the position of the metal ions can be altered without destroying the crystal structure resulting in a plastically deformable solid.

Some metals are used as passive substitutes for hard tissue replacement such as total hip and knee joints, for fracture healing aids as bone plates and screws, spinal fixation devices, and dental implants because of their excellent mechanical properties and **corrosion** resistance. Some metallic alloys are used for more active roles in devices such as vascular stents, catheter guide wires, orthodontic archwires, and cochlea implants.

The first metal alloy developed specifically for human use was the "vanadium steel" which was used to manufacture bone fracture plates (Sherman plates) and screws. Most metals such as iron (Fe), chromium (Cr), cobalt (Co), nickel (Ni), titanium (Ti), tantalum (Ta), niobium (Nb), molybdenum (Mo), and tungsten (W), that were used to make alloys for manufacturing implants can only be tolerated by the body in minute amounts. Sometimes those metallic elements, in naturally occurring forms, are essential in red blood cell functions (Fe) or synthesis of a vitamin B_{12} (Co), but cannot be tolerated in large amounts in the body [Black, 1992]. The biocompatibility of the metallic implant is of considerable concern because these implants can corrode in an *in vivo* environment [Williams, 1982]. The consequences of corrosion are the disintegration of the implant material per se, which will weaken the implant, and the harmful effect of corrosion products on the surrounding tissues and organs.

1.2 Stainless Steels

The first stainless steel utilized for implant fabrication was the 18-8 (type 302 in modern classification), which is stronger and more resistant to corrosion than the vanadium steel. Vanadium steel is no longer used in implants since its corrosion resistance is inadequate *in vivo*. Later 18-8sMo stainless steel was introduced which contains a small percentage of molybdenum to improve the corrosion resistance in chloride solution (salt water). This alloy became known as *type 316 stainless steel*. In the 1950s the carbon content of 316 stainless steel was reduced from 0.08 to a maximum amount of 0.03% (all are weight percent unless specified) for better corrosion resistance to chloride solution and to minimize the sensitization and hence, became known as type *316L stainless steel*. The minimum effective concentration of chromium is 11% to impart corrosion resistance in stainless steels. The chromium is a reactive element, but it and its alloys can be *passivated* by 30% nitric acid to give excellent corrosion resistance.

The *austenitic stainless steels*, especially type 316 and 316L, are most widely used for implant fabrication. These cannot be hardened by heat treatment but can be hardened by cold-working. This group of stainless steels is nonmagnetic and possesses better corrosion resistance than any others. The inclusion of molybdenum enhances resistance to *pitting corrosion* in salt water. The American Society of Testing and Materials (ASTM) recommends type 316L rather than 316 for implant fabrication. The specifications for 316L stainless steel are given in Table 1.1. The only difference in composition between the 316L and 316 stainless steel is the maximum content of carbon, that is, 0.03 and 0.08%, respectively, as noted earlier.

The nickel stabilizes the austenitic phase [γ, face centered cubic crystal (fcc) structure], at room temperature and enhances corrosion resistance. The austenitic phase formation can be influenced by both the Ni and Cr contents as shown in Figure 1.1 for 0.10% carbon stainless steels. The minimum amount of Ni for maintaining austenitic phase is approximately 10%.

Table 1.2 gives the mechanical properties of 316L stainless steel. A wide range of properties exists depending on the heat treatment (annealing to obtain softer materials) or cold working (for greater strength and hardness). Figure 1.2 shows the effect of cold working on the yield and ultimate tensile strength of 18-8 stainless steels. The engineer must consequently be careful when selecting materials of this type. Even the 316L stainless steels may corrode inside the body under certain circumstances in a highly stressed and oxygen depleted region, such as the contacts under the screws of the bone fracture

TABLE 1.1 Compositions of
316L Stainless Steel

Element	Composition (%)
Carbon	0.03 max.
Manganese	2.00 max.
Phosphorus	0.03 max.
Sulfur	0.03 max.
Silicon	0.75 max.
Chromium	17.00–20.00
Nickel	12.00–14.00
Molybdenum	2.00–4.00

Source: American Society for Testing and Materials, F139–86, p. 61, 1992.

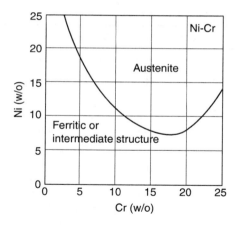

FIGURE 1.1 The effect of Ni and Cr contents on the austenitic phase of stainless steels containing 0.1% C [Keating, 1956].

TABLE 1.2 Mechanical Properties of 316L Stainless Steel for Implants

Condition	Ultimate tensile strength, min. (MPa)	Yield strength (0.2% offset), min. (MPa)	Elongation 2 in. (50.8 mm) min. %	Rockwell hardness
Annealed	485	172	40	95 HRB
Cold-worked	860	690	12	—

Source: American Society for Testing and Materials, F139–86, p. 61, 1992.

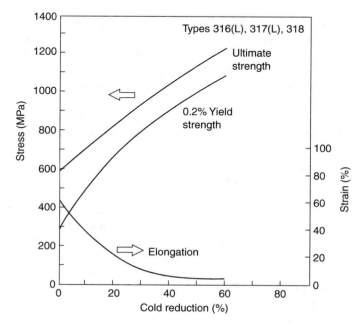

FIGURE 1.2 Effect of cold-work on the yield and ultimate tensile strength of 18-8 stainless steel [ASTM, 1980].

TABLE 1.3 Chemical Compositions of Co–Cr Alloys

Element	CoCrMo (F75)		CoCrWNi (F90)		CoNiCrMo (F562)		CoNiCrMoWFe (F563)	
	Min.	Max.	Min.	Max.	Min.	Max.	Min.	Max.
Cr	27.0	30.0	19.0	21.0	19.0	21.0	18.00	22.00
Mo	5.0	7.0	—	—	9.0	10.5	3.00	4.00
Ni	—	2.5	9.0	11.0	33.0	37.0	15.00	25.00
Fe	—	0.75	—	3.0	—	1.0	4.00	6.00
C	—	0.35	0.05	0.15	—	0.025	—	0.05
Si	—	1.00	—	1.00	—	0.15	—	0.50
Mn	—	1.00	—	2.00	—	0.15	—	1.00
W	—	—	14.0	16.0	—	—	3.00	4.00
P	—	—	—	—	—	0.015	—	—
S	—	—	—	—	—	0.010	—	0.010
Ti	—	—	—	—	—	1.0	0.50	3.50
Co			Balance					

Source: American Society for Testing and Materials, F75–87, p. 42; F90–87, p. 47; F562–84, p. 150, 1992.

plate. Thus, these stainless steels are suitable to use only in temporary implant devices such as fracture plates, screws, and hip nails. Surface modification methods such as anodization, **passivation**, and glow-discharge nitrogen-implantation, are widely used in order to improve corrosion resistance, wear resistance, and fatigue strength of 316L stainless steel [Bordiji et al., 1996].

1.3 CoCr Alloys

There are basically two types of cobalt–chromium alloys (1) the castable CoCrMo alloy and (2) the CoNiCrMo alloy which is usually *wrought* by (hot) *forging*. The castable CoCrMo alloy has been used for many decades in dentistry and, relatively recently, in making artificial joints. The wrought CoNiCrMo alloy is relatively new, now used for making the stems of prostheses for heavily loaded joints such as the knee and hip.

The ASTM lists four types of CoCr alloys which are recommended for surgical implant applications (1) cast CoCrMo alloy (F75), (2) wrought CoCrWNi alloy (F90), (3) wrought CoNiCrMo alloy (F562), and (4) wrought CoNiCrMoWFe alloy (F563). The chemical compositions of each are summarized in Table 1.3. At the present time only two of the four alloys are used extensively in implant fabrications, the castable CoCrMo and the wrought CoNiCrMo alloy. As can be noticed from Table 1.3, the compositions are quite different from each other.

The two basic elements of the CoCr alloys form a solid solution of up to 65% Co. The molybdenum is added to produce finer grains which results in higher strengths after casting or forging. The chromium enhances corrosion resistance as well as solid solution strengthening of the alloy.

The CoNiCrMo alloy originally called MP35N (Standard Pressed Steel Co.) contains approximately 35% Co and Ni each. The alloy is highly corrosion resistant to seawater (containing chloride ions) under stress. Cold working can increase the strength of the alloy considerably as shown in Figure 1.3. However, there is a considerable difficulty of cold working on this alloy, especially when making large devices such as hip joint stems. Only hot-forging can be used to fabricate a large implant with the alloy.

The abrasive wear properties of the wrought CoNiCrMo alloy are similar to the cast CoCrMo alloy (about 0.14 mm/yr in joint simulation tests with ultra high molecular weight polyethylene acetabular cup); however, the former is not recommended for the bearing surfaces of joint prosthesis because of its poor frictional properties with itself or other materials. The superior fatigue and ultimate tensile strength of the wrought CoNiCrMo alloy make it suitable for the applications which require long service life without fracture or stress fatigue. Such is the case for the stems of the hip joint prostheses. This advantage is better appreciated when the implant has to be replaced, since it is quite difficult to remove the failed

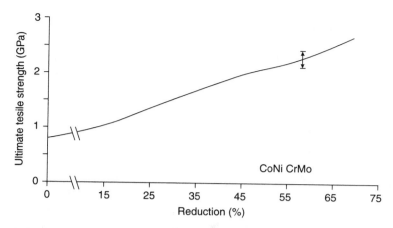

FIGURE 1.3 Relationship between ultimate tensile and the amount of cold-work for CoNiCrMo alloy [Devine and Wulff, 1975].

TABLE 1.4 Mechanical Property Requirements of Co-Cr Alloys

Property	Cast CoCrMo (F75)	Wrought CoCrWNi (F90)	Wrought CoNiCrMo (F562) Solution annealed	Wrought CoNiCrMo (F562) Cold-worked and aged
Tensile strength (MPa)	655	860	793–1000	1793 min.
Yield strength (0.2% offset) (MPa)	450	310	240–655	1585
Elongation (%)	8	10	50.0	8.0
Reduction of area (%)	8	—	65.0	35.0
Fatigue strength (MPa)[a]	310	—	—	—

[a] From Semlitch, M. (1980). *Eng. Med.* 9, 201–207.
Note: ASTM F76, F90, F562.
Source: American Society for Testing and Materials, F75–87, p. 42; F90–87, p. 47; F562–84, p. 150, 1992.

piece of implant embedded deep in the femoral medullary canal. Furthermore, the revision arthroplasty is usually inferior to the primary surgery in terms of its function due to poorer fixation of the implant.

The mechanical properties required for CoCr alloys are given in Table 1.4. As with the other alloys, the increased strength is accompanied by decreased ductility. Both the cast and wrought alloys have excellent corrosion resistance.

Experimental determination of the rate of nickel release from the CoNiCrMo alloy and 316L stainless steel in 37°C Ringer's solution showed an interesting result. Although the cobalt alloy has more initial release of nickel ions into the solution, the rate of release was about the same (3×10^{-10} g/cm^2/day) for both alloys [Richards-Mfg-Company, 1980]. This is rather surprising since the nickel content of the CoNiCrMo alloy is about three times that of 316L stainless steel.

The metallic products released from the prosthesis because of wear, corrosion, and fretting may impair organs and local tissues. *In vitro* studies have indicated that particulate Co is toxic to human osteoblast-like cell lines and inhibits synthesis of type-I collagen, osteocalcin and alkaline phosphatase in the culture medium. However, particulate Cr and CoCr alloy are well tolerated by cell lines with no significant toxicity. The toxicity of metal extracts *in vitro* have indicated that Co and Ni extracts at 50% concentration appear to be highly toxic since all viability parameters were altered after 24 h. However, Cr extract seems to be less toxic than Ni and Co [Granchi et al., 1996].

The modulus of elasticity for the CoCr alloys does not change with the changes in their ultimate tensile strength. The values range from 220 to 234 GPa which are higher than other materials such as stainless steels. This may have some implications of different load transfer modes to the bone in artificial joint replacements, although the effect of the increased modulus on the fixation and longevity of implants is not clear. Low wear (average linear wear on the MeKee-Farrar component was 4.2 μm/yr) has been recognized as an advantage of metal-on-metal hip articulations because of its hardness and toughness [Schmalzried et al., 1996].

1.4 Ti Alloys

1.4.1 Pure Ti and Ti6Al4V

Attempts to use titanium for implant fabrication dates to the late 1930s. It was found that titanium was tolerated in cat femurs, as was stainless steel and Vitallium® (CoCrMo alloy). Titanium's lightness (4.5 g/cm^3, see Table 1.5) and good mechanochemical properties are salient features for implant application.

There are four grades of unalloyed commercially pure (cp) titanium for surgical implant applications as given in Table 1.6. The impurity contents separate them; oxygen, iron, and nitrogen should be controlled carefully. Oxygen in particular has a great influence on the ductility and strength.

One titanium alloy (Ti6Al4V) is widely used to manufacture implants and its chemical requirements are given in Table 1.7. The main alloying elements of the alloy are aluminum (5.5–6.5%) and vanadium

TABLE 1.5 Specific Gravities of Some Metallic Implant Alloys

Alloys	Density (g/cm^3)
Ti and its allloys	4.5
316 Stainless steel	7.9
CoCrMo	8.3
CoNiCrMo	9.2
NiTi	6.7

TABLE 1.6 Chemical Compositions of Titanium and its Alloy

Element	Grade 1	Grade 2	Grade 3	Grade 4	Ti6Al4V[a]
Nitrogen	0.03	0.03	0.05	0.05	0.05
Carbon	0.10	0.10	0.10	0.10	0.08
Hydrogen	0.015	0.015	0.015	0.015	0.0125
Iron	0.20	0.30	0.30	0.50	0.25
Oxygen	0.18	0.25	0.35	0.40	0.13
Titanium			Balance		

[a] Aluminum 6.00% (5.50–6.50), vanadium 4.00% (3.50–4.50), and other elements 0.1% maximum or 0.4% total.
All are maximum allowable weight percent.
Source: American Society for Testing and Materials, F67–89, p. 39; F136–84, p. 55, 1992.

TABLE 1.7 Mechanical Properties of Ti and its Alloys (ASTM F136)

Properties	Grade 1	Grade 2	Grade 3	Grade 4	Ti6Al4V	Ti13Nb13Zr
Tensile strength (MPa)	240	345	450	550	860	1030
Yield strength (0.2% offset) (MPa)	170	275	380	485	795	900
Elongation (%)	24	20	18	15	10	15
Reduction of area (%)	30	30	30	25	25	45

Source: American Society for Testing and Materials, F67–89, p. 39; F136–84, p. 55, 1992 and Davidson et al., 1994.

(3.5~4.5%). The Ti6Al4V alloy has approximately the same fatigue strength (550 MPa) of CoCr alloy after rotary bending fatigue tests [Imam et al., 1983]. Titanium is an allotropic material, which exists as a hexagonal close packed structure (hcp, α-Ti) up to 882°C and body-centered cubic structure (bcc, β-Ti) above that temperature. Titanium alloys can be strengthened and mechanical properties varied by controlled composition and thermomechanical processing techniques. The addition of alloying elements to titanium enables it to have a wide range of properties: (1) Aluminum tends to stabilize the α-phase, that is increase the transformation temperature from α- to β-phase (Figure 1.4). (2) Vanadium stabilizes the β-phase by lowering the temperature of the transformation from α to β.

The α-alloy has a single-phase microstructure (Figure 1.5a) which promotes good weldability. The stabilizing effect of the high aluminum content of these groups of alloys makes excellent strength

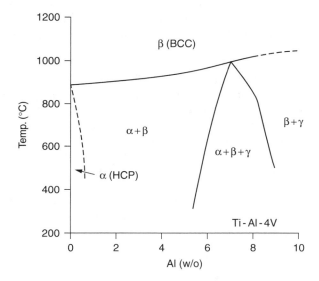

FIGURE 1.4 Part of Phase-diagram of Ti–Al–V at 4 w/o V [Smith and Hughes, 1966].

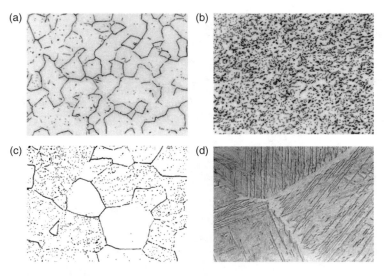

FIGURE 1.5 Microstructure of Ti alloys (all are 500X) [Hille, 1966]. (a) Annealed α-alloy. (b) Ti6Al4V, α–β alloy, annealed. (c) β-alloy, annealed. (d) Ti6Al4V, heat-treated at 1650°C and quenched [Imam et al., 1983].

characteristics and oxidation resistance at high temperature (300 to 600°C). These alloys cannot be heat-treated for precipitation hardening since they are single-phased.

The addition of controlled amounts of β-stabilizers causes the higher strength β-phase to persist below the transformation temperature which results in the two-phase system. The precipitates of β-phase will appear by heat treatment in the solid solution temperature and subsequent quenching, followed by aging at a somewhat lower temperature. The aging cycle causes the coherent precipitation of some fine α particles from the metastable β, imparting α structure may produce local strain field capable of absorbing deformation energy. Cracks are stopped or deterred at the α particles, so that the hardness is higher than for the solid solution (Figure 1.5b).

The higher percentage of β-stabilizing elements (13%V in Ti13V11Cr3Al alloy) results in a microstructure that is substantially β which can be strengthened by heat-treatment (Figure 1.5c). Another Ti alloy (Ti13Nb13Zr) with13%Nb and 13%Zr showed **martensite** structure after water quenched and aged, which showed high corrosion resistant with low modulus ($E = 79$ MPa) [Davidson et al., 1994]. Formation of plates of martensite induces considerable elastic distortion in the parent crystal structure and increases strength (Figure 1.5d).

The mechanical properties of the commercially pure titanium and its alloys are given in Table 1.7. The modulus of elasticity of these materials is about 110 GPa except 13Nb13Zr alloy. From Table 1.7 one can see that the higher impurity content of the cp-Ti leads to higher strength and reduced ductility. The strength of the material varies from a value much lower than that of 316 stainless steel or the CoCr alloys to a value about equal to that of annealed 316 stainless steel of the cast CoCrMo alloy. However, when compared by the specific strength (strength per density) the titanium alloys exceed any other implant materials as shown in Figure 1.6. Titanium, nevertheless, has poor shear strength making it less desirable for bone screws, plates, and similar applications. It also tends to gall or seize when in sliding contact with itself or another metal.

Titanium derives its resistance to corrosion by the formation of a solid oxide layer to a depth of 10 nm. Under *in vivo* conditions the oxide (TiO_2) is the only stable reaction product. However, micromotion at the cement-prosthesis and cement-bone are inevitable and consequently, titanium oxide and titanium

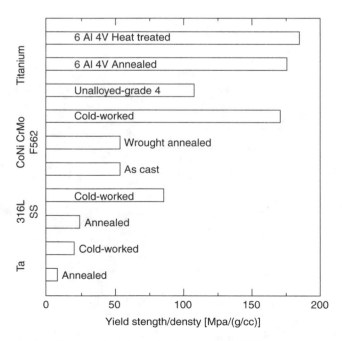

FIGURE 1.6   Yield strength-to-density ratios of some implant materials [Hille, 1966].

alloy particles are released in cemented joint prosthesis. Sometimes this wear debris accumulates as periprosthetic fluid collections and triggers giant cell response around the implants. This cystic collection continued to enlarge and aspiration revealed "dark" heavily stained fluid containing titanium wear particles and histiocytic cells. Histological examination of the stained soft tissue showed "fibrin necrotic debris" and collagenous, fibrous tissue containing a histiocytic and foreign body giant cell infiltrate. The metallosis, black staining of the periprosthetic tissues, has been implicated in knee implant [Breen and Stoker, 1993].

The titanium implant surface consists of a thin oxide layer and the biological fluid of water molecules, dissolved ions, and biomolecules (proteins with surrounding water shell) as shown in Figure 1.7. The microarchitecture (microgeometry, roughness, etc.) of the surface and its chemical compositions are important due to the following reasons:

1. Physical nature of the surface either at the atomic, molecular, or higher level relative to the dimensions of the biological units may cause different contact areas with biomolecules, cells, etc. The different contact areas, in turn, may produce different perturbations and types of bonding of the biological units, which may influence their conformation and function.
2. Chemical composition of the surface may produce different types of bonding to the biomolecules, which may then also affect their properties and function. Metals undergo chemical reactions at the surface depending on the environment which cause the difficulties of understanding the exact nature of the interactions.

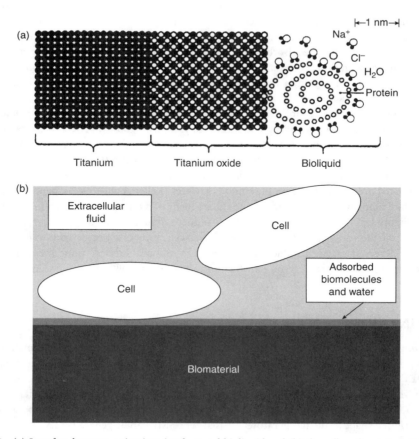

FIGURE 1.7 (a) Interface between a titanium implant and bioliquid and (b) the cell surface interaction [Kasemo and Lausma, 1988].

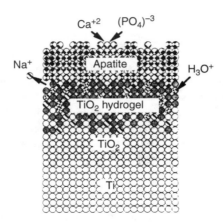

FIGURE 1.8   Chemical change of titanium implant surface of alkali following heat treatment [Kim et al., 1996].

The surface-tissue interaction is dynamic rather than static, i.e., it will develop into new stages as time passes, especially during the initial period after implantation. During the initial few seconds after implantation, there will be only water, dissolved ions, and free biomolecules in the closest proximity of the surface but no cells. The composition of biofluid will then change continuously as inflammatory and healing processes proceed, which in turn also probably causes changes in the composition of the adsorbed layer of biomolecules on the implant surface until quasiequilibrium sets in. Eventually, cells and tissues will approach the surface and, depending on the nature of the adsorbed layer, they will respond in specific ways that may further modify the adsorbed biomolecules. The type of cells closest to the surface and their activities will change with time. For example, depending on the type of initial interaction, the final results may be fibrous capsule formation or tissue integration [Kasemo and Lausma, 1988; Hazan et al., 1993; Takatsuka et al., 1995; Takeshita et al., 1997; Yan et al., 1997].

Osseointegration is defined as direct contact without intervening soft tissue between viable remodeled bone and an implant. Surface roughness of titanium alloys have a significant effect on the bone apposition to the implant and on the bone implant interfacial pull out strength. The average roughness increased from 0.5 to 5.9 μm and the interfacial shear strength increased from 0.48 to 3.5 MPa [Feighan et al., 1995]. Highest levels of osteoblast cell attachment are obtained with rough sand blast surfaces where cells differentiated more than those on the smooth surfaces [Keller et al., 1994]. Chemical changes of the titanium surface following heat treatment is thought to form a TiO_2 hydrogel layer on top of the TiO_2 layer as shown in Figure 1.8. The TiO_2 hydrogel layer may induce the apatite crystal formation [Kim et al., 1996].

In general, on the rougher surfaces there are lower cell numbers, decreased rate of cellular proliferation, and increased matrix production compared to smooth surface. Bone formation appears to be strongly related to the presence of transforming growth factor β_1 in the bone matrix [Kieswetter et al., 1996].

1.4.2   TiNi Alloys

The *titanium–nickel* alloys show unusual properties, that is, after it is deformed the material can snap back to its previous shape following heating of the material. This phenomenon is called **shape memory effect (SME)**. The SME of TiNi alloy was first observed by Buehler and Wiley at the U.S. Naval Ordnance Laboratory [Buehler et al., 1963]. The equiatomic TiNi or NiTi alloy (Nitinol) exhibits an exceptional SME near room temperature: if it is plastically deformed below the transformation temperature, it reverts back to its original shape as the temperature is raised. The SME can be generally related to a diffusionless martensitic phase transformation which is also thermoelastic in nature, the thermoelasticity being attributed to the ordering in the parent and martensitic phases [Wayman and Shimizu, 1972]. Another unusual

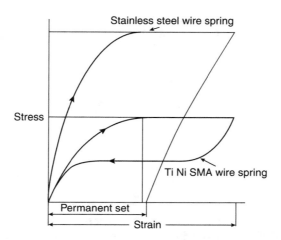

FIGURE 1.9   Schematic illustration of the stainless steel wire and TiNi SMA wire springs for orthodontic archwire behavior. (Modified from Wayman, C.M. and Duerig, T.W. (1990). London: Butterworth-Heinemann, pp. 3–20.)

property is the **superelasticity**, which is shown schematically in Figure 1.9. As can be seen, the stress does not increase with increased strain after the initial elastic stress region and upon release of the stress or strain the metal springs back to its original shape in contrast to other metals such as stainless steel. The superlastic property is utilized in orthodontic archwires since the conventional stainless steel wires are too stiff and harsh for the tooth. In addition, the shape memory effect can also be utilized.

Some possible applications of shape memory alloys are orthodontic dental archwire, intracranial aneurysm clip, *vena cava* filter, contractile artificial muscles for an artificial heart, vascular stent, catheter guide wire, and orthopedic staple [Duerig et al., 1990].

In order to develop such devices, it is necessary to understand fully the mechanical and thermal behavior associated with the martensitic phase transformation. A widely known NiTi alloy is 55-Nitinol (55 weight% or 50 atomic % Ni), which has a single phase and the mechanical memory plus other properties, for example, high acoustic damping, direct conversion of heat energy into mechanical energy, good fatigue properties, and low temperature ductility. Deviation from the 55-Nitinol (near stoichiometric NiTi) in the Ni-rich direction yields a second group of alloys which are also completely nonmagnetic but differ from 55-Nitinol in their ability to be thermally hardened to higher hardness levels. Shape recovery capability decreases and heat treatability increases rapidly as the Ni content approaches 60%. Both 55 and 60-Nitinols have relatively low modulus of elasticity and can be tougher and more resilient than stainless steel, NiCr, or CoCr alloys.

Efficiency of 55-Nitinol shape recovery can be controlled by changing the final annealing temperatures during preparation of the alloy device [Lee et al., 1988]. For the most efficient recovery, the shape is fixed by constraining the specimen in a desired configuration and heating to 482 to 510°C. If the annealed wire is deformed at a temperature below the shape recovery temperature, shape recovery will occur upon heating, provided the deformation has not exceeded crystallographic strain limits (~8% strain in tension). The NiTi alloys also exhibit good biocompatibility and corrosion resistance *in vivo*.

There is no significant difference between titanium and NiTi in the inhibition of mitosis in human fibroblasts. NiTi showed lower percentage bone and bone contact area than titanium and the Ti6Al4V alloy [Takeshita et al., 1997].

The mechanical properties of NiTi alloys are especially sensitive to the stoichiometry of composition (typical composition is given in Table 1.8) and the individual thermal and mechanical history. Although much is known about the processing, mechanical behavior, and properties relating to the shape memory effect, considerably less is known about the thermomechanical and physical metallurgy of the alloy.

TABLE 1.8 Chemical
Composition of Ni–Ti Alloy Wire

Element	Composition (%)
Ni	54.01
Co	0.64
Cr	0.76
Mn	0.64
Fe	0.66
Ti	Balance

1.5 Dental Metals

Dental **amalgam** is an alloy made of liquid mercury and other solid metal particulate alloys made of silver, tin, copper, etc. The solid alloy is mixed with (liquid) mercury in a mechanical vibrating mixer and the resulting material is packed into the prepared cavity. One of the solid alloys is composed of at least 65% silver, and not more than 29% tin, 6% copper, 2% zinc, and 3% mercury. The reaction during setting is thought to be

$$\gamma + Hg \rightarrow \gamma + \gamma_1 + \gamma_2 \tag{1.1}$$

in which the γ phase is Ag_3Sn, the γ_1 phase is Ag_2Hg_3, and the γ_2 phase is Sn_7Hg. The phase diagram for the Ag-Sn-Hg system shows that over a wide compositional range all three phases are present. The final composition of dental amalgams typically contain 45% to 55% mercury, 35% to 45% silver, and about 15% tin after fully set in about one day.

Gold and gold alloys are useful metals in dentistry as a result of their durability, stability, and corrosion resistance [Nielsen, 1986]. Gold fillings are introduced by two methods: casting and malleting. *Cast* restorations are made by taking a wax impression of the prepared cavity, making a mold from this impression in a material such as gypsum silica, which tolerates high temperature, and casting molten gold in the mold. The patient is given a temporary filling for the intervening time. Gold *alloys* are used for cast restorations, since they have mechanical properties which are superior to those of pure gold. Corrosion resistance is retained in these alloys provided they contain 75% or more of gold and other **noble** metals. Copper, alloyed with gold, significantly increases its strength. Platinum also improves the strength, but no more than about 4% can be added, or the melting point of the alloy is elevated excessively. Silver compensates for the color of copper. A small amount of zinc may be added to lower the melting point and to scavenge oxides formed during melting. Gold alloys of different composition are available. Softer alloys containing more than 83% gold are used for inlays which are not subjected to much stress. Harder alloys containing less gold are chosen for crowns and cusps which are more heavily stressed.

Malleted restorations are built up in the cavity from layers of *pure* gold foil. The foils are welded together by pressure at ambient temperature. In this type of welding, the metal layers are joined by thermal diffusion of atoms from one layer to the other. Since intimate contact is required in this procedure, it is particularly important to avoid contamination. The pure gold is relatively soft, so this type of restoration is limited to areas not subjected to much stress.

1.6 Other Metals

Several other metals have been used for a variety of specialized implant applications. *Tantalum* has been subjected to animal implant studies and has been shown very biocompatible. Due to its poor mechanical properties (Table 1.9) and its high density (16.6 g/cm^3) it is restricted to few applications such as wire sutures for plastic surgeons and neurosurgeons and a radioisotope for bladder tumors.

TABLE 1.9 Mechanical Properties of Tantalum

Properties	Annealed	Cold-worked
Tensile strength (MPa)	207	517
Yield strength (0.2% offset) (MPa)	138	345
Elongation (%)	20–30	2
Young's modulus (GPa)	—	190

Source: American Society for Testing and Materials, F560–86, p. 143, 1992.

Platinum group metals (PGM) such as Pt, Pd, Rh, Ir, Ru, and Os are extremely corrosion resistant but have poor mechanical properties [Wynblatt, 1986]. They are mainly used as alloys for electrodes such as pacemaker tips because of their high resistance to corrosion and low threshold potentials for electrical conductivity.

Thermoseeds made of 70% Ni and 30% Cu have been produced which possess Curie points in the therapeutic **hyperthermia** range, approximately 40 to 50°C [Ferguson et al., 1992]. Upon the application of an alternating magnetic field, eddy currents are induced, which will provide a continuous heat source through resistive heating of the material. As the temperature of a ferromagnetic substance nears its Curie point, however, there is a loss of ferromagnetic properties and a resulting loss of heat output. Thus, self-regulation of temperature is achieved and can be used to deliver a constant hyperthermic temperature extracorporeally at any time and duration.

Surface modifications of metal alloys such as coatings by plasma spray, physical or chemical vapor deposition, ion implantaion, and fluidized bed deposition have been used in industry [Smith, 1993]. Coating implants with tissue compatible materials such as hydroxyapatite, oxide ceramics, Bioglass®, and pyrolytic carbon are typical applications in implants. Such efforts have been largely ineffective if the implants are permanent and particularly if the implants are subjected to a large loading. The main problem is the delamination of the coating or eventual wear of the coating. The added cost of coating or ion implanting hinders the use of such techniques unless the technique shows unequivocal superiority compared to the non-treated implants.

1.7 Corrosion of Metallic Implants

Corrosion is the unwanted chemical reaction of a metal with its environment, resulting in its continued degradation to oxides, hydroxides, or other compounds. Tissue fluid in the human body contains water, dissolved oxygen, proteins, and various ions such as chloride and hydroxide. As a result, the human body presents a very aggressive environment for metals used for implantation. Corrosion resistance of a metallic implant material is consequently an important aspect of its biocompatibility.

1.7.1 Electrochemical Aspects

The lowest free energy state of many metals in an oxygenated and hydrated environment is that of the oxide. Corrosion occurs when metal atoms become ionized and go into solution, or combine with oxygen or other species in solution to form a compound which flakes off or dissolves. The body environment is very aggressive in terms of corrosion since it is not only aqueous but also contains chloride ions and proteins. A variety of chemical reactions occur when a metal is exposed to an aqueous environment, as shown in Figure 1.10. The electrolyte, which contains ions in solution, serves to complete the electric circuit. In the human body, the required ions are plentiful in the body fluids. Anions are negative ions which migrate toward the **anode**, and cations are positive ions which migrate toward the **cathode**. At the anode, or positive electrode, the metal oxidizes by losing valence electrons as in the following:

$$M \rightarrow M^{+n} + ne^-$$

(1.2)

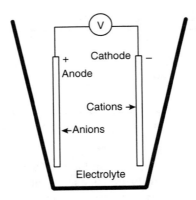

FIGURE 1.10   Electrochemical cell.

TABLE 1.10   Standard Electrochemical Series

Reaction	ΔE_0 [volts]
$Li \leftrightarrow Li^+$	−3.05
$Na \leftrightarrow Na^+$	−2.71
$Al \leftrightarrow Al^{+++}$	−1.66
$Ti \leftrightarrow Ti^{+++}$	−1.63
$Cr \leftrightarrow Cr^{++}$	−0.56
$Fe \leftrightarrow Fe^{++}$	−0.44
$Cu \leftrightarrow Cu^{++}$	−0.34
$Co \leftrightarrow Co^{++}$	−0.28
$Ni \leftrightarrow Ni^{++}$	−0.23
$H_2 \leftrightarrow 2H^+$	−0.00
$Ag \leftrightarrow Ag^+$	+0.80
$Au \leftrightarrow Au^+$	+1.68

At the cathode, or negative electrode, the following reduction reactions are important:

$$M^{+n} + ne^- \rightarrow M \tag{1.3}$$

$$M^{++} + OH^- + 2e^- \rightarrow MOH \tag{1.4}$$

$$2H_3O^+ + 2e^- \rightarrow H_2 \uparrow + 2H_2O \tag{1.5}$$

$$1/2O_2 + H_2O + 2e^- \rightarrow 2OH^- \tag{1.6}$$

The tendency of metals to corrode is expressed most simply in the standard electrochemical series of **Nernst potentials**, shown in Table 1.10. These potentials are obtained in electrochemical measurements in which one electrode is a standard hydrogen electrode formed by bubbling hydrogen through a layer of finely divided platinum black. The potential of this reference electrode is defined to be zero. Noble metals are those which have a potential higher than that of a standard hydrogen electrode; base metals have lower potentials.

If two dissimilar metals are present in the same environment, the one which is most negative in the **galvanic series** will become the anode, and bimetallic (or galvanic) corrosion will occur. **Galvanic corrosion** can be much more rapid than the corrosion of a single metal. Consequently, implantation of dissimilar metals (mixed metals) is to be avoided. Galvanic action can also result in corrosion within a single metal, if there is inhomogeneity in the metal or in its environment, as shown in Figure 1.11.

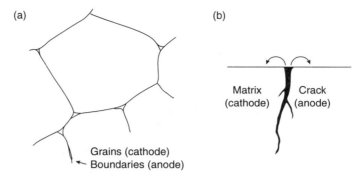

FIGURE 1.11 Micro-corrosion cells. (a) Grain boundaries are anodic with respect to the grain interior. (b) Crevice corrosion due to oxygen-deficient zone in metal's environment.

The potential difference, E, actually observed depends on the concentration of the metal ions in solution according to the Nernst equation,

$$E = E_o + (RT/nF)\ln[M^{+n}] \tag{1.7}$$

in which R is the gas constant, E_o is the standard electrochemical potential, T is the absolute temperature, F' is Faraday's constant (96,487 C/mol), and n is the number of moles of ions.

The order of nobility observed in actual practice may differ from that predicted thermodynamically. The reasons are that some metals become covered with a *passivating* film of reaction products which protects the metal from further attack. The dissolution reaction may be strongly irreversible so that a potential barrier must be overcome. In this case, corrosion may be inhibited even though it remains energetically favorable. The kinetics of corrosion reactions are not determined by the thermodynamics alone.

1.7.2 Pourbaix Diagrams in Corrosion

The **Pourbaix diagram** is a plot of regions of *corrosion*, **passivity**, and **immunity** as they depend on electrode potential and pH [Pourbaix, 1974]. The Pourbaix diagrams are derived from the Nernst equation and from the solubility of the degradation products and the equilibrium constants of the reaction. For the sake of definition, the *corrosion region* is set arbitrarily at a concentration of greater than 10^{-6} g atom/l (molar) or more of metal in the solution at equilibrium. This corresponds to about 0.06 mg/l for metals such as iron and copper, and 0.03 mg/l for aluminum. *Immunity* is defined as equilibrium between metal and its ions at less than 10^{-6} M. In the region of immunity, the corrosion is energetically impossible. Immunity is also referred to as cathodic protection. In the passivation domain, the stable solid constituent is an oxide, hydroxide, hydride, or a salt of the metal. *Passivity* is defined as equilibrium between a metal and its reaction products (oxides, hydroxides, etc.) at a concentration of 10^{-6} M or less. This situation is useful if reaction products are adherent. In the biomaterials setting, passivity may or may not be adequate; disruption of a passive layer may cause an increase in corrosion. The equilibrium state may not occur if reaction products are removed by the tissue fluid. Materials differ in their propensity to re-establish a passive layer which has been damaged. This layer of material may protect the underlying metal if it is firmly adherent and nonporous; in that case further corrosion is prevented. Passivation can also result from a concentration polarization due to a buildup of ions near the electrodes. This is not likely to occur in the body since the ions are continually replenished. Cathodic depolarization reactions can aid in the passivation of a metal by virtue of an energy barrier which hinders the kinetics. Equation 1.5 and Equation 1.6 are examples.

There are two diagonal lines in the diagrams shown in Figure 1.12. The top oxygen line represents the upper limit of the stability of water and is associated with oxygen rich solutions or electrolytes near

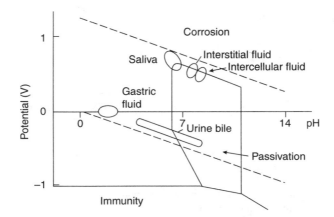

FIGURE 1.12   Pourbaix diagram for chromium, showing regions associated with various body fluids. (Modified from Black, J. (1992). *Biological Performance of Materials, 2nd ed.* New York: M. Dekker, Inc.)

oxidizing materials. In the region above this line, oxygen is evolved according to $2H_2O \rightarrow O_2 \uparrow + 4H^+ + 4e^-$. In the human body, saliva, intracellular fluid, and interstitial fluid occupy regions near the oxygen line, since they are saturated with oxygen. The lower hydrogen diagonal line represents the lower limit of the stability of water. Hydrogen gas is evolved according to Equation 1.5. Aqueous corrosion occurs in the region between these diagonal lines on the Pourbaix diagram. In the human body, urine, bile, the lower gastrointestinal tract, and secretions of ductless glands, occupy a region somewhat above the hydrogen line.

The significance of the Pourbaix diagram is as follows. Different parts of the body have different pH values and oxygen concentrations. Consequently, a metal which performs well (is immune or passive) in one part of the body may suffer an unacceptable amount of corrosion in another part. Moreover, pH can change dramatically in tissue that has been injured or infected. In particular, normal tissue fluid has a pH of about 7.4, but in a wound it can be as low as 3.5, and in an infected wound the pH can increase to 9.0.

Pourbaix diagrams are useful, but do not tell the whole story; there are some limitations. Diagrams are made considering equilibrium among metal, water, and reaction products. The presence of other ions, for example, chloride, may result in very much different behavior and large molecules in the body may also change the situation. Prediction of passivity may in some cases be optimistic, since reaction rates are not considered.

1.7.3   Rate of Corrosion and Polarization Curves

The regions in the Pourbaix diagram specify whether corrosion will take place, but they do not determine the rate. The rate, expressed as an electric current density (current per unit area), depends upon electrode potential as shown in the polarization curves shown in Figure 1.13. From such curves, it is possible to calculate the number of ions per unit time liberated into the tissue, as well as the depth of metal removed by corrosion in a given time. An alternative experiment is one in which the weight loss of a specimen of metal due to corrosion is measured as a function of time.

The rate of corrosion also depends on the presence of synergistic factors, such as those of mechanical origin (uneven distribution of mechanical stress). The stressed alloy failures occur due to the propagation of cracks in corrosive environments. For example, in corrosion fatigue (stress corrosion cracking), repetitive deformation of a metal in a corrosive environment results in acceleration of both the corrosion and the fatigue microdamage. Since the body environment involves both repeated mechanical loading and a chemically aggressive environment, fatigue testing of implant materials should always be performed under physiological environmental conditions; under Ringer's solution at body temperature. In *fretting corrosion*, rubbing of one part on another disrupts the passivation layer, resulting in accelerated corrosion. In

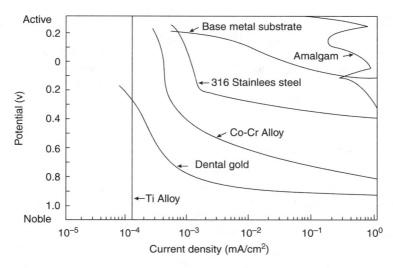

FIGURE 1.13　Potential-current density curves for some biomaterials [Greener et al., 1972].

pitting, the corrosion rate is accelerated in a local region. Stainless steel is vulnerable to pitting. Localized corrosion can occur if there is inhomogeneity in the metal or in the environment. *Grain boundaries* in the metal may be susceptible to the initiation of corrosion, as a result of their higher energy level. *Crevices* are also vulnerable to corrosion, since the chemical environment in the crevice may differ from that in the surrounding medium. The area of contact between a screw and a bone plate, for example, can suffer **crevice corrosion**.

1.7.4　Corrosion of Available Metals

Choosing a metal for implantation should take into account the corrosion properties discussed above. Metals which are in current use as biomaterials include gold, cobalt chromium alloys, type 316 stainless steel, cp-titanium, titanium alloys, nickel–titanium alloys, and silver-tin-mercury amalgam.

The noble metals are immune to corrosion and would be ideal materials if corrosion resistance were the only concern. Gold is widely used in dental restorations and in that setting it offers superior performance and longevity. Gold is not, however, used in orthopaedic applications as a result of its high density, insufficient strength, and high cost.

Titanium is a base metal in the context of the electrochemical series, however, it forms a robust passivating layer and remains passive under physiological conditions. Corrosion currents in normal saline are very low: 10^{-8} A/cm^2. Titanium implants remain virtually unchanged in appearance. Ti offers superior corrosion resistance but is not as stiff or strong as steel or Co–Cr alloys.

Cobalt–chromium alloys, like titanium, are passive in the human body. They are widely in use in orthopedic applications. They do not exhibit pitting corrosion.

Stainless steels contain enough chromium to confer corrosion resistance by passivity. The passive layer is not as robust as in the case of titanium or the cobalt chrome alloys. Only the most corrosion resistant of the stainless steels are suitable for implants. These are the austenitic types — 316, 316L, and 317, which contain molybdenum. Even these types of stainless steel are vulnerable to pitting and to crevice corrosion around screws.

The phases of dental amalgam are passive at neutral pH, the transpassive potential for the γ_2 phase is easily exceeded, due to interphase galvanic couples or potentials due to differential aeration under dental plaque. Amalgam, therefore, often corrodes and is the most active (corrosion prone) material used in dentistry.

Corrosion of an implant in the clinical setting can result in symptoms such as local pain and swelling in the region of the implant, with no evidence of infection; cracking or flaking of the implant as seen on x-ray films, and excretion of excess metal ions. At surgery, gray or black discoloration of the surrounding tissue may be seen and flakes of metal may be found in the tissue. Corrosion also plays a role in the mechanical failures of orthopaedic implants. Most of these failures are due to fatigue, and the presence of a saline environment certainly exacerbates fatigue. The extent to which corrosion influences fatigue in the body is not precisely known.

1.7.5 Stress Corrosion Cracking

When an implant is subjected to stress, the corrosion process could be accelerated due to the mechanical energy. If the mechanical stress is repeated then fatigue stress corrosion takes place such as in the femoral stem of the hip joint and hip nails made of stainless steels [Dobbs and Scales, 1979; Sloter and Piehler, 1979]. However, other mechanisms of corrosion such as fretting may also be involved at point of contact such as in the counter-sink of the hip nail or bone fracture plate for the screws.

1.8   Manufacturing of Implants

1.8.1 Stainless Steels

The austenitic stainless steels work-harden very rapidly as shown in Figure 1.2 and therefore, cannot be cold-worked without intermediate heat treatments. The heat treatments should not induce, however, the formation of chromium carbide (CCr_4) in the grain boundaries; this may cause corrosion. For the same reason, the austenitic stainless steel implants are not usually welded.

The distortion of components by the heat treatments can occur but this problem can be solved by controlling the uniformity of heating. Another undesirable effect of the heat treatment is the formation of surface oxide scales which have to be removed either chemically (acid) or mechanically (sand-blasting). After the scales are removed the surface of the component is polished to a mirror or mat finish. The surface is then cleaned, degreased, and passivated in nitric acid (ASTM Standard F86). The component is washed and cleaned again before packaging and sterilizing.

1.8.2 Co–Cr Alloys

The CoCrMo alloy is particularly susceptible to work-hardening so that the normal fabrication procedure used with other metals cannot be employed. Instead, the alloy is cast by a lost wax (or investment casting) method which involves making a wax pattern of the desired component. The pattern is coated with a refractory material, first by a thin coating with a slurry (suspension of silica in ethyl silicate solution) followed by complete investing after drying (1) the wax is then melted out in a furnace (100–150°C), (2) the mold is heated to a high temperature burning out any traces of wax or gas forming materials, (3) molten alloy is poured with gravitational or centrifugal force, and (4) the mold is broken after cooled. The mold temperature is about 800–1000°C and the alloy is at 1350–1400°C.

Controlling the mold temperature will have an effect on the grain size of the final cast; coarse ones are formed at higher temperatures which will decrease the strength. However, high processing temperature will result in larger carbide precipitates with greater distances between them resulting in a less brittle material. Again there is a complementary (trade off) relationship between strength and toughness.

1.8.3 Ti and Its Alloys

Titanium is very reactive at high temperature and burns readily in the presence of oxygen. Therefore, it requires an inert atmosphere for high temperature processing or is processed by vacuum melting. Oxygen diffuses readily in titanium and the dissolved oxygen embrittles the metal. As a result, any hot working or forging operation should be carried out below 925°C. Machining at room temperature is not the solution to all the problems since the material also tends to gall or seize the cutting tools. Very sharp tools with slow speeds and large feeds are used to minimize this effect. Electrochemical machining is an attractive means.

Defining Terms

Amalgam: An alloy obtained by mixing silver tin alloy with mercury.

Anode: Positive electrode in an electrochemical cell.

Cathode: Negative electrode in an electrochemical cell.

Corrosion: Unwanted reaction of metal with environment. In a Pourbaix diagram, it is the region in which the metal ions are present at a concentration of more than 10^{-6} M.

Crevice corrosion: A form of localized corrosion in which concentration gradients around pre-existing crevices in the material drive corrosion processes.

Curie temperature: Transition temperature of a material from ferromagnetic to paramagnetic.

Galvanic corrosion: Dissolution of metal driven by macroscopic differences in electrochemical potential, usually as a result of dissimilar metals in proximity.

Galvanic series: Table of electrochemical potentials (voltage) associated with the ionization of metal atoms. These are called Nernst potentials.

Hyperthermia: Application of high enough thermal energy (heat) to suppress the cancerous cell activities. Above 41.5°C (but below 60°C) is needed to have any effect.

Immunity: Resistance to corrosion by an energetic barrier. In a Pourbaix diagram, it is the region in which the metal is in equilibrium with its ions at a concentration of less than 10^{-6} M. Noble metals resist corrosion by immunity.

Martensite: A metastable structure formed by quenching of austenite (g) structure in alloys such as steel and Ti alloys. They are brittle and hard, and therefore, are further treated with heat to make tougher.

Nernst potential: Standard electrochemical potential measured with respect to a standard hydrogen electrode.

Noble: Type of metal with a positive standard electrochemical potential.

Passivation: Production of corrosion resistance by a surface layer of reaction products (Normally oxide layer which is impervious to gas and water.)

Passivity: Resistance to corrosion by a surface layer of reaction products. In a Pourbaix diagram, it is the region in which the metal is in equilibrium with its reaction products at a concentration of less than 10^{-6} molar.

Pitting: A form of localized corrosion in which pits form on the metal surface.

Pourbaix diagram: Plot of electrical potential vs. pH for a material in which the regions of corrosion, passivity, and immunity are identified.

Shape memory effect (SME): Thermoelastic behavior of some alloys which can revert back to their original shape when the temperature is greater than the phase transformation temperature of the alloy.

Superelasticity: Minimal stress increase beyond the initial strain region resulting in very low modulus in the region for some shape memory alloys.

References

ASTM (1980). Annual Book of *ASTM Standards* (Philadelphia, PA).

Black, J. (1992). *Biological Performance of Materials*, 2nd ed. New York: M. Dekker, Inc.

Bordiji, K., Jouzeau, J., Mainard, D., Payan, E., Delagoutte, J., and Netter, P. (1996). Evaluation of the effect of three surface treatments on the biocompatibility of 316L stainless steel using human differentiated cells. *Biomaterials* 17, 491–500.

Breen, D.J. and Stoker, D.J. (1993). Titanium lines: a manifestation of metallosis and tissue response to titanium alloy megaprostheses at the knee. *Clin. Radiol.* 43, 274–277.

Buehler, W.J., Gilfrich, J.V., and Wiley, R.C. (1963). Effect of low-temperature phase changes on the mechanical properties of alloys near composition Ti–Ni. *J. Appl. Phys.* 34, 1475–1477.

Davidson, J.A., Mishra, A.K., Kovacs, P., and Poggie, R.A. (1994). New surface hardened, low-modulus, corrosion-resistant Ti–13Nb–13Zr alloy for total hip arthroplasty. *Biomed. Mater. Eng.* 4, 231–243.

Devine, T.M. and Wulff, J. (1975). Cast vs. wrought cobalt–chromium surgical implant alloys. *J. Biomed. Mater. Res.* 9, 151–167.

Dobbs, H.S. and Scales, J.T. (1979). Fracture and corrosion in stainless steel hip replacement stems. In *Corrosion and Degradation of Implant Materials*, Syrett, B.C. and Acharya, A. Eds. Philadelphia: American Society for Testing and Materials, pp. 245–258.

Duerig, T.W., Melton, K.N., Stockel, D., and Wayman, C.M. (1990). *Engineering Aspects of Shape Memory Alloys*. London: Butterworth-Heinemann.

Feighan, J.E., Goldberg, V.M., Davy, D., Parr, J.A., and Stevenson, S. (1995). The influence of surface-blasting on the incorporation of titanium–alloy implants in a rabbit intramedullary model. 77A, 1380–1395.

Ferguson, S.D., Paulus, J.A., Tucker, R.D., Loening, S.A., and Park, J.B. (1992). Effect of thermal treatment on heating characteristics of Ni–Cu Alloy for hyperthermie: preliminary studies. *J. Appl. Biomater.* 4, 55–60.

Granchi, D., Ciapetti, G., Savarino, L., Cavedagna, D., Donati, M.E., and Pizzoferrato, A. (1996). Assessment of metal extract toxicity on human lymphocytes cultured *in vitro*. *J. Biomed. Mater. Res.* 31, 183–191.

Greener, E.H., Harcourt, J.K., and Lautenschlager, E.P. (1972). *Materials Science in Dentistry*. Baltimore, MD: Williams and Wilkins.

Hazan, R., Brener, R., and Oron, U. (1993). Bone growth to metal implants is regulated by their surface chemical properties. *Biomaterials* 570–574.

Hille, G.H. (1966). Titanium for surgical implants. *J. Mater.* 1, 373–383.

Imam, M.A., Fraker, A.C., Harris, J.S., and Gilmore, C.M. (1983). Influence of heat treatment on the fatigue lives of Ti–6Al–4V and Ti–4.5Al–5Mo–1.5CR. In *Titanium Alloys in Surgical Implants*, Luckey, H.A. and Kubli, F.E. Eds. Philadelphia, PA: ASTM special technical publication 796, pp. 105–119.

Kasemo, B. and Lausma, J. (1988). Biomaterial and implant surface: A surface science approach. *Int. J. Oral Maxillofac Implant.* 3, 247–259.

Keating, F.H. (1956). *Chromium-Nickel Autentic Steels*. London: Butterworths.

Keller, J.C., Stanford, C.M., Wightman, J.P., Draughn, R.A., and Zaharias, R. (1994). Characterizations of titanium implant surfaces. III. *J. Biomed. Mater. Res.* 28, 939–946.

Kieswetter, K., Schwartz, Z., Hummert, T.W., Cochran, D.L., Simpson, J., and Boyan, B.D. (1996). Surface roughness modulates the local production of growth factors and cytokines by osteoblast-like MG-63 cells. *J. Biomed. Mater. Res.* 32, 55–63.

Kim, H., Miyaji, F., Kokubo, T., and Nakamura, T. (1996). Preparation of bioactive Ti and its alloys via simple chemical surface treatment. *J. Biomed. Mater. Res.* 32, 409–417.

Lee, J.H., Park, J.B., Andreasen, G.F., and Lakes, R.S. (1988). Thermomechanical study of Ni–Ti alloys. *J. Biomed. Mater. Res.* 22, 573–588.

Nielsen, J.P. (1986). Dental noble-metal casting alloys: composition and properties. In *Encyclopedia of Materials Science and Engineering*, Bever, M.B. Ed. Oxford, Cambridge: Pergamon Press, pp. 1093–1095.

Pourbaix, M. (1974). *Atlas of Electrochemical Equilibria in Aqueous Solutions*, 2nd ed. Houston/CEBELCOR, Brussels: NACE.

Richards-Mfg-Company. (1980). Biophase implant material, technical information publication 3846 Memphis, TN.

Schmalzried, T.P., Peters, P.C., Maurer, B.T., Bragdon, C.R., and Harris, W.H. (1996). Long-duration metal-on-metal total hip arthroplasties with low wear of the articulating surfaces. *J. Arthroplasty* 11, 322–331.

Semlitch, M. (1980). Properties of wrought CoNiCrMo alloy Protasul-10, a highly corrosion and fatigue resistant implant material for joint endoprostheses. *Eng. Med.* 9, 201–207.

Sloter, L.E. and Piehler, H.R. (1979). Corrosion-fatigue performance of stainless steel hip nails — Jewett type. In *Corrosion and Degradation of Implant Materials*, Syrett, B.C. and Acharya, A. Eds. Philadelphia, PA: American Society for Testing and Materials, pp. 173–195.

Smith, C.J.E. and Hughes, A.N. (1966). The corrosion-fatigue behavior of a titanium-6w/o aluminum-4w/o vanadium alloy. *Eng. Med.* 7, 158–171.

Smith, W.F. (1993). *Structure and Properties of Engineering Alloys*, 2nd ed. New York: McGraw-Hill.

Takatsuka, K., Yamamuro, T., Nakamura, T., and Kokubo, T. (1995). Bone-bonding behavior of titanium alloy evaluated mechanically with detaching failure load. *J. Biomed. Mater. Res.* 29, 157–163.

Takeshita, F., Ayukawa, Y., Iyama, S., Murai, K., and Suetsugu, T. (1997). Long-term evaluation of bone-titanium interface in rat tibiae using light microscopy, transmission electron microscopy, and image processing. *J. Biomed. Mater. Res.* 37, 235–242.

Wayman, C.M. and Duerig, T.W. (1990). An introduction of martensite and shape memory. In *Engineering Aspects of Shape Memory Alloys*, Duerig, T.W., Melton, K.N., Stockel, D., and Wayman, C.M. Eds. London: Butterworth-Heinemann, pp. 3–20.

Wayman, C.M. and Shimizu, K. (1972). The shape memory ('Marmem') effect in alloys. *Metal Sci. J.* 6, 175–183.

Williams, D.F. (1982). *Biocompatibility in Clinical Practice*. Boca Raton, FL: CRC Press.

Wynblatt, P. (1986). Platinum Group Metals and Alloys. In *Encyclopedia of Materials Science and Engineering*, Bever, M.B. Ed. Oxford, Cambridge: Pergamon Press, pp. 3576–3579.

Yan, W., Nakamura, T., Kobayashi, M., Kim, H., Miyaji, F., and Kokubo, T. (1997). Bonding of chemically treated titanium implants to bone. *J. Biomed. Mater. Res.* 37, 267–275.

Further Reading

American Society for Testing and Materials. 1992. *Annual Book of ASTM Standards*, vol. 13, *Medical Devices and Services*, American Society for Testing and Materials, Philadelphia, PA.

Bardos, D.I. 1977. Stainless steels in medical devices, in *Handbook of Stainless Steels*. Peckner, D. and Bernstein, I.M., Eds. pp. 1–10, McGraw-Hill, New York.

Bechtol, C.O., Ferguson, A.B., Jr., and Laing, P.G. 1959. *Metals and Engineering in Bone and Joint Surgery*, Williams and Wilkins, Baltimore, MD.

Comte, T.W. 1984. Metallurgical observations of biomaterials, in *Contemporary Biomaterials*, Boretos, J.W. and Eden, M., Eds., pp. 66–91, Noyes, Park Ridge, NJ.

Duerig, T.W., Melton, K.N., Stockel, D., and Wayman, C.M., Eds. 1990. *Engineering Aspects of Shape Memory Alloys*, Butterworth-Heinemann, London.

Dumbleton, J.H. and Black, J. 1975. *An Introduction to Orthopaedic Materials*, C. Thomas, Springfield, IL.

Fontana, M.G. and Greene, N.O. 1967. *Corrosion Engineering*, pp. 163–168, McGraw-Hill, New York.

Greener, E.H., Harcourt, J.K., and Lautenschlager, E.P. 1972. *Materials Science in Dentistry*, Williams and Wilkins, Baltimore, MD.

Hildebrand, H.F. and Champy, M., Eds. 1988. *Biocompatibility of Co–Cr–Ni Alloys*. Plenum Press, New York.

Levine, S.N., Ed. 1968. *Materials in Biomedical Engineering*, Annals of New York Academy of Science, vol. 146. New York.

Luckey H.A., Ed. 1983. *Titanium Alloys in Surgical Implants*, ASTM Special Technical Publication 796, Philadelphia, PA

Mears, D.C. 1979. *Materials and Orthopaedic Surgery*, Williams and Wilkins, Baltimore, MD.

Park, J.B. 1984. *Biomaterials Science and Engineering*, Plenum Pub., New York.

Perkins, J., Ed. 1975. *Shape Memory Effects in Alloys*, Plenum Press, New York.

Puckering, F.B., Ed. 1979. *The Metallurgical Evolution of Stainless Steels*, 1–42, American Society for Metals and the Metals Society, Metals Park, OH.

Smith, W.F. 1993. *Structure and Properties of Engineering Alloys*, 2nd ed., McGraw-Hill, New York.

Weinstein, A., Horowitz, E., and Ruff, A.W., Eds. 1977. *Retrieval and Analysis of Orthopaedic Implants*, NBS, U.S. Department of Commerce, Washington, D.C.

Williams, D.F. and Roaf, R. 1973. Implants in *Surgery*, W.B. Sauders Co., LTD, London.

2

Ceramic Biomaterials

W.C. Billotte
University of Dayton

2.1 Introduction

Ceramics are defined as the art and science of making and using solid articles that have as their essential component, inorganic nonmetallic materials [Kingery et al., 1976]. Ceramics are refractory, polycrystal line compounds, usually inorganic, including silicates, metallic oxides, carbides and various refractory hydrides, sulfides, and selenides. Oxides such as Al_2O_3, MgO, SiO_2, and ZrO_2 contain metallic and nonmetallic elements and ionic salts, such as $NaCl$, $CsCl$, and ZnS [Park and Lakes, 1992]. Exceptions to the preceding include covalently bonded ceramics such as diamond and carbonaceous structures like graphite and pyrolized carbons [Park and Lakes, 1992].

Ceramics in the form of pottery have been used by humans for thousands of years. Until recently, their use was somewhat limited because of their inherent brittleness, susceptibility to notches or micro-cracks, low tensile strength, and low impact strength. However, within the last 100 years, innovative techniques for fabricating ceramics have led to their use as "high tech" materials. In recent years, humans have realized that ceramics and their composites can also be used to augment or replace various parts of the body, particularly bone. Thus, the ceramics used for the latter purposes are classified as *bioceramics*. Their relative inertness to the body fluids, high compressive strength, and aesthetically pleasing appearance led to the use of ceramics in dentistry as dental crowns. Some carbons have found use as implants especially for blood interfacing applications such as heart valves. Due to their high specific strength as fibers and their biocompatibility, ceramics are also being used as reinforcing components of composite implant materials and for tensile loading applications such as artificial tendon and ligaments [Park and Lakes, 1992].

Unlike metals and polymers, ceramics are difficult to shear plastically due to the (ionic) nature of the bonding and minimum number of slip systems. These characteristics make the ceramics nonductile and are responsible for almost zero creep at room temperature [Park and Lakes, 1992]. Consequently, ceramics are very susceptible to notches or microcracks because instead of undergoing plastic deformation (or yield) they will fracture elastically on initiation of a crack. At the crack tip the stress could be many times higher than the stress in the material away from the tip, resulting in a *stress concentration* which weakens the material considerably. The latter makes it difficult to predict the tensile strength of the material (ceramic). This is also the reason ceramics have low tensile strength compared to compressive strength. If a ceramic is flawless, it is very strong even when subjected to tension. Flawless glass fibers have twice the tensile strengths of high strength steel ($\sim$7 GPa) [Park and Lakes, 1992].

Ceramics are generally hard; in fact, the measurement of hardness is calibrated against ceramic materials. Diamond is the hardest, with a hardness index of 10 on Moh's scale, and talc ($Mg_3Si_3O_{10}COH$) is the softest ceramic (Moh's hardness 1), while ceramics such as **alumina** (Al_2O_3; hardness 9), quartz (SiO_2; hardness 8), and apatite ($Ca_5P_3O_{12}F$; hardness 5) are in the middle range. Other characteristics of ceramic materials are (1) their high melting temperatures and (2) low conductivity of electricity and heat. These characteristics are due to the chemical bonding within ceramics.

In order to be classified as a bioceramic, the ceramic material must meet or exceed the properties listed in Table 2.1. The number of specific ceramics currently in use or under investigation cannot be accounted for in the space available for bioceramics in this book. Thus, this chapter will focus on a general overview of the relatively bioinert, bioactive or surface reactive ceramics, and biodegradable or resorbable bioceramics.

Ceramics used in fabricating implants can be classified as nonabsorbable (relatively inert), bioactive or surface reactive (semi-inert) [Hench, 1991, 1993] and biodegradable or resorbable (non-inert) [Hentrich et al., 1971; Graves et al., 1972]. Alumina, zirconia, silicone nitrides, and carbons are inert bioceramics. Certain **glass ceramics** and dense **hydroxyapatites** are semi-inert (bioreactive) and **calcium phosphates** and calcium aluminates are resorbable ceramics [Park and Lakes, 1992].

2.2 Nonabsorbable or Relatively Bioinert Bioceramics

2.2.1 Relatively Bioinert Ceramics

Relatively bioinert ceramics maintain their physical and mechanical properties while in the host. They resist corrosion and wear and have all the properties listed for bioceramics in Table 2.1. Examples of relatively bioinert ceramics are dense and porous aluminum oxides, zirconia ceramics, and single phase calcium aluminates (Table 2.2). Relatively bioinert ceramics are typically used as structural-support implants.

TABLE 2.1 Desired Properties of Implantable Bioceramics

1. Should be nontoxic
2. Should be noncarcinogenic
3. Should be nonallergic
4. Should be noninflammatory
5. Should be biocompatible
6. Should be biofunctional for its lifetime in the host

Some of these are bone plates, bone screws, and femoral heads (Table 2.3). Examples of nonstructural support uses are ventilation tubes, sterilization devices [Feenstra and de Groot, 1983] and drug delivery devices (see Table 2.3).

2.2.2 Alumina (Al$_2$O$_3$)

The main source of high purity alumina (aluminum oxide, Al$_2$O$_3$) is bauxite and native corundum. The commonly available alumina (alpha, α) can be prepared by calcining alumina trihydrate. The chemical composition and density of commercially available "pure" calcined alumina are given in Table 2.4. The American Society for Testing and Materials (ASTM) specifies that alumina for implant use should contain 99.5% pure alumina and less than 0.1% combined SiO$_2$ and alkali oxides (mostly Na$_2$O) (F603-78).

Alpha alumina has a rhombohedral crystal structure ($a = 4.758$ Å and $c = 12.991$ Å). Natural alumina is known as sapphire or ruby, depending on the types of impurities which give rise to color. The single

TABLE 2.2 Examples of Relatively Bioinert Bioceramics

Bioinert Ceramics	References
1. Pyrolitic carbon coated devices	Adams and Williams, 1978
	Bokros et al., 1972
	Bokros, 1972
	Chandy and Sharma, 1991
	Dellsperger and Chandran, 1991
	Kaae, 1971
	More and Silver, 1990
	Shimm and Haubold, 1980
	Shobert, 1964
2. Dense and nonporous aluminum oxides	Hench, 1991
	Hentrich et al., 1971
	Krainess and Knapp, 1978
	Park, 1991
	Ritter et al., 1979
	Shackelford, 1988
3. Porous aluminum oxides	Hench, 1991
	Hentrich et al., 1971
	Park, 1991
	Ritter et al., 1979
	Shackelford, 1988
4. Zirconia ceramics	Barinov and Baschenko, 1992
	Drennan and Steele, 1991
	Hench, 1991
	Kumar et al., 1989
5. Dense hydroxyapatites	Bajpai, 1990
	Cotell et al., 1992
	Fulmer et al., 1992
	Huaxia et al., 1992
	Kijima and Tsutsumi, 1979
	Knowles et al., 1993
	Meenan et al., 1992
	Niwa et al., 1980
	Posner et al., 1958
	Schwartz et al., 1993
	Valiathan et al., 1993
	Whitehead et al., 1993
6. Calcium aluminates	Hammer et al., 1972
	Hentrich et al., 1971
	Hulbert and Klawitter, 1971

TABLE 2.3   Uses of Bioinert Bioceramics

Bioinert Ceramics	References
1. In reconstruction of acetabular cavities	Boutin, 1981
	Dorlot et al., 1986
2. As bone plates and screws	Zimmermann et al., 1991
3. In the form of ceramic–ceramic composites	Boutin, 1981
	Chignier et al., 1987
	Sedel et al., 1991
	Terry et al., 1989
4. In the form of ceramic–polymer composites	Hulbert, 1992
5. As drug delivery devices	Buykx et al., 1992
6. As femoral heads	Boutin, 1981
	Dörre, 1991
	Ohashi et al., 1988
	Oonishi, 1992
7. As middle ear ossicles	Grote, 1987
8. In the reconstruction of orbital rims	Heimke, 1992
9. As components of total and partial hips	Feenstra and de Groot, 1983
10. In the form of sterilization tubes	Feenstra and de Groot, 1983
11. As ventilation tubes	Feenstra and de Groot, 1983
12. In the repair of the cardiovascular area	Chignier et al., 1987
	Ely and Haubold, 1993

TABLE 2.4   Chemical Composition of Calcined Alumina

Chemicals	Composition (Weight %)
Al_2O_3	99.6
SiO_2	0.12
Fe_2O_3	0.03
Na_2O	0.04

Source: Park, J.B., and Lakes, R.S. 1992. *Ceramic Implant Materials*. In: *Biomaterials An Introduction*, 2nd ed., p. 121 Plenum Press, New York.

crystal form of alumina has been used successfully to make implants [Kawahara, 1989; Park 1991]. Single crystal alumina can be made by feeding fine alumina powders onto the surface of a seed crystal which is slowly withdrawn from an electric arc or oxy-hydrogen flame as the fused powder builds up. Single crystals of alumina up to 10 cm in diameter have been grown by this method [Park and Lakes, 1992].

The strength of polycrystalline alumina depends on its grain size and porosity. Generally, the smaller the grains, the lower the porosity and the higher the strength [Park and Lakes, 1992]. The ASTM standards (F603-78) requires a flexural strength greater than 400 MPa and elastic modulus of 380 GPa (Table 2.5).

Aluminum oxide has been used in the area of orthopaedics for more than 25 years [Hench, 1991]. Single crystal alumina has been used in orthopaedics and dental surgery for almost 20 years. Alumina is usually a quite hard material, its hardness varies from 20 to 30 GPa. This high hardness permits its use as an abrasive (emery) and as bearings for watch movements [Park and Lakes, 1992]. Both polycrystalline and single crystal alumina have been used clinically. The high hardness is accompanied by low friction and wear and inertness to the *in vivo* environment. These properties make alumina an ideal material for use in joint replacements [Park and Lakes, 1992]. Aluminum oxide implants in bones of rhesus monkeys have shown no signs of rejection or toxicity for 350 days [Graves et al., 1972; Hentrich et al., 1971]. One of the most popular uses for aluminum oxide is in total hip protheses. Aluminum oxide hip protheses

TABLE 2.5   Physical Property Requirements of
Alumina and Partially Stabilized Zirconia

Properties	Alumina	Zirconia
Elastic Modulus (GPa)	380	190
Flexural strength (GPa)	>0.4	1.0
Hardness, Mohs	9	6.5
Density (g/cm^3)	3.8–3.9	5.95
Grain size (μm)	4.0	0.6

Note: Both the ceramics contain 3 mole %Y_2O_3.
Soruce: Park, J.B. personal communication, 1993.

with an ultra-high molecular weight polyethylene (UHMWPE) socket have been claimed to be a better device than a metal prostheses with a UHMWPE socket [Oonishi, 1992]. However, the key for success of any implant, besides the correct surgical implantation, is the highest possible quality control during fabrication of the material and the production of the implant [Hench, 1991].

2.2.3   Zirconia (ZrO_2)

Pure zirconia can be obtained from chemical conversion of zircon ($ZrSiO_4$), which is an abundant mineral deposit [Park and Lakes, 1992]. Zirconia has a high melting temperature ($T_m = 2953$ K) and chemical stability with $a = 5.145$ Å, $b = 0.521$ Å, $c = 5.311$ Å, and $\beta = 99°14$ in [Park and Lakes, 1992]. It undergoes a large volume change during phase changes at high temperature in pure form; therefore, a dopant oxide such as Y_2O_3 is used to stabilize the high temperature (cubic) phase. We have used 6 mole% Y_2O_3 as dopant to make zirconia for implantation in bone [Hentrich et al., 1971]. Zirconia produced in this manner is referred to as *partially stabilized* zirconia [Drennan and Steele, 1991]. However, the physical properties of zirconia are somewhat inferior to that of alumina (Table 2.5).

High density zirconia oxide showed excellent compatibility with autogenous rhesus monkey bone and was completely nonreactive to the body environment for the duration of the 350 day study [Hentrich et al., 1971]. Zirconia has shown excellent biocompatibility and good wear and friction when combined with ultra-high molecular weight polyethylene [Kumar et al., 1989; Murakami and Ohtsuki, 1989].

2.2.4   Carbons

Carbons can be made in many allotropic forms: crystalline diamond, graphite, noncrystalline glassy carbon, and quasicrystalline pyrolitic carbon. Among these, only pyrolitic carbon is widely utilized for implant fabrication; it is normally used as a surface coating. It is also possible to coat surfaces with diamond. Although the techniques of coating with diamond have the potential to revolutionize medical device manufacturing, it is not yet commercially available [Park and Lakes, 1992].

The crystalline structure of carbon, as used in implants, is similar to the graphite structure shown in Figure 2.1. The planar hexagonal arrays are formed by strong covalent bonds in which one of the valence electrons or atoms is free to move, resulting in high but anisotropic electric conductivity. Since the bonding between the layers is stronger than the van der Waals force, it has been suggested that the layers are *cross-linked*. However, the remarkable lubricating property of graphite cannot be attained unless the cross-links are eliminated [Park and Lakes, 1992].

The poorly crystalline carbons are thought to contain unassociated or unoriented carbon atoms. The hexagonal layers are not perfectly arranged, as shown in Figure 2.2. Properties of individual crystallites seem to be highly anisotropic. However, if the crystallites are randomly dispersed, the aggregate becomes isotropic [Park and Lakes, 1992].

The mechanical properties of carbon, especially pyrolitic carbon, are largely dependent on its density, as shown in Figure 2.3 and Figure 2.4. The increased mechanical properties are directly related to increased

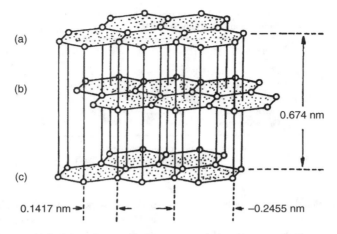

FIGURE 2.1 Crystal structure of graphite. (From Shobert, E.I. 1964. *Carbon and Graphite*. New York, Academic Press. With permission.)

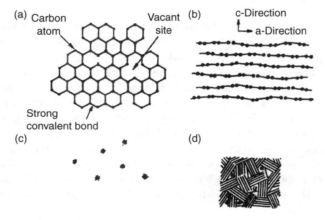

FIGURE 2.2 Schematic presentation of poorly crystalline carbon. (a) Single-layer plane, (b) parallel layers in a crystallite, (c) unassociated carbon, (d) an aggregate of crystallites, single layers and unassociated carbon. (From Bokros, J.C. 1972. *Chem. Phys. Carbon*. 5: 70–81, New York, Marcel Dekker. With permission.)

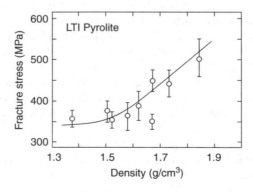

FIGURE 2.3 Fracture stress vs. density for unalloyed LTI pyrolite carbons. (From Kaae, J.L. 1971. *J. Nucl. Mater.* 38: 42–50. With permission.)

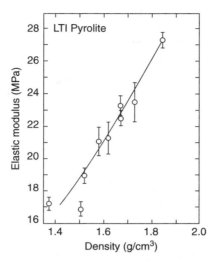

FIGURE 2.4   Elastic modulus vs. density for unalloyed LTI pyrolite carbons. (From Kaae, J.L. 1971. *J. Nucl. Mater.* 38:42–50. With permission.)

TABLE 2.6   Properties of Various Types of Carbon

	Types of Carbon		
Properties	Graphite	Glassy	Pyrolitica[a]
Density (g/cm3)	1.5–1.9	1.5	1.5–2.0
Elastic modulus (GPa)	24	24	28
Compressive strength (MPa)	138	172	517 (575[a])
Toughness (Mn/cm^3)[b]	6.3	0.6	4.8

[a] 1.0 w/o Si-alloyed pyrolitic carbon, Pyrolite ™(Carbomedics, Austin, TX).
[b] 1 m-N/cm^3 = 1.45×10^{-3} in.-lb/in.^{3}.
Source: Park, J.B., and Lakes, R.S. 1992. *Biomaterials An Introduction*, 2nd ed, p. 133. Plenum Press, New York.

density, which indicates that the properties of pyrolitic carbon depend mainly on the aggregate structure of the material [Park and Lakes, 1992].

Graphite and glassy carbon have a much lower mechanical strength than pyrolitic carbon (Table 2.6). However, the average modulus of elasticity is almost the same for all carbons. The strength of pyrolitic carbon is quite high compared to graphite and glassy carbon. Again, this is due to the fewer number of flaws and unassociated carbons in the aggregate.

A composite carbon which is reinforced with carbon fiber has been considered for making implants. However, the carbon–carbon composite is highly anisotropic, and its density is in the range of 1.4 to 1.45 g/cm^3 with a porosity of 35 to 38% (Table 2.7).

Carbons exhibit excellent compatibility with tissue. Compatibility of pyrolitic carbon-coated devices with blood have resulted in extensive use of these devices for repairing diseased heart valves and blood vessels [Park and Lakes, 1992].

Pyrolitic carbons can be deposited onto finished implants from hydrocarbon gas in a *fluidized bed* at a controlled temperature and pressure. The anisotropy, density, crystallite size and structure of the deposited carbon can be controlled by temperature, composition of the fluidized gas, the bed geometry, and the residence time (velocity) of the gas molecules in the bed. The microstructure of deposited carbon should be highly controlled, since the formation of growth features associated with uneven crystallization

TABLE 2.7   Mechanical Properties of Carbon Fiber-Reinforced Carbon

Property	Fiber lay-up	
	Unidirectional	0–90° Crossply
Flexural Modulus (GPa)		
Longitudinal	140	60
Transverse	7	60
Flexural Strength (MPa)		
Longitudinal	1,200	500
Transverse	15	500
Interlaminar shear strength (MPa)	18	18

Source: Adams, D. and Williams, D.F. 1978. *J. Biomed. Mater. Res.* 12: 38.

can result in a weaker material (Figure 2.5). It is also possible to introduce various elements into the fluidized gas and co-deposit them with carbon. Usually silicon (10 to 20 w/o) is co-deposited (or alloyed) to increase hardness for applications requiring resistance to abrasion, such as heart valve discs.

Recently, success was achieved in depositing pyrolitic carbon onto the surfaces of blood vessel implants made of polymers. This type of carbon is called ultra low temperature isotropic (ULTI) carbon instead of low temperature isotropic (**LTI**) **carbon**. The deposited carbon has excellent compatibility with blood and is thin enough not to interfere with the flexibility of the grafts [Park and Lakes, 1992].

The vitreous or glassy carbon is made by controlled pyrolysis of polymers such as phenolformaldehyde, Rayon (cellulose), and polyacrylnitrite at a high temperature in a controlled environment. This process is particularly useful for making carbon fibers and textiles which can be used alone or as components of composites.

2.3   Biodegradable or Resorbable Ceramics

Although Plaster of Paris was used in 1892 as a bone substitute [Peltier, 1961], the concept of using synthetic resorbable ceramics as bone substitutes was introduced in 1969 [Hentrich et al., 1969; Graves et al., 1972]. *Resorbable ceramics*, as the name implies, degrade upon implantation in the host. The resorbed material is replaced by endogenous tissues. The rate of degradation varies from material to material. Almost all bioresorbable ceramics except Biocoral and Plaster of Paris (calcium sulfate dihydrate) are variations of calcium phosphate (Table 2.8). Examples of resorbable ceramics are aluminum calcium phosphate, coralline, Plaster of Paris, hydroxyapatite, and tricalcium phosphate (Table 2.8).

2.3.1   Calcium Phosphate

Calcium phosphate has been used in the form of artificial bone. This material has been synthesized and used for manufacturing various forms of implants, as well as for solid or porous coatings on other implants (Table 2.9).

Calcium phosphate can be crystallized into salts such as hydroxyapatite and β-whitlockite depending on the Ca : P ratio, presence of water, impurities, and temperature. In a wet environment and at lower temperatures ($<900°C$), it is more likely that hydroxyl- or hydroxyapatite will form, while in a dry atmosphere and at a higher temperature, β-whitlockite will be formed [Park and Lakes 1992]. Both forms are very tissue compatible and are used as bone substitutes in a granular form or a solid block. The apatite form of calcium phosphate is considered to be closely related to the mineral phase of bone and teeth.

The mineral part of bone and teeth is made of a crystalline form of calcium phosphate similar to hydroxyapatite $[Ca_{10}(PO_4)_6(OH)_2]$. The apatite family of mineral $[A_{10}(BO_4)_6X_2]$ crystallizes into

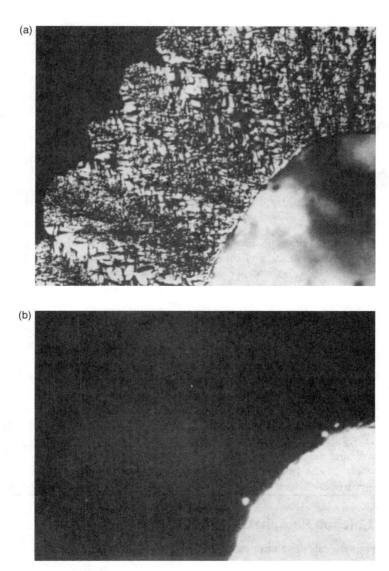

FIGURE 2.5 Microstructure of carbons deposited in a fluidized bed. (a) A granular carbon with distinct growth features. (b) An isotropic carbon without growth features. Both under polarized light. 240×. (From Bokros, J.C., LaGrange, L.D., and Schoen, G.J. 1972. *Chem. Phys. Carbon.* 9: 103–171. New York, Marcel Dekker. With permission.)

hexagonal rhombic prisms and has unit cell dimensions $a = 9.432$Å and $c = 6.881$ Å. The atomic structure of hydroxyapatite projected down the c-axis onto the basal plane is shown in Figure 2.6. Note that the hydroxyl ions lie on the corners of the projected basal plane and they occur at equidistant intervals (3.44 Å) along the columns perpendicular to the basal plane and parallel to the c-axis. Six of the ten calcium ions in the unit cell are associated with the hydroxyls in these columns, resulting in strong interactions among them [Park and Lakes, 1992].

The ideal Ca : P ratio of hydroxyapatite is 10 : 6 and the calculated density is 3.219 g/cm^3. Substitution of OH with fluoride gives the apatite greater chemical stability due to the closer coordination of fluoride (symmetric shape) as compared to the hydroxyl (asymmetric, two atoms) by the nearest calcium. This is why fluoridation of drinking water helps in resisting caries of the teeth [Park and Lakes, 1992].

TABLE 2.8   Examples of Biodegradable Bioceramics

Biodegradable or Resorbable Bioceramics	References
1. Aluminum–Calcium–Phosphorous Oxides	Bajpai et al., 1985
	Mattie and Bajpai, 1988
	Wyatt et al., 1976
2. Glass Fibers and their composites	Alexander et al., 1987
	Zimmermann et al., 1991
3. Corals	Bajpai, 1983
	Guillemin et al., 1989
	Khavari and Bajpai, 1993
	Sartoris et al., 1986
	Wolford et al., 1987
4. Calcium Sulfates, including Plaster of Paris	Bajpai, 1983
	Peltier, 1961
	Scheidler and Bajpai, 1992
5. Ferric Calcium Phosphorous Oxides	Fuski et al., 1993
	Larrabee et al., 1993
	Stricker et al., 1992
6. Hydroxyapatites	Bajpai and Fuchs, 1985
	Bajpai, 1983
	Jenei et al., 1986
	Ricci et al., 1986
7. Tricalcium Phosphate	Bajpai, 1983
	Bajpai et al., 1988
	Lemons et al., 1988
	Morris and Bajpai, 1989
8. Zinc–Calcium–Phosphorous Oxides	Arar et al., 1989
	Bajpai, US. Patent No. 4778471
	Binzer and Bajpai, 1987
	Gromofsky et al., 1988
9. Zinc-Sulfate–Calcium–Phosphorous Oxides	Scheidler and Bajpai, 1992

TABLE 2.9   Uses of Biodegradable Bioceramics

Biodegradable or resorbable ceramics	References
1. As drug delivery devices	Abrams and Bajpai, 1994
	Bajpai, 1992
	Bajpai, 1994
	Benghuzzi et al., 1991
	Moldovan and Bajpai, 1994
	Nagy and Bajpai, 1994
2. For repairing damaged bone due to disease or trauma	Bajpai, 1990
	Gromofsky et al., 1988
	Khavari and Bajpai, 1993
	Morris and Bajpai, 1987
	Scheidler and Bajpai, 1992
3. For filling space vacated by bone screws, donor bone, excised tumors, and diseased bone loss	Bajpai and Fuchs, 1985
	Ricci et al., 1986
4. For repairing and fusion of spinal and lumbo-sacral vertebrae	Bajpai et al., 1984
	Yamamuro et al., 1988
5. For repairing herniated discs	Bajpai et al., 1984
6. For repairing maxillofacial and dental defects	Freeman et al., 1981
7. Hydroxyapatite Ocular Implants	De Potter et al., 1994
	Shields et al., 1993

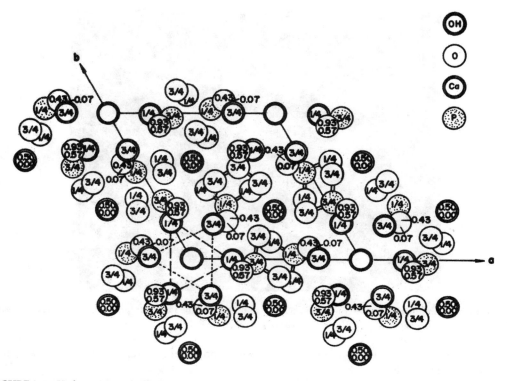

FIGURE 2.6 Hydroxyapatite structure projected down the *c*-axis onto the basal plane. (From Posner A.S., Perloff A., and Diorio A.D. 1958. *Acta. Cryst.* 11: 308–309.)

TABLE 2.10 Physical Properties of Calcium Phosphate

Properties	Values
Elastic modulus (GPa)	4.0–117
Compressive strength (MPa)	294
Bending strength (MPa)	147
Hardness (Vickers, GPa)	3.43
Poisson's ratio	0.27
Density (theoretical, g/cm^3)	3.16

Source: Park J.B. and Lakes R.S. 1992. *Biomaterials: An Introduction*, 2nd ed., p. 125. Plenum Press, New York.

The mechanical properties of synthetic calcium phosphates vary considerably (Table 2.10). The wide variations in properties of polycrystalline calcium phosphates are due to the variations in the structure and manufacturing processes. Depending on the final firing conditions, the calcium phosphate can be calcium hydroxyapatite or β-whitlockite. In many instances, both types of structures exist in the same final product [Park and Lakes, 1992].

Polycrystalline hydroxyapatite has a high elastic modulus (40 to 117 GPa). Hard tissue such as bone, dentin, and dental enamel are natural composites which contain hydroxyapatite (or a similar mineral), as well as protein, other organic materials, and water. Enamel is the stiffest hard tissue, with an elastic modulus of 74 GPa, and contains the most mineral. Dentin (E = 21 GPa) and compact bone (E = 12 to

FIGURE 2.7   Scanning electron micrograph ($\times 500$) of a set and hardened hydroxyaptite (HA)-cysteine composite. The small white cysteine particles can be seen on the larger HA particles.

18 GPa) contain comparatively less mineral. The Poisson's ratio for the mineral or synthetic hydroxyapatite is about 0.27 which is close to that of bone (≈ 0.3) [Park and Lakes, 1992].

Hontsu et al. [1997] were able to deposit an amorphous HA film on Ti, α-Al$_2$O$_3$, SiO//Si(100), and SrTiO$_3$ using a pulsed ArF excimer laser. Upon heat treatment the amorphous film was converted to the crystalline form of HA. The HA film's electrical properties were measured for the first time (Table 2.14).

Among the most important properties of hydroxyapatite as a biomaterial is its excellent biocompatibility. Hydroxyapatite appears to form a direct chemical bond with hard tissues [Piattelli and Trisi, 1994]. On implantation of hydroxyapatite particles or porous blocks in bone, new lamellar cancellous bone forms within 4 to 8 weeks [Bajpai and Fuchs, 1985]. Scanning electron micrograph ($500\times$) of a set and hardened hydroxyapatite-cysteine composite is shown in Figure 2.7. The composite sets and hardens on addition of water.

Many different methods have been developed to make precipitates of hydroxyapatite from an aqueous solution of Ca(NO$_3$)$_2$ and NaH$_2$PO$_4$. There has been successful use of modifications to Jarcho and colleagues wet precipitation procedure for synthesizing hydroxyapatites for use as bone implants [Jarcho et al., 1979; Bajpai and Fuchs, 1985], and drug delivery devices [Bajpai, 1992, 1994; Parker and Bajpai, 1993; Abrams and Bajpai, 1994]. The dried, filtered precipitate is placed in a high-temperature furnace and calcined at 1150°C for 1 h. The calcined powder is then ground in a ball mill, and the particles are separated by an automatic sieve shaker and sieves. The sized particles are then pressed in a die and sintered at 1200°C for 36 h for making drug delivery devices [Bajpai, 1989, 1992; Abrams and Bajpai, 1994]. Above 1250°C, hydroxyapatite shows a second precipitation phase along the grain boundaries [Park and Lakes 1992].

2.3.2   Aluminum–Calcium–Phosphate (ALCAP) Ceramics

Initially we fabricated a calcium aluminate ceramic containing phosphorous pentoxide [Hentrich et al., 1969, 1971; Graves et al., 1972]. Aluminum–calcium–phosphorous oxide ceramic (ALCAP) was developed later [Bajpai and Graves, 1980]. ALCAP has insulating dielectric properties but no magnetic or piezoelectric properties [Allaire et al., 1989]. ALCAP ceramics are unique because they provide a multipurpose crystallographic system where one phase of the ceramic on implantation can be more rapidly resorbed than the others [Wyatt et al., 1976; Bajpai, 1983; Mattie and Bajpai, 1988]. ALCAP is prepared from stock powders of aluminum oxide, calcium oxide, and phosphorous pentoxide. A ratio of $50 : 34 : 16$ by weight of $AlO_2 : CaO : P_2O_5$ is used to obtain the starting mixture for calcining at $1350°C$ in a high temperature furnace for 12 h. The calcined material is ground in a ball mill and sieved by an automatic siever to obtain particles of the desired size. The particulate powder is then pressed into solid blocks or hollow cylinders (green shape) and sintered at $1400°C$ for 36 h to increase the mechanical strength. ALCAP ceramic implants have given excellent results in terms of biocompatibility and gradual replacement of the ceramic material with endogenous bone [Bajpai, 1982; Mattie and Bajpai, 1988]. A scanning electron micrograph $(1000\times)$ of sintered porous ALCAP is shown in Figure 2.8.

2.3.3 Coralline

Coral is a natural substance made by marine invertebrates. According to Holmes et al. [1984], the marine invertebrates live in the limestone exostructure, or coral. The porous structure of the coral is unique for each species of marine invertebrate [Holmes et al., 1984]. Corals for use as bone implants are selected on the basis of structural similarity to bone [Holmes et al., 1984]. Coral provides an excellent structure for the ingrowth of bone, and the main component, calcium carbonate, is gradually resorbed by the body [Khavari and Bajpai, 1993]. Corals can also be converted to hydroxyapatite by a hydrothermal exchange process. Interpore 200, a coral hydroxyapatite, resembles cancellous bone [Sartoris et al., 1986]. Both pure coral (Biocoral) and coral transformed to hydroxyapatite are currently used to repair traumatized bone, replace diseased bone, and correct various bone defects.

Biocoral is composed of crystalline calcium carbonate or aragonite, the metastable form of calcium carbonate. The compressive strength of Biocoral varies from 26 (50% porous) to 395 MPa (dense) and depends on the porosity of the ceramic. Likewise, the modulus of elasticity (Young's Modulus) of Biocoral varies from 8 (50% porous) to 100 GPa (dense) [Biocoral, 1989].

2.3.4 Tricalcium Phosphate (TCP) Ceramics

A multicrystalline porous form of β-tricalcium phosphate $[\beta\text{-}Ca_3(PO_4)_2]$ has been used successfully to correct periodontal defects and augment bony contours [Metsger et al., 1982]. X-ray diffraction

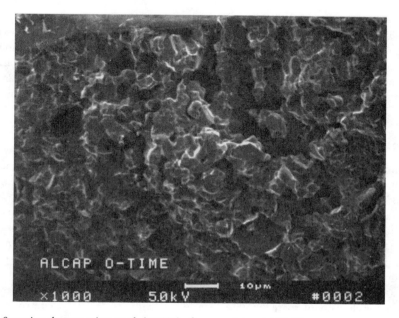

FIGURE 2.8 Scanning electron micrograph ($\times 1000$) of porous sintered ALCAP.

FIGURE 2.9   Scanning electron micrograph (×500) of a set and hardened TCP–cysteine composite. The small white cysteine particles can be seen on the larger TCP particles.

of β-tricalcium phosphate shows an average interconnected porosity of over 100 μm [Lemons et al., 1979]. Often tribasic calcium phosphate is mistaken for β-tricalcium phosphate. According to Metsger et al. [1982], tribasic calcium phosphate is a nonstoichiometric compound often bearing the formula of hydroxyapatite [$Ca_10(PO_4)_6(OH)_2$].

β-Tricalcium phosphate is prepared by a wet precipitation procedure from an aqueous solution of $Ca(NO_3)_2$ and NaH_2PO_4 [Bajpai et al., 1988]. The precipitate is calcined at 1150°C for 1 h, ground, and sieved to obtain the desired size particles for use as bone substitutes [Bajpai et al., 1988; Bajpai, 1990] and for making ceramic matrix drug delivery systems [Morris and Bajpai, 1989; Nagy and Bajpai, 1994; Moldovan and Bajpai, 1994]. These particles are used as such or pressed into cylindrical shapes and sintered at 1150 to 1200°C for 36 h to achieve the appropriate mechanical strength for use as drug delivery devices [Bajpai, 1989, 1992, 1994; Benghuzzi et al., 1991]. A scanning electron micrograph (500×) of a set and hardened TCP-cysteine composite is shown in Figure 2.9. The composite sets and hardens on the addition of water. TCP is usually more soluble than synthetic hydroxyapatite and, on implantation, allows good bone ingrowth and eventually is replaced by endogenous bone.

2.3.5   Zinc–Calcium–Phosphorous Oxide (ZCAP) Ceramics

Zinc is essential for human metabolism and is a component of at least 30 metalloenzymes [Pories and Strain, 1970]. In addition, zinc may also be involved in the process of wound healing [Pories and Strain 1970]. Thus zinc–calcium–phosphorous oxide polyphasic ceramics (ZCAP) were synthesized to repair bone defects and deliver drugs [Binzer and Bajpai, 1987; Bajpai, 1988, 1993; Arar and Bajpai, 1992]. ZCAP is prepared by a thermal mixing of zinc oxide, calcium oxide, and phosphorous pentoxide powders [Bajpai, 1988]. ZCAP, like ALCAP, has insulating dielectric properties but no magnetic or piezoelectric properties [Allaire et al., 1989]. Various ratios of these powders have been used to produce the desired material [Bajpai, 1988]. The oxide powders are mixed in a ball mill and subsequently calcined at 800°C for 24 h. The calcined ceramic is then ground and sieved to obtain the desired size particles. Scanning electron micrograph (500×) of a set and hardened ZCAP–cysteine composite is shown in Figure 2.10. The composite sets and hardens on addition of water. To date, ZCAP ceramics have been

FIGURE 2.10 Scanning electron micrograph ($\times 500$) of a set and hardened ZCAP–cysteine composite. The small white cysteine particles have blended with the ZCAP particles.

used to repair experimentally induced defects in bone and for delivering drugs [Binzer and Bajpai, 1987; Bajpai, 1993].

2.3.6 Zinc–Sulfate–Calcium–Phosphate (ZSCAP) Ceramics

Zinc–sulfate–calcium–phosphate polyphasic ceramics (ZSCAP) are prepared from stock powders of zinc sulfate, zinc oxide, calcium oxide, and phosphorous pentoxide [Bajpai, 1988]. A ratio of $15:30:30:25$ by weight of $ZnSO_4 : ZnO : CaO : P_2O_5$ is mixed in a crucible and allowed to cool for 30 min after the exothermal reaction has subsided. The cooled mixture is calcined in a crucible at $650°C$ for 24 h. The calcined ceramic is ground in a ball mill and the particles of the desired size are separated by sieving in an automatic siever. Scanning electron micrograph ($2000\times$) of set and hardened ZSCAP particles (45 to 63 μm) is shown in Figure 2.11. ZSCAP sets and hardens on addition of water. ZSCAP particles, on implantation in bone, set and harden on contact with blood and have been used to repair experimentally induced defects in bone [Scheidler and Bajpai, 1992].

2.3.7 Ferric–Calcium–Phosphorous–Oxide (FECAP) Ceramics

Ferric–Calcium–Phosphorous–Oxide polyphasic ceramic (FECAP) is prepared from powders of ferric (III) oxide, calcium oxide, and phosphorous pentoxide [Stricker et al.,1992; Fuski et al., 1993; Larrabee et al., 1993]. The powders are combined in various ratios by weight and mixed in a blender. Blocks of the mixture are then pressed in a die by means of a hydraulic press and calcined at $1100°C$ for 12 h. The calcined ceramic blocks are crushed and ground in a ball mill. The calcined ceramic is ground in a ball mill and the particles of the desired size are separated by sieving in an automatic siever. A scanning electron micrograph ($1000\times$) of a set and hardened FECAP-α ketoglutaric acid composite is shown in Figure 2.12. The composite sets and hardens on the addition of water. Studies conducted to date suggest complete resorption of FECAP particles implanted in bone within 60 days [Larrabee et al., 1993]. This particular ceramic could be used in patients suffering from anemia and similar diseases [Fuski et al., 1993].

FIGURE 2.11   Scanning electron micrograph ($\times$2000) of a set and hardened ZSCAP particles (45–63 μm). Sulfate is hardly visible between the cube-shaped ZCAP particles.

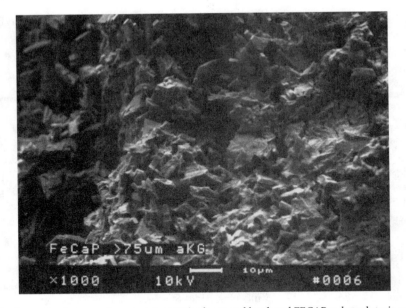

FIGURE 2.12   Scanning electron micrograph ($\times$1000) of a set and hardened FECAP-α-ketoglutaric acid composite. Plate-shaped FECAP particles have been aggregated by the acid.

2.4   Bioactive or Surface-Reactive Ceramics

Upon implantation in the host, surface reactive ceramics form strong bonds with adjacent tissue. Examples of surface reactive ceramics are dense nonporous glasses, Bioglass and Ceravital, and hydroxyapatites (Table 2.11). One of their many uses is the coating of metal prostheses. This coating provides a stronger

TABLE 2.11   Examples of Surface Reactive Bioceramics

Surface reactive bioceramics	References
1. Bioglasses and Ceravital™	Ducheyne, 1985
	Gheyson et al., 1983
	Hench, 1991
	Hench, 1993
	Ogino et al., 1980
	Ritter et al., 1979
2. Dense and non-porous glasses	Andersson et al., 1992
	Blencke et al., 1978
	Li et al., 1991
	Ohtsuki et al., 1992
	Ohura et al., 1992
	Schepers et al., 1993
	Takatsuko et al., 1993
3. Hydroxyapatite	Bagambisa et al., 1993
	Bajpai, 1990
	Fredette et al., 1989
	Huaxia et al., 1992
	Knowles and Bonfield, 1993
	Niwa et al., 1980
	Park and Lakes, 1992
	Posner et al., 1958
	Schwartz et al., 1993
	Whitehead et al., 1993

bonding to the adjacent tissues, which is very important for protheses. A list of the uses of surface-reactive ceramics is shown in Table 2.12.

2.4.1 Glass Ceramics

Several variations of Bioglass and Ceravital glass ceramics have been used by various workers within the last decade. Glass ceramics used for implantation are silicon oxide based systems with or without phosphorous pentoxide.

Glass ceramics are polycrystalline ceramics made by controlled crystallization of glasses developed by S.D. Stookey of Corning Glass Works in the early 1960s [Park and Lakes, 1992]. Glass ceramics were first utilized in photosensitive glasses, in which small amounts of copper, silver, and gold are precipitated by ultraviolet light irradiation. These metallic precipitates help to nucleate and crystallize the glass into a fine grained ceramic which possesses excellent mechanical and thermal properties. Both Bioglass and Ceravital glass ceramics have been used as implants [Yamamuro et al., 1990].

The formation of glass ceramics is influenced by the nucleation and growth of small (<1 μm diameter) crystals as well as the size distribution of these crystals. It is estimated that about 10^{12} to 10^{15} nuclei/cm^3 are required to achieve such small crystals. In addition to the metallic agents already mentioned, Pt groups, TiO_2, ZrO_2, and P_2O_5 are widely used for nucleation and crystallization. The nucleation of glass is carried out at temperatures much lower than the melting temperature. During processing the melt viscosity is kept in the range of 10^{11} and 10^{12} Poise for 1 to 2 h. In order to obtain a larger fraction of the microcrystalline phase, the material is further heated to an appropriate temperature for maximum crystal growth. Deformation of the product, phase transformation within the crystalline phases, or redissolution of some of the phases should be avoided. The crystallization is usually more than 90% complete with grain sizes 0.1 to 1 μm. These grains are much smaller than those of conventional ceramics. Figure 2.13 shows a schematic representation of temperature–time cycle for a glass ceramic [Park and Lakes, 1992].

TABLE 2.12   Uses of Surface Reactive Bioceramics

Surface Reactive Bioceramics	References
1. For coating of metal prostheses	Cotell et al., 1992
	Huaxia et al., 1992
	Ritter et al., 1979
	Takatsuko et al., 1993
	Whitehead et al., 1993
2. In reconstruction of dental defects	Hulbert et al., 1987
	Gheysen et al., 1983
	Schepers et al., 1988
	Schepers et al., 1989
3. For filling space vacated by bone screws, donor bone, excised tumors, and diseased bone loss	Hulbert et al., 1987
	Schepers et al., 1993
	Terry et al., 1989
4. As bone plates and screws	Doyle, 1990
	Ducheyne and McGuckin, 1990
	Yamamuro et al., 1988
5. As replacements of middle ear ossicles	Feenstra and de Groot, 1983
	Grote, 1987
	Hench, 1991
	Hench, 1993
	Reck et al., 1988
6. For lengthening of rami	Feenstra and de Groot, 1983
7. For correcting periodontal defects	Feenstra and de Groot, 1983
8. In replacing subperiosteal teeth	Hulbert, 1992

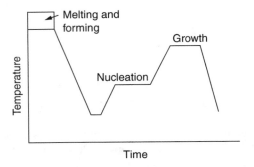

FIGURE 2.13   Temperature–time cycle for a glass-ceramic. (From Kingery W.D., Bowen H.K., and Uhlmann D.R. 1976. *Introduction to Ceramics*, 2nd ed., p. 368. New York, John Wiley & Sons, With permission.)

The glass ceramics developed for implantation are SiO_2–CaO–Na_2O–P_2O_5 and Li_2O–ZnO–SiO_2 systems. Two major groups are experimenting with the SiO_2–CaO–Na_2O–P_2O_5 glass ceramic. One group varied the compositions (except for P_2O_5) in order to obtain the best glass ceramic composition for inducing direct bonding with bone (Table 2.13). The bonding to bone is related to the simultaneous formation of a calcium phosphate and SiO_2-rich film layer on the surface, as exhibited by the 46S5.2 type Bioglass. If a SiO_2-rich layer forms first and a calcium phosphate film develops later (46 to 55 mol % SiO_2 samples) or no phosphate film is formed (60 mol % SiO_2) then direct bonding with bone does not occur [Park and Lakes, 1992]. The approximate region of the SiO_2–CaO–Na_2O system for the tissue–glass–ceramic reaction is shown in Figure 2.14. As can be seen, the best region (region A) for good tissue bonding is the composition given for 46S5.2 type Bioglass (see Table 2.13) [Park and Lakes, 1992].

TABLE 2.13   Compositions of Bioglass and Ceravital Glass Ceramics

Type	Code	SiO$_2$	CaO	Na$_2$O	P$_2$O$_5$	MgO	K$_2$O
Bioglass	42S5.6	42.1	29.0	26.3	2.6	—	—
	(45S5)46S5.2	46.1	26.9	24.4	2.6	—	—
	49S4.9	49.1	25.3	23.8	2.6	—	—
	52S4.6	52.1	23.8	21.5	2.6	—	—
	55S4.3	55.1	22.2	20.1	2.6	—	—
	60S3.8	60.1	19.6	17.7	2.6	—	—
Ceravital	Bioactive[a]	40–50	30–35	5–10	10–15	2.5–5	0.5–3
	Nonbioactive[b]	30–35	25–30	3.5–7.5	7.5–12	1–2.5	0.5–2

[a] The Ceravital® composition is in weight % while the Bioglass® compositions are in mol %.
[b] In addition Al$_2$O$_3$ (5.0–15.0), TiO$_2$(1.0–5.0) and Ta$_2$O$_5$ (5–15) are added.
Source: Park J.B. and Lakes R.S. 1992. *Biomaterials: An Introduction*, 2nd ed., p. 127. Plenum Press, New York. With permission.

TABLE 2.14   Electrical Properties of an HA Film

Dielectric constant (ε_r)	5.7 (25°C 1 MHz)
Loss tangent (tan δ)	<2%
Breakdown electric field	10^4 V cm^{-1}

Source: Hontsu S. et al. 1997. Electrical properties of hydroxyapatite thin films grown by pulsed laser deposition. *Thin Solid Films* 295: 214–217.

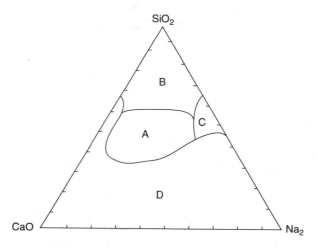

FIGURE 2.14   Approximate regions of the tissue-glass-ceramic bonding for the SiO$_2$–CaO–Na$_2$O system. A: Bonding within 30 days. B: Nonbonding; reactivity is too low. D: Bonding does not form glass. (From Hench L.L. and Ethridge E.C. 1982. *Biomaterials: An Interfacial Approach.* p. 147, New York, Academic Press. With permission.)

2.4.2 Ceravital

The composition of Ceravital is similar to that of Bioglass in SiO$_2$ content but differs somewhat in other components (see Table 2.13). In order to control the dissolution rate, Al$_2$O$_3$, TiO$_2$, and Ta$_2$O$_5$ are added in Ceravital glass ceramic. The mixtures, after melting in a platinum crucible at 1500°C for three h, are annealed and cooled. The nucleation and crystallization temperatures are 680 and 750°C, respectively,

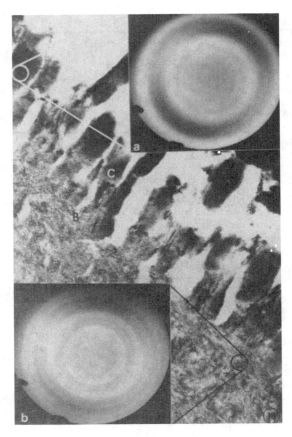

FIGURE 2.15 Transmission electron micrograph of well-mineralized bone (b) juxtaposed to the glass-ceramic (c) which fractured during sectioning. ×51,500. Insert a is the diffraction pattern from ceramic area and b is from bone area. (From Beckham C.A., Greenlee T.K. Jr, and Crebo A.R. 1971. *Calc. Tiss. Res.* 8: 165–171. With permission.)

each for 24 h. When the size of crystallites reaches approximately 4 Å and the characteristic needle structure is not formed, the process is stopped to obtain a fine grain structured glass ceramic [Park and Lakes, 1993].

Glass ceramics have several desirable properties compared to glasses and ceramics. The thermal coefficient of expansion is very low, typically 10^{-7} to $10^{-5} °C^{-1}$, and in some cases it can even be made negative. Due to the controlled grain size and improved resistance to surface damage, the tensile strength of these materials can be increased by at least a factor of two, from about 100 to 200 MPa. The resistance to scratching and abrasion of glass ceramics is similar to that of sapphire [Park and Lakes, 1992].

A transmission electron micrograph of Bioglass glass ceramic implanted in the femur of rats for six weeks showed intimate contacts between the mineralized bone and the Bioglass (Figure 2.15). The mechanical strength of the interfacial bond between bone and Bioglass ceramic is on the same order of magnitude as the strength of the bulk glass ceramic (850 kg/cm² or 83.3 MPa), which is about three-fourths that of the host bone strength [Park and Lakes, 1992].

A negative characteristic of the glass ceramic is its brittleness. In addition, limitations on the compositions used for producing a biocompatibile (or osteoconductive) glass ceramic hinders the production of a glass ceramic which has substantially higher mechanical strength. Thus glass ceramics cannot be used for making major load-bearing implants such as joint implants. However, they can be used as fillers for bone cement, dental restorative composites, and coating material (see Table 2.12). A glass ceramic containing 36 wt% of magnetite in a β-wollastonite- and $CaOSiO_2$-based glassy matrix has been synthesized for treating bone tumors by hyperthermia [Kokubo et al., 1992].

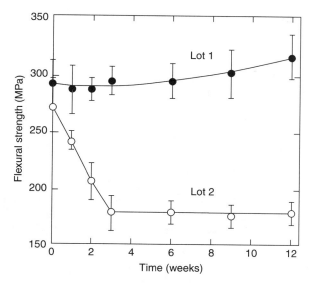

FIGURE 2.16   Flexural strength of dense alumina rods after aging under stress in Ringer's solution. Lot 1 and 2 are from different batches of production. (From Krainess F.E. and Knapp W.J. 1978. *J. Biomed. Mater. Res.* 12: 245. With permission.)

2.5   Deterioration of Ceramics

It is of great interest to know whether the inert ceramics such as alumina undergo significant static or dynamic fatigue. Even for the biodegradable ceramics, the rate of degradation *in vivo* is of paramount importance. Controlled degradation of an implant with time on implantation is desirable. Above a critical stress level, the fatigue strength of alumina is reduced by the presence of water. This is due to the delayed crack growth, which is accelerated by the water molecules [Park and Lakes, 1992]. Reduction in strength occurs if water penetrates the ceramic. Decrease in strength was not observed in samples which did not show water marks on the fractured surface (Figure 2.16). The presence of a small amount of silica in one sample lot may have contributed to the permeation of water molecules that is detrimental to the strength [Park and Lakes, 1992]. It is not clear whether the static fatigue mechanism operates in single crystal alumina. It is reasonable to assume, that static fatigue will occur if the ceramic contains flaws or impurities, because these will act as the source of crack initiation and growth under stress [Park and Lakes, 1992].

Studies of the fatigue behavior of vapor-deposited pyrolitic carbon fibers (4000 to 5000 Å thick) onto a stainless steel substrate showed that the film does not break unless the substrate undergoes plastic deformation at 1.3×10^{-2} strain and up to one million cycles of loading. Therefore, the fatigue is closely related to the substrate, as shown in Figure 2.17. Similar substrate-carbon adherence is the basis for the pyrolitic carbon deposited polymer arterial grafts [Park and Lakes, 1992].

The fatigue life of ceramics can be predicted by assuming that the fatigue fracture is due to the slow growth of preexisting flaws. Generally, the strength distribution, σ_i, of ceramics in an inert environment can be correlated with the probability of failure F by the following equation:

$$\mathrm{LnLn}\left(\frac{1}{1-F}\right) = m\mathrm{Ln}\left(\frac{s_i}{s_o}\right) \qquad (2.1)$$

Both m and s_o are constants in the equation. Figure 2.18 shows a good fit for Bioglass coated alumina [Park and Lakes, 1992].

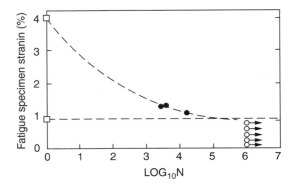

FIGURE 2.17 Strain vs. number of cycles to failure (o = absence of fatigue cracks in carbon film; • = fracture of carbon film due to fatigue failure of substrates; □ = data from substrate determined in single-cycle tensile test). (From Shimm H.S. and Haubold A.D. 1980. *Biomater. Med. Dev. Art. Org.* 8: 333–344. With permission.)

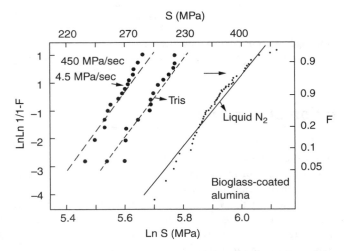

FIGURE 2.18 Plot of Ln Ln $[1/(1 - F)]$ vs. Ln S for Bioglass-coated alumina in a tris-hydroxyaminomethane buffer and liquid nitrogen. F is the probability of failure and S is strength. (From Ritter J.E. Jr, Greenspan D.C., Palmer R.A., and Hench L.L. 1979. *J. Biomed. Mater. Res.* 13: 260. With permission.)

A minimum service life (t_{min}) of a specimen can be predicted by means of a proof test wherein it is subjected to stresses that are greater than those expected in service. Proof tests also eliminate the weaker pieces. This minimum life can be predicted from the following equation:

$$t_{min} = B\sigma_p^{N-2}\sigma_a^{-N} \tag{2.2}$$

Here σ_p is the proof test stress, σ_a is the applied stress, and B and N are constants.

Equation 2.2 after rearrangement, reads as follows:

$$t_{min}\sigma_a^2 = B(\sigma_p/\sigma_a)^{N-2} \tag{2.3}$$

Figure 2.19 shows a plot of Equation 2.3 for alumina on a logarithmic scale [Park and Lakes, 1992].

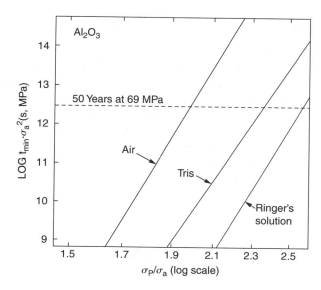

FIGURE 2.19 Plot of Equation 2.3 for alumina after proof testing. $N = 43.85$, $m = 13.21$, and $\sigma_o = 55728$ psi. (From Ritter J.E. Jr, Greenspan D.C., Palmer R.A., and Hench L.L. 1979. *J. Biomed. Mater. Res.* 13: 261. With permission.)

2.6 Bioceramic Manufacturing Techniques

In order to fabricate bioceramics in more and more complex shapes, scientists are investigating the use of old and new manufacturing techniques. These techniques range from the adaptation of an age old pottery technique to the latest manufacturing methods for high temperature ceramic parts for airplane engines. No matter where the technique is perfected, the ultimate goal is the fabrication of bioceramic particles or devices in a desired shape in a consistent manner with the desired properties. The technique used to fabricate the bioceramic device will depend greatly on the ultimate application of the device, whether it is for hard tissue replacement or the integration of the device within the surrounding tissue.

2.6.1 Hard Tissue Replacement

Hard tissue replacement implies that the bioceramic device will be used for load bearing applications. Although it is desirable to have a device with a sufficient porosity for the surrounding tissue to infiltrate and attach to the device, the most important and immediate property is the strength of the device. In order to accomplish this, one must manufacture a bioceramic implant with a density and strength sufficient to mimic that of bone. However, if the bioceramic part is significantly stronger than the surrounding bone, one runs into the common problem seen with metals called *stress shielding*. The density of the bioceramic greatly determines its overall strength. As the density increases so does the overall strength of the bioceramic. Some of the techniques used to manufacture dense bioceramics are injection molding, gel casting, bicontinuous microemulsion, inverse microemulsion, emulsion, and additives.

Injection molding is a common technique used to form plastic parts for many commercial applications such as automobile parts. Briefly, the process involves forcing a heated material into a die and then ejecting the formed piece from the die. Injection molding allows for making complex shapes. Cihlar and Trunec [1996] found that by calcining (1273 K, 3 h) and milling the hydroxyapatite (HA) prior to mixing with a binder, an ethylene vinyl acetate copolymer (EVA)/HA mixture of 63% HA, they achieved 98% relative density with only 16% shrinkage using injection molding. The maximum flexural strength was 60 MPa for HA products sintered at 1473 K. However, this is still not strong enough for load bearing applications.

They also observed that HA decomposed to α-TCP at temperatures greater than 1573 K [Cihlar and Trunec, 1996].

In gel casting, HA is formed using the standard chemical precipitation. The calcium phosphate precipitant (30% w/v) is then mixed with glycerol and filtered. The "gel cake" is sintered at 1200°C for 2 h. This yielded a density of >99% and a highly uniform microstructure [Varma and Sivakumar, 1996].

Bicontinuous microemulsion, inverse microemulsion, and emulsion are all wet chemistry based methods to produce nanometer size HA powders. All three methods yield >97% relative density upon sintering at 1200°C for 2 h. The biocontinuous and inverse microemulsion resulted in the two smallest HA particle sizes, 22 and 24 nm respectively [Lim et al., 1997].

Another strategy to increase the density of ceramics is to use additives or impurities in small weight percents during sintering. The main disadvantages to this technique include (1) the possible decomposition of the original pure bioceramic and (2) results may end in all or portions of the bioceramic being nonbiocompatible.

Suchanek et al. [1997] studied the addition of several different additives to HA in 5 wt% amounts. The additives studied were K_2CO_3, Na_2CO_3, H_3BO_3, KF, $CaCl_2$, KCl, KH_2PO_4, $(KPO_3)_n$, $Na_2Si_2O_5$, $Na_2P_2O_7$, Na_3PO_4, $(NaPO_3)_n$, $Na_5P_3O_{10}$, and β-$NaCaPO_4$. HA has a fracture toughness of 1 MPa·$m^{1/2}$, whereas human bone has a fracture toughness of 2 to 12 MPa · $m^{1/2}$. One of the ways to improve the mechanical properties is to improve the densification of HA. Suchanek et al. [1997] found that the following additives (5 wt.%) did not improve the densification of HA: H_3BO_3, $CaCl_2$, KCl, KH_2PO_4, $(KPO_3)_n$, and $Na_2Si_2O_5$. The densification of HA was improved through the addition (5 wt.%) of K_2CO_3, Na_2CO_3, KF, $Na_2P_2O_7$, Na_3PO_4, $(NaPO_3)_n$, $Na_5P_3O_{10}$, and β-$NaCaPO_4$. However, H_3BO_3, $CaCl_2$, KH_2PO_4, $(KPO_3)_n$, $Na_2Si_2O_5$, K_2CO_3, Na_2CO_3, and KF produced the formation of β-TCP or CaO. The sodium phosphates used in this study were added to HA without the formation of β-TCP or CaO. The only compound that improved densification, did not cause formation of β-TCP or CaO, and provided a weak interface for HA was β-$NaCaPO_4$.

Another additive that has been investigated to improve the performance of HA is lithium (Li). The addition of lithium can increase the microhardness and produces a fine microstructure in HA. Fanovich et al. [1998] found that the addition of 0.2 wt% of Li to HA produced the maximum microhardness (5.9 GPa). However, the addition of high amounts of Li to HA results in abnormal grain growth and large pores. Furthermore, Li addition to HA results in the formation of β-TCP upon sintering.

Zirconia has been used as an additive to HA in order to improve its mechanical strength. Kawashima et al. [1997] found that addition of partially stabilized zirconia (PSZ) to HA can be used to increase the fracture toughness to 2.8 MPa · $m^{1/2}$. Bone has a fracture toughness of 2-12 MPa · $m^{1/2}$ [Suchanek et al., 1997]. PSZ was added to HA in different percentages (17, 33, 50 wt%) and it was found that 50 wt% PSZ had the highest fracture toughness. The surface energy of the PSZ-HA was not significantly different from HA alone. This suggests that the PSZ-HA composite could be biocompatible because of the similarity of the surface with HA [Kawashima et al., 1997].

2.6.2 Tissue Integration

The porosity is a critical factor for growth and integration of a tissue into the bioceramic implant. In particular the open porosity, that which is connected to the outside surface, is critical to the integration of tissue into the ceramic especially if the bioceramic is inert. Several methods have been developed to form porous ceramics, two of these are starch consolidation and drip casting.

In starch consolidation (Table 2.15), starch powders of a specific size are mixed with a bioceramic slurry at a predetermined weight percent. Upon heating, the starch will uptake water from the slurry mixture and swell. Upon sintering of the starch-bioceramic mixture, the starch is burned out and the pores are left in their place. Starch consolidation has been used to form complex shapes in alumina with ultimate porosities between 23 and 70 vol %. By controlling the starch content, one can control the ultimate porosity and resulting pore sizes. Large pores formed using starch consolidation in alumina were in the size range 10 to 80 μm whereas small pores varied between 0.5 and 9.5 μm [Lyckfeldt and Ferreira, 1998].

TABLE 2.15 Bioceramic Manufacturing Techniques for Hard Tissue Replacement or Tissue Integration

Manufacturing Technique	References
Hard Tissue Replacement	
Injection Molding	Cihlar and Trunec, 1996
Bicontinuous microemulsion	Lim et al., 1997
Inverse microemulsion	
Emulsion	
Additives	Fanovich et al., 1998; Kawashima et al., 1997; Suchanek et al., 1997
Tissue Integration	
Drip casting	Liu, 1996; Lyckfeldt and Ferreira, 1998
Starch consolidation	Lyckfeldt and Ferreira, 1998
Polymeric sponge method	
Foaming method	
Organic additives	
Gel casting	
Slip casting	
Direct coagulation consolidation	
Hydrolysis assisted solidification	
Freezing	

Liu [1996] used a drip casting technique (Table 2.15) to form porous HA granules with pore sizes from 95 to 400 μm. The granules had a total porosity from 24 to 76 vol %. The HA was made into a slurry using water and poly(vinyl butyral) powders. The slurry was then dripped onto a spherical mold surface. This technique is similar to that of drip casting by dripping an HA slurry into a liquid nitrogen bath. In both instances, calcining and sintering procedures are used to produce the final product [Liu, 1996].

2.6.3 Hydroxyapatite Synthesis Method

Prepare the following solutions:

Solution 1: Dissolve 157.6 g calcium nitrate tetrahydrate [$Ca(NO_3)_2 4H_2O$] in 500 ml DI water. Bring the solution to a pH of 11 by adding $\approx$70 ml ammonium hydroxide [NH_4OH]. Bring the solution to 800 ml with DI water.

Solution 2: Dissolve 52.8 g ammonium phosphate dibasic [$(NH_4)_2HPO_4$] in 500 ml DI water. Bring the solution to a pH of 11 by adding $\approx$150 ml ammonium hydroxide [NH_4OH]. Add DI water until the precipitate is completely dissolved, 250–350 ml.

2.6.3.1 Special Note

If you use calcium nitrate [$Ca(NO_3)_2 nH_2O$] instead of calcium nitrate tetrahydrate you need to recalculate the amount of calcium nitrate to add to make Solution 1 on the basis of the absence of the extra four waters. If you do not, you will have to add a large amount of ammonium hydroxide to pH the solution:

1. Add one-half of Solution 1 to a 2 l separatory funnel
2. Add one-half of Solution 2 to a 2 l separatory funnel
3. Titrate both solutions into a 4 l beaker with heat and constant stirring
4. Boil gently for 30 min
5. Repeat steps 1–4 for the rest of Solutions 1 and 2
6. Let cool completely allowing precipitate to settle to bottom of beaker
7. Pour contents of beaker into 250 ml polypropylene bottles
8. Centrifuge bottles for 10 min at (10,000 rpm) 16,000 g

9. Collect precipitate from the six bottles into two and resuspend with DI water
10. Fill the four empty bottles with the reaction mixture and centrifuge all six as before
11. Collect precipitant from the two bottles that were resuspended with DI water
12. Combine remaining four bottles into two bottles and resuspend with DI water
13. Repeat steps 10–12 as necessary
14. Dry precipitate for 24–48 h at 70°C
15. Calcine the precipitate for one hour at 1140°C
16. Grind and sieve product as desired

Defining Terms

Alumina: Aluminum oxide (Al_2O_3) which is very hard (Mohs hardness is 9) and strong. Single crystals are called sapphire or ruby depending on color. Alumina is used to fabricate hip joint socket components or dental root implants.

Calcium phosphate: A family of calcium phosphate ceramics including aluminum calcium phosphate, ferric calcium phosphate, hydroxyapatite and tricalcium phosphate (TCP), and zinc calcium phosphate which are used to substitute or augment bony structures and deliver drugs.

Glass-ceramics: A glass crystallized by heat treatment. Some of those have the ability to form chemical bonds with hard and soft tissues. Bioglass and Ceravital are well known examples.

Hydroxyapatite: A calcium phosphate ceramic with a calcium to phosphorus ratio of 5/3 and nominal composition $Ca_{10}(PO_4)_6(OH)_2$. It has good mechanical properties and excellent biocompatibility. Hydroxyapatite is the mineral constituent of bone.

LTI carbon: A silicon alloyed pyrolitic carbon deposited onto a substrate at low temperature with isotropic crystal morphology. It is highly compatible with blood and used for cardiovascular implant fabrication such as artificial heart valve.

Maximum radius ratio: The ratio of atomic radii computed by assuming the largest atom or ion which can be placed in a crystal's unit cell structure without deforming the structure.

Mohs scale: A hardness scale in which 10 (diamond) is the hardest and 1 (talc) is the softest.

Acknowledgments

The author is grateful to Dr. Joon B. Park for inviting him to write this chapter and providing the basic shell from his book to expand upon. The author would like to thank his wife, Zoe, for her patience and understanding. The author would also like to thank Dr. M.C. Hofmann for her support and help.

In Memory Of

In memory of Dr. P.K. Bajpai who left us early in 1998, I would like to share our lab's recipe for hydroxyapatite. Dr. Bajpai's lab and research at the University of Dayton have ended after 30 plus years, but I felt it would be fitting to share this recipe as a way to encourage other scientists to continue exploring the possibilities of bioceramics.

References

Abrams L. and Bajpai P.K. 1994. Hydroxyapatite ceramics for continuous delivery of heparin. *Biomed. Sci. Instrum.* 30: 169–174.

Adams D. and Williams D.F. 1978. Carbon fiber-reinforced carbon as a potential implant material. *J. Biomed. Mater. Res.* 12: 35–42.

Alexander H., Parsons J.R., Ricci J.L., Bajpai P.K., and Weiss A.B. 1987. Calcium-based ceramics and composites in bone reconstruction. *Crit. Rev.* 4: 43–47.

Allaire M., Reynolds D., and Bajpai P.K. 1989. Electrical properties of biocompatible ALCAP and ZCAP ceramics. *Biomed. Sci. Instrum.* 25: 163–168.

Andersson O.H., Guizhi L., Kangasniemi K., and Juhanoja J. 1992. Evaluation of the acceptance of glass in bone. *J. Mat. Sci.: Mater. Med.* 3: 145–150.

Annual Book of ASTM Standards, part 46, F603-78, American Society for Testing and Materials, Philadelphia, 1980.

Arar H.A. and Bajpai P.K. 1992. Insulin delivery by zinc calcium phosphate (ZCAP) ceramics. *Biomed. Sci. Instrum.* 28: 172–178.

Bagambisa F.B., Joos U., and Schilli W. 1993. Mechanisms and structure of the bond between bone and hydroxyapatite ceramics. *J. Biomed. Mater. Res.* 27: 1047–1055.

Bajpai P.K., Fuchs C.M., and Strnat M.A.P. 1985. *Development of Alumino-Calcium Phosphorous Oxide* (ALCAP) *Ceramic Cements.* In: *Biomedical Engineering IV Recent Developments.* Proceedings of the Fourth Southern Biomedical Engineering Conference, Jackson, M.S. and B. Sauer (Ed.), pp. 22–25, Pergamon Press, New York, NY.

Bajpai P.K. 1994. Ceramic drug delivery systems. In: *Biomedical Materials Research in The Far East* (I). Xingdong Zhang and Yoshito Ikada (Eds.), pp. 41–42. Kobunshi Kankokai Inc., Kyoto, Japan.

Bajpai P.K. 1993. Zinc based ceramic cysteine composite for repairing vertebral defects. *J. Instrum. Sci.* 6: 346.

Bajpai P.K. 1992. Ceramics: a novel device for sustained long term delivery of drugs. In: *Bioceramics* Vol. 3, J.A. Hulbert and S.F. Hulbert (Eds.), pp. 87–99. Rose-Hulman Institute of Technology, Terra Haute, IN.

Bajpai P.K. 1990. Ceramic amino acid composites for repairing traumatized hard tissues. In: *Handbook of Bioactive Ceramics,* Vol. II: *Calcium Phosphate and Hydroxylapatite Ceramics.* T. Yamamuro, L.L. Hench, and J. Wilson-Hench (Eds.), pp. 255–270. CRC Press, Baton Raton, FL.

Bajpai P.K. 1989. Ceramic implantable drug delivery system. *T.I.B. & A.O.* 3: 203–211.

Bajpai P.K. 1988. ZCAP Ceramics. US. Patent No. 4778471.

Bajpai P.K. 1983. Biodegradable scaffolds in orthopedic, oral, and maxillo facial surgery. In: *Biomaterials in Reconstructive Surgery.* L.R. Rubin, Ed. pp. 312–328. C.V. Mosby Co., St. Louis, MO.

Bajpai P.K., Fuchs C.M., and McCullum D.E. 1988. *Development of tricalcium phosphate ceramic cements.* In: *Quantitative Characterization and Performance of Porous Implants for Hard Tissue Applications,* ASTM STP 953, J.E. Lemons (Ed.), pp. 377–388. American Society for Testing and Materials, Philadelphia, PA.

Bajpai P.K. and Fuchs C.M. 1985. Development of a hydroxyapatite bone grout. In: *Proceedings of the First Annual Scientific Session of the Academy of Surgical Research.* San Antonio, Texas. C.W. Hall, (Ed.), pp. 50–54. Pergamon Press, New York, NY.

Bajpai P.K., Graves G.A. Jr, Wilcox L.G., and Freeman M.J. 1984. Use of resorbable alumino-calcium-phosphorous-oxide ceramics (ALCAP) in health care. *Trans. Soc. Biomater.* 7: 353.

Bajpai P.K. and Graves G.A. Jr. 1980. Porous Ceramic Carriers for Controlled Release of Proteins, Polypeptide Hormones and other Substances within Human and/or Mammalian Species. US. Patent No. 4218255.

Barinov S.M. and Bashenko YuV. 1992. Application of ceramic composites as implants: result and problem. In: *Bioceramics and the Human Body.* A. Ravaglioli and A. Krajewski (Eds.), pp. 206–210. Elsevier Applied Science, London.

Beckham C.A., Greenlee T.K. Jr, and Crebo A.R. 1971. Bone formation at a ceramic implant interface. *Calc. Tiss. Res.* 8: 165–171.

Benghuzzi H.A., Giffin B.F., Bajpai P.K., and England B.G. 1991. Successful antidote of multiple lethal infections with sustained delivery of difluoromethylornithine by means of tricalcium phosphate drug delivery devices. *Trans. Soc. Biomater.* 24: 53.

Binzer T.J. and Bajpai P.K. 1987. The use of zinc–calcium-phosphorous oxide (ZCAP) ceramics in reconstructive bone surgery. *Digest of Papers, Sixth Southern Biomedical Engineering Conference.* Dallas, Texas. R.C. Eberhart (Ed.), pp. 182–185. McGregor and Werner, Washington, D.C.

Biocoral. 1989. From coral to biocoral, p. 46. *Innoteb*, Paris, France.

Blencke B.A., Bromer H., Deutscher K.K. 1978. Compatibility and long-term stability of glass-ceramic implants. *J. Biomed. Mater. Res.* 12: 307–318.

Bokros J.C. 1972. Deposition structure and properties of pyrolitic carbon. *Chem. Phys. Carbon.* 5: 70–81.

Bokros J.C., LaGrange L.D., and Schoen G.J. 1972. Control of structure of carbon for use in bioengineering. *Chem. Phys. Carbon.* 9: 103–171.

Boutin P. 1981. T.H.R. Using Alumina–Alumina Sliding and a Metallic Stem: 1330 Cases and an 11-Year Follow-up. In: *Orthopaedic Ceramic Implants*, Vol. 1. H. Oonishi and H.Y. Ooi (Eds.), Tokyo, Japanese Society of Orthopaedic Ceramic Implants.

Buykx W.J., Drabarek E., Reeve K.D., Anderson N., Mathivanar R., and Skalsky M. 1992. Development of porous ceramics for drug release and other applications. In: *Bioceramics*, Vol. 3. J.E. Hulbert and S.F. Hulbert (Eds.), pp. 349–354. Rose Hulman Institute of Technology, Terre Haute, Indiana.

Chandy T. and Sharma C.P. 1991. Biocomaptibility and toxicological screening of materials. In: *Blood Compatible Materials and Devices*. C.P. Sharma and M. Szycher (Eds.), pp. 153–166. Technomic Publishing Co., Lancaster, PA.

Chignier E., Monties J.R., Butazzoni B., Dureau G., and Eloy R. 1987. Haemocompatibility and biological course of carbonaceous composites for cardiovascular devices. *Biomaterials* 8: 18–23.

Cihlar J. and Trunec M. 1996. Injection moulded hydroxyapatite ceramics. *Biomaterials* 17: 1905–1911.

Cotell C.M., Chrisey D.B., Grabowski K.S., Sprague J.A., and Gossett C.R. 1992. Pulsed laser deposition of hydroxyapatite thin films on Ti–6Al–4V, *J. Appl. Biomater.* 3: 87–93.

de Groot K. 1983. *Bioceramics of Calcium Phosphate.* CRC Press, Boca Raton, FL.

De Potter P., Shields C.L., Shields J.L., and Singh A.D. 1994. Use of the hydroxyapatite ocular implant in the pediatric population. *Arch. Opthalmol.* 112: 208–212.

Dellsperger K.C. and Chandran K.B. 1991. Prosthetic heart valves. In: *Blood Compatible Materials and Devices*. C.P. Sharma and M. Szycher (Eds.), Technomic Publishing Co., Lancaster, PA.

Dorlot J.M., Christel P., and Meunier A. 1988. Alumina hip prostheses: long term behaviors. In: *Bioceramics. Proceedings of 1st International Symposium on Ceramics in Medicine*. H. Oonishi, H. Aoki, and K. Sawai (Eds.), pp. 236–301. Ishiyaku EuroAmerica, Inc. Tokyo.

Dörre E. 1991. Problems concerning the industrial production of alumina ceramic components for hip prosthesis. In: *Bioceramics and the Human Body*. A. Ravaglioli and A. Krajewski (Eds.), pp. 454–460. Elsevier Applied Science, London and New York.

Doyle C. 1990. Composite bioactive ceramic-metal materials. In: *Handbook of Bioactive Ceramics*. T. Yamamuro, L.L. Hench, and J. Wilson (Eds.), pp. 195–208. CRC Press, Boca Raton, FL.

Drennan J. and Steele B.C.H. 1991. Zirconia and hafnia, In: *Concise Encyclopedia of Advanced Ceramic Materials*. R.J. Brook (Ed.), pp. 525–528. Pergamon Press, Oxford, NY.

Ducheyne P. and McGuckin, J.F. Jr. 1990. Composite Bioactive Ceramic-Metal Materials. In: *Handbook of Bioactive Ceramics*. T. Yamamuro, L.L. Hench, and J. Wilson (Eds.), pp. 75–86. CRC Press, Boca Raton, FL.

Ducheyne P. 1985. Bioglass coatings and bioglass composites as implant materials. *J. Biomed. Mater. Res.* 19: 273–291.

Ely J.L. and Haubald A.O. 1993. Static fatigue and stress corrosion in pyrolitic carbon. In: *Bioceramics*, Vol. 6. P. Ducheyne and D. Christiansen (Eds.), pp. 199–204. Butterworth-Heinemann, Ltd. Boston. MA.

Fanovich M.A., Castro M.S., and Porto Lopez J.M. 1998. Improvement of the microstructure and microhardness of hydroxyapatite ceramics by addition of lithium. *Mat. Lett.* 33: 269–272.

Feenstra L and de Groot K. 1983. Medical use of calcium phosphate ceramics. In: *Bioceramics of calcium phosphate*. K. de Groot (Ed.), pp. 131–141, CRC Press, Boca Raton, FL.

Freeman M.J., McCullum D.E., and Bajpai P.K. 1981. Use of ALCAP ceramics for rebuilding maxillo-facial defects. *Trans. Soc. Biomater.* 4: 109.

Fredette S.A., Hanker J.S., Terry B.C., and Beverly L. 1989. Comparison of dense versus porous hydroxyapatite (HA) particles for rat mandibular defect repair. *Mat. Res. Soc. Symp. Proc.* 110: 233–238.

Fulmer M.T., Martin R.I., and Brown P.W. 1992. Formation of calcium deficient hydroxyapatite at near-physiological temperature. *J. Mat. Sci.: Mater. Med.* 3: 299–305.

Fuski M.P., Larrabee R.A., and Bajpai P.K. 1993. Effect of ferric calcium phosphorous oxide ceramic implant in bone on some parameters of blood. *T.I.B. & A.O.* 7: 16–19.

Gheysen G., Ducheyne P., Hench L.L., and de Meester P. 1983. Bioglass composites: a potential material for dental application. *Biomaterials* 4: 81–84.

Graves G.A. Jr, Hentrich R.L. Jr, Stein H.G., and Bajpai P.K. 1972. Resorbable Ceramic implants in bioceramics. In: *Engineering and Medicine (Part I)*. C.W. Hall, S.F. Hulbert, S.N. Levine, and F.A. Young (Eds.), pp. 91–115. Interscience Publishers, New York, NY.

Grenoble D.E., Katz J.L., Dunn K.L., Gilmore R.S., and Murty K.L. 1972. The elastic properties of hard tissues and apatites. *J. Biomed. Mater. Res.* 6: 221–233.

Gromofsky J.R., Arar H., and Bajpai P.K. 1988. Development of zinc calcium phosphorous oxide ceramic–organic acid composites for repairing traumatized hard tissue. In: *Digest of Papers, Seventh Southern Biomedical Engineering Conference*. Greenville, S.C. and D.D. Moyle (Eds.), pp. 20–23. Mcgregor and Werner, Washington, D.C.

Grote J.J. 1987. Reconstruction of the ossicular chain with hydroxyapatite prostheses. *Am. J. Otol.* 8: 396–401.

Guillemin G., Meunier A., Dallant P., Christel P., Pouliquen J.C., and Sedel L. 1989. Comparison of coral resorption and bone apposition with two natural corals of different porosities. *J. Biomed. Mater. Res.* 23: 765–779.

Hammer J. III, Reed O., and Greulich R. 1972. Ceramic root implantation in baboons. *J. Biomed. Mater. Res.* 6: 1–13.

Heimke G. 1992. Use of alumina ceramics in medicine. In: *Bioceramics*, Vol. 3. J.E. Hulbert and S.F. Hulbert (Eds.), pp. 19–30. Rose Hulman Institute of Technology, Terre Haute, IN.

Hench L.L. 1991. Bioceramics: From concept to clinic. *J. Am. Ceram. Soc.* 74: 1487–1510.

Hench L.L. 1993. Bioceramics: From concept to clinic. *Am. Ceram. Soc. Bull.* 72: 93–98.

Hench L.L. and Ethridge E.C. 1982. *Biomaterials: An Interfacial Approach*. p.147, Academic Press, NY.

Hentrich R.L. Jr, Graves G.A. Jr, Stein H.G., and Bajpai P.K. 1971. An evaluation of inert and resorbable ceramics for future clinical applications. *J. Biomed. Mater. Res.* 5: 25–51.

Hentrich R.L. Jr, Graves G.A. Jr, Stein H.G., and Bajpai P.K. 1969. An evaluation of inert and resorbable ceramics for future clinical applications. Fall Meeting, Ceramics-Metals Systems, Division of the American Ceramic Society, Cleveland, Ohio.

Holmes R., Mooney V., Bucholz R., and Tencer A. 1984. A coralline hydroxyapatite bone graft substitute. *Clin. Orthopaed. Relat. Res.* 188: 252–262.

Hontsu S., Matsumoto T., Ishii J., Nakamori M., Tabata H., and Kawai T. 1997. Electrical properties of hydroxyapatite thin films grown by pulsed laser deposition. *Thin Solid Films* 295: 214–217.

Huaxia J.I., Ponton C.B., and Marquis P.M. 1992. Microstructural characterization of hydroxyapatite coating on titanium. *J. Mat. Sci.: Mater. Med.* 3: 283–287.

Hulbert S.F. 1992. Use of ceramics in medicine. In: *Bioceramics*, Vol. 3. J.E. Hulbert and S.F. Hulbert (Eds.), pp. 1–18. Rose Hulman Institute of Technology, Terre Haute, IN.

Hulbert S.F. and Klawitter J.J. 1971. Application of porous ceramics for the development of load-bearing internal orthopedic applications. *Biomed. Mater. Symp.* pp. 161–229.

Hulbert S.F., Bokros J.C., Hench L.L., Wilson J., and Heimke G. 1987. Ceramics in clinical applications: past, present, and future. In: *High Tech Ceramics*. P. Vincezini (Ed.), pp. 189–213. Elsevier, Amsterdam, Netherlands.

Jarcho M., Salsbury R.L., Thomas M.B., and Doremus R.H. 1979. Synthesis and fabrication of β-tricalcium phosphate (whitlockite) ceramics for potential prosthetic applications. *J. Mater. Sci.* 14: 142–150.

Jenei S.R., Bajpai P.K., and Salsbury R.L. Resorbability of commercial hydroxyapatite in lactate buffer. *Proceedings of the Second Annual Scientific Session of the Academy of Surgical Research.* S.C. Clemson and D.N. Powers (Ed.), pp. 13–16. Clemson University Press, Clemson, SC.

Kaae J.L. 1971. Structure and mechanical properties of isotropic pyrolitic carbon deposited below 1600°C. *J. Nucl. Mater.* 38: 42–50.

Kawahara H. Ed. 1989. *Oral Implantology and Biomaterials*, Elsevier, Amsterdam, Netherlands.

Kawashima N., Soetanto K., Watanabe K., Ono K., and Matsuno T. 1997. The surface characteristics of the sintered body of hydroxyapatite–zirconia composite particles. *Coll. Surf. B: Biointerf.* 10: 23–27.

Khavari F. and Bajpai P.K. 1993. Coralline-sulfate bone substitutes. *Biomed. Sci. Instrum.* 29: 65–69.

Kijima T. and Tsutsumi M. 1979. Preparation and thermal properties of dense polycrystalline oxyhydroxyapatite. *J. Am. Cer. Soc.* 62: 954–960.

Kingery W.D., Bowen H.K., and Uhlmann D.R. 1976. *Introduction to Ceramics*, 2nd ed., p. 368, John Wiley & New York.

Knowles J.C. and Bonfield W. 1993. Development of a glass reinforced hydroxyapatite with enhanced mechanical properties. the effect of glass composition on mechanical properties and its relationship to phase changes. *J. Biomed. Mater. Res.* 27: 1591–1598.

Kokubo T., Kushitani H., Ohtsuki C., Sakka S., and Yamamuro T. 1992. Chemical reaction of bioactive glass and glass-ceramics with a simulated body fluid. *J. Mat. Sci.: Mater. Med.* 3: 79–83.

Krainess F.E. and Knapp W.J. 1978. Strength of a dense alumina ceramic after aging *in vitro*. *J. Biomed. Mater. Res.* 12: 241–246.

Kumar P., Shimizu K., Oka M., Kotoura Y., Nakayama Y., Yamamuro T., Yanagida T., and Makinouchi K. 1989. Biological reaction of zirconia ceramics. In: *Bioceramics. Proceedings of 1st International Symposium on Ceramics in Medicine.* H. Oonishi, H. Aoki, and K. Sawai (Eds.), pp. 341–346, Ishiyaku Euroamerica, Inc., Tokyo.

Larrabee R.A., Fuski M.P., and Bajpai P.K. 1993. A ferric–calcium–phosphorous–oxide ceramic for rebuilding bone. *Biomed. Sci. Instrum.* 29: 59–64.

Lemons J.E., Bajpai P.K., Patka P., Bonel G., Starling L.B., Rosenstiel T., Muschler G, Kampnier S., and Timmermans T. 1988. Significance of the porosity and physical chemistry of calcium phosphate ceramics orthopaedic uses. In: *Bioceramics: Material Characteristics Versus In Vivo Behavior.* Annals of New York Academy of Sciences, Vol. 523, pp. 190–197.

Lemons J.E. and Niemann K.M.W. 1979. *Porous Tricalcium Phosphate Ceramic for Bone Replacement.* 25th Annual O.R.S., Meetings, San Francisco, CA, February 20–22, p. 162.

Li R., Clark A.E., and Hench L.L. 1991. An investigation of bioactive glass powders by sol–gel processing, *J. Appl. Biomater.* 2: 231–239.

Lim G.K., Wang J., Ng S.C., Chew C.H., and Gan L.M. 1997. Processing of hydroxyapatite via microemulsion and emulsion routes. *Biomaterials* 18: 1433–1439.

Liu D. 1996. Fabrication and characterization of porous hydroxyapatite granules. *Biomaterials* 17: 1955–1957.

Lyckfeldt O. and Ferreira J.M.F. 1998. Processing of porous ceramics by starch consolidation. *J. Europ. Cer. Soc.* 18: 131–140.

Mattie D.R. and Bajpai P.K. 1988. Analysis of the biocompatibility of ALCAP ceramics in rat femurs. *J. Biomed. Mater. Res.*, 22: 1101–1126.

Meenen N.M., Osborn J.F., Dallek M., and Donath K. 1992. Hydroxyapatite-ceramic for juxta-articular implantation. *J. Mat. Sci.: Mater. Med.* 3: 345–351.

Metsger S., Driskell T.D., and Paulsrud J.R. 1982. Tricalcium phosphate ceramic — A resorbable bone implant: review and current status. *JADA* 105: 1035–1038.

Moldovan K. and Bajpai P.K. 1994. A ceramic system for continuous release of aspirin. *Biomed. Sci. Instrum.* 30: 175–180.

More R.B. and Silver M.D. 1990. Pyrolitic carbon prosthetic heart valve occluder wear: *in vitro* results for the Bjork–Shiiey prosthesis. *J. Appl. Biomater.* 1: 267–278.

Morris L.M. and Bajpai P.K. 1989. Development of a resorbable tricalcium phosphate (TCP) amine antibiotic composite. *Mat. Res. Soc. Symp.* 110: 293–300.

Murakami T. and Ohtsuki N. 1989. Friction and wear characteristics of sliding pairs of bioceramics and polyethylene. In: *Bioceramics. Proceedings of 1st International Symposium on Ceramics in Medicine.* H. Oonishi, H. Aoki, and K. Sawai (Eds.), pp. 225–230. Ishiyaku Euroamerica, Inc., Tokyo.

Nagy E.A. and Bajpai P.K. 1994. Development of a ceramic matrix system for continuous delivery of azidothymidine. *Biomed. Sci. Instrum.* 30: 181–186.

Niwa S., Sawai K., Takahashie S., Tagai H., Ono M., and Fukuda Y. 1980. *Experimental Studies on the Implantation of Hydroxyapatite in the Medullary Canal of Rabbits, Trans. First World Biomaterials Congress,* Baden, Austria. p. 4.10.4.

Ogino M., Ohuchi F., and Hench L.L. 1980. Compositional dependence of the formation of calcium phosphate film on bioglass, *J. Biomed. Mater. Res.* 12: 55–64.

Ohashi T., Inoue S., Kajikawa K., Ibaragi K., Tada T., Oguchi M., Arai T., and Kondo K. 1988. The clinical wear rate of acetabular component accompanied with alumina ceramic head. In: *Bioceramics. Proceedings of 1st International Symposium on Ceramics in Medicine.* H. Oonishi, H. Aoki, and K. Sawai (Eds.), pp. 278–283. Ishiyaku EuroAmerica, Inc. Tokyo.

Ohtsuki C., Kokubo T., and Yamamuro T. 1992. Compositional dependence of bioactivity of glasses in the system $CaO-SiO_2-Al_2O_3$: its *in vitro* evaluation. *J. Mat. Sci.: Mater. Med.* 3: 119–125.

Ohura K., Nakamura T., Yamamuro T., Ebisawa Y., Kokubo T., Kotoura Y., and Oka M. 1992. Bioactivity of $Cao-SiO_2$ glasses added with various ions. *J. Mat. Sci.: Mater. Med.* 3: 95–100.

Oonishi H. 1992. Bioceramic in orthopaedic surgery — our clinical experiences. In: *Bioceramics,* Vol. 3. J.E. Hulbert and S.F. Hulbert (Eds.), pp. 31–42. Rose Hulman Institute of Technology, Terre Haute, IN.

Park J.B. and Lakes R.S. 1992. *Biomaterials — An Introduction,* 2nd ed., Plenum Press, New York.

Park J.B. 1991. Aluminum oxides: biomedical applications. In: *Concise Encyclopedia of Advanced Ceramic Materials,* R.J. Brook (Ed.), pp.13–16. Pergamon Press, Oxford.

Parker D.R. and Bajpai P.K. 1993. Effect of locally delivered testosterone on bone healing. *Trans. Soc. Biomater.* 26: 293.

Peltier L.F. 1961, The use of plaster of Paris to fill defects in bone. *Clin. Orthop.* 21: 1–29.

Piattelli A. and Trisi P. 1994. A light and laser scanning microscopy study of bone/hydroxyapatite-coated titanium implants interface: histochemical evidence of unmineralized material in humans. *J. Biomed. Mater. Res.* 28: 529–536.

Pories W.J. and Strain W.H. 1970. Zinc and wound healing. In: *Zinc Metabolism,* A.S. Prasad (Ed.), pp. 378–394. Thomas, Springfield, IL.

Posner A.S., Perloff A., and Diorio A.D. 1958. Refinement of hydroxyapatite structure. *Acta. Cryst.* 11: 308–309.

Reck R., Störkel S, and Meyer A. 1988. Bioactive glass-ceramics in middle ear surgery: an eight year review. In: *Bioceramics: Material Characteristics Versus In Vivo Behavior. Ann. NY Acad. Sci.* 253: 100–106.

Ricci J.L., Bajpai P.K., Berkman A., Alexander H., and Parsons J.R. 1986. Development of a fast-setting ceramic based grout material for filling bone defects. In: *Biomedical Engineering V Recent Developments. Proceedings of the Fifth Southern Biomedical Engineering Conference.* Shreveport L.A. and S. Saha (Eds.), pp. 475–481. Pergamon Press, New York.

Ritter J.E. Jr, Greenspan D.C., Palmer R.A., and Hench L.L. 1979. Use of fracture of an alumina and bioglass coated alumina, *J. Biomed. Mater. Res.* 13: 251–263.

Sartoris D.J., Gershuni D.H., Akeson W.H., Holmes R.E., and Resnick D. 1986. Coralline hydroxyapatite bone graft substitutes: Preliminary report of radiographic evaluation. *Radiology* 159: 133–137.

Scheidler P.A. and Bajpai P.K. 1992. Zinc sulfate calcium phosphate (ZSCAP) composite for repairing traumatized bone. *Biomed. Sci. Instrum.* 28: 183–188.

Schepers E., Ducheyne P., and De Clercq M. 1989. Interfacial analysis of fiber-reinforced bioactive dental root implants. *J. Biomed. Mater. Res.* 23: 735–752.

Schepers E., De Clercq M., and Ducheyne P. 1988. Interfacial behavior of bulk bioactive glass and fiber-reinforced bioactive glass dental root implants. *Ann. NY Acad. Sci.* 523: 178–189.

Schepers E.J.G, Ducheyne P., Barbier L., and Schepers S. 1993. Bioactive glass particles of narrow size range: A new material for the repair of bone defects. *Impl. Dent.* 2: 151–156.

Schwartz Z., Braun G., Kohave D., Brooks B., Amir D., Sela J., and Boyan B. 1993. Effects of hydroxyapatite implants on primary mineralization during rat tibial healing: Biochemical and morphometric analysis. *J. Biomed. Mater. Res.* 27: 1029–1038.

Sedel L., Meunier A., Nizard R.S., and Witvoet J. 1991. Ten year survivorship of cemented ceramic–ceramic total hip replacement. In: *Bioceramics* Volume 4. *Proceedings of the 4th International Symposium on Ceramics in Medicine.* W. Bonfield, G.W. Hastings, and K.E. Tanner (Eds.), pp.27–37. Butterworth-Heinemann Ltd., London, UK.

Shackelford J.F. 1988. *Introduction to Materials Science for Engineers*, 2nd ed., Macmillan Publishing Co., New York.

Shimm H.S. and Haubold A.D. 1980. The fatigue behavior of vapor deposited carbon films. *Biomater. Med. Dev. Art. Org.* 8: 333–344.

Shields J.A., Shields C.L., and De Potter P. 1993. Hydroxyapatite orbital implant after enucleation-experience with 200 cases. *Mayo Clinic Proc.* 68: 1191–1195.

Shobert E.I. II. 1964. *Carbon and Graphite*, Academic Press, N.Y.

Stricker N.J., Larrabee R.A., and Bajpai P.K. 1992. Biocompatibility of ferric calcium phosphorous oxide ceramics. *Biomed. Sci. Instrum.* 28: 123–128.

Suchanek W., Yashima M., Kakihana M., and Yoshimura M. 1997. Hydroxyapatite ceramics with selected sintering additives. *Biomaterials* 18: 923–933.

Sudanese A., Toni A., Cattaneo G.L., Ciaroni D., Greggi T., Dallart D., Galli G., and Giunti A. 1989. In *Bioceramics. Proceedings of 1st International Symposium on Ceramics in Medicine.* H. Oonishi, H. Aoki, and K. Sawai (Eds.), pp. 237–240, Ishiyaku Euroamerica, Inc., Tokyo.

Takatsuko K., Yamamuro T., Kitsugi T., Nakamura T., Shibuya T., and Goto T. 1993. A new bioactive glass-ceramic as a coating material on titanium alloy. *J. Appl. Biomater.* 4: 317–329.

Terry B.C., Baker R.D., Tucker M.R., and Hanker J.S. 1989. Alveolar ridge augmentation with composite implants of hydroxylapatite and plaster for correction of bony defects, deficiencies and related contour abnormalities. *Mat. Res. Soc. Symp.* 110: 187–198.

Valiathan A., Randhawa G.S., and Randhawa A. 1993. Biomaterial aspects of calcium hydroxyapatite, *T.I.B. & A.O.* 7: 1–7.

Varma H.K. and Sivakumar R. 1996. Dense hydroxyapatite ceramics through gel casting technique. *Mater. Lett.* 29: 57–61.

Whitehead R.Y., Lacefield W.R., and Lucas L.C. 1993. Structure and integrity of a plasma sprayed hydroxyapatite coating on titanium, *J. Biomed. Mater. Res.* 27: 1501–1507.

Wolford L.M., Wardrop R.W., and Hartog J.M. 1987. Coralline porous hydroxylapatite as a bone graft substitute in orthognathic surgery. *J. Oral. Maxillofacial. Surg.* 45: 1034–1042.

Wyatt D.F., Bajpai P.K., Graves G.A. Jr, and Stull P.A. 1976. Remodelling of calcium aluminate phosphorous pentoxide ceramic implants in bone. *IRCS. Med. Sci.* 4: 421.

Yamamuro T., Hench L.L., and Wilson J. 1990. *Handbook of Bioactive Ceramics I and II.* CRC Press, Boca Raton, FL.

Yamamuro T., Shikata J., Kakutani Y., Yoshii S., Kitsugi T., and Ono K. 1988. Novel methods for clinical applications of bioactive ceramics. In: *Bioceramics: Material Characteristics Versus In Vivo Behavior.* *Ann. NY Acad. Sci.* 523: 107–114.

Zimmerman M.C., Alexander H., Parsons J.R., and Bajpai P.K. 1991. The design and analysis of laminated degradable composite bone plates for fracture fixation. In: *High-Tech Textiles*, T.L. Vigo and A.F. Turbak (Eds.), pp. 132–148. ACS Symposium Series 457, American Chemical Society, Washington, D.C.

Further Information

Bajpai P.K. 1988. *ZCAP Ceramics*. US. Patent No. 4778471.

Bajpai P.K. 1987. *Surgical Cements*. US. Patent No. 4668295.

Bonfield W., Hastings G.W., and Tanner K.E. 1991. *Bioceramics, Vol. 4. Proceedings of the 4th International Symposium on Ceramics in Medicine*. Butterworth-Heinemann Ltd., London, UK.

Brook J. 1991. *Concise Encyclopedia of Advanced Ceramic Materials*. Pergamon Press, Oxford.

de Groot K. 1983. *Bioceramics of Calcium Phosphate*. CRC Press, Boca Raton, FL.

Ducheyne P. and Lemons J.E. 1988. Bioceramics: material characteristics versus *in vivo* behavior. *Ann. NY Acad. Sci.*, New York, NY.

Ducheyne P. and Christiansen D. 1993. *Bioceramics*, Vol. 6. Butterworth-Heinemann Ltd., Boston.

Filgueiras M.R.T., LaTorre G., and Hench L.L. 1993. Solution effects on the surface reactions of three bioactive glass compositions. *J. Biomed. Mater. Res.* 27: 1485–1493.

Frank R.M., Wiedemann P., Hemmerle J., and Freymann M. 1991. Pulp capping with synthetic hydroxyapatite in human premolars. *J. Appl. Biomater.* 2: 243–250.

Fulmer M.T. and Brown P.W. 1993. Effects of Na_2HPO_4 and NaH_2PO_4 on hydroxyapatite formation. *J. Biomed. Mater. Res.* 27: 1095–1102.

Garcia R. and Doremus R.H. 1992. Electron microscopy of the bone-hydroxyapatite interface from a human dental implant. *J. Mater. Sci.: Mater. Med.* 3: 154–156.

Hall C.W., Hulbert S.F., Levine S.N., and Young F.A. 1972. *Engineering and Medicine*. Interscience Publishers, New York.

Hench L.L. 1991. Bioceramics: From concept to clinic. *J. Am. Ceram. Soc.* 74: 1487–1510.

Hench L.L. and Ethridge E.C. 1982. *Biomaterials: An Interfacial Approach*. Academic Press, New York.

Hulbert J.A. and Hulbert S.F. 1992. *Bioceramics*, Vol. 3. *Proceedings of the 3rd International Symposium on Ceramics in Medicine*, Rose-Hulman Institute of Technology, Terra Haute, IN.

Kawahara H. Ed. 1989. *Oral Implantology and Biomaterials*. Elsevier, Amsterdam, Netherlands.

Kingery W.D., Bowen H.K., and Uhlmann D.R. 1976. *Introduction to Ceramics*, 2nd ed., p. 368, John Wiley & Sons, New York.

Lemons J.E. 1988. *Quantitative Characterization and Performance of Porous Implants for Hard Tissue Applications*, ASTM STP 953. American Society for Testing and Materials, Philadelphia, PA.

Mattie D.R. and Bajpai P.K. 1986. Biocompatibility testing of ALCAP ceramics. *IRCS. Med. Sci.* 14: 641–643.

Neo M., Nakamura T., Ohtsuki C., Kokubo T., and Yamamuro T. 1993. Apatite formation on three kinds of bioactive material at an early stage *in vivo*: A comparative study by transmission electron microscopy. *J. Biomed. Mater. Res.* 27: 999–1006.

Oonishi H., Aoki H., and Sawai K. 1988. *Bioceramics*, Vol. 1. *Proceedings of 1st International Symposium on Ceramics in Medicine*. Ishiyaku EuroAmerica, Inc. Tokyo.

Oonishi H. and Ooi Y. 1981. *Orthopaedic ceramic implants*, Vol. I. *Proceedings of Japanese Society of Orthopaedic Ceramic Implants*.

Park J.B. and Lakes R.S. 1992. *Biomaterials: An Introduction*, 2nd ed., Plenum Press, New York, NY.

Ravaglioli A. and Krajewski A. 1992. *Bioceramics and the Human Body*. Elsevier Applied Science, London and New York.

Rubin L.R. 1983. *Biomaterials in Reconstructive Surgery*. C.V. Mosby Co., St. Louis, MO.

Sharma C.P. and Szycher M. 1991. *Blood Compatible Materials and Devices: Perspectives Toward 21st Century*. Technomic Publishing Co., Lancaster, PA.

Signs S.A., Pantano C.G., Driskell T.D., and Bajpai P.K. 1979. In vitro dissolution of synthos ceramics in an acellular physiological environment. *Biomater. Med. Dev. Art. Org.* 7: 183–190.

Stea S., Tarabusi C., Ciapetti G., Pizzoferrato A., Toni A., and Sudanese A. 1992. Microhardness evaluations of the bone growing into porous implants. *J. Mater. Sci.: Mater. Med.* 3: 252–254.

van Blitterswijk C.A. and Grote J.J. 1989. Biological performance of ceramics during inflammation and infection. *Crit. Rev. Biocompatib.* 5: 13–43.

Wilson J. and Low S.B. 1992. Bioactive ceramics for periodontal treatment: Comparative studies in the patus monkey. *J. Appl. Biomater.* 3: 123–129.

Yamamuro T., Hench L.L., and Wilson J. 1990. *Handbook of Bioactive Ceramics.* CRC Press, Boca Raton, FL.

Zhang X. and Ikada Y. 1994. *Biomedical Materials Research in The Far East (I).* Kobunshi Kankokai Inc., Kyoto, Japan.

3

Polymeric Biomaterials

Hai Bang Lee
Korea Research Institute of Chemical Technology

Gilson Khang
Chonbuk National University

Jin Ho Lee
Hannam University

3.1 Introduction

Synthetic polymeric materials have been widely used in medical disposable supply, prosthetic materials, dental materials, implants, dressings, extracorporeal devices, encapsulants, polymeric drug delivery systems, tissue engineered products, and orthodoses as that of metal and ceramics substituents [Lee, 1989]. The main advantages of the polymeric **biomaterials** compared to metal or ceramic materials are ease of manufacturability to produce various shapes (latex, film, sheet, fibers, etc.), ease of secondary processability, reasonable cost, and availability with desired mechanical and physical properties. The required properties of polymeric biomaterials are similar to other biomaterials, that is, **biocompatibility**, sterilizability, adequate mechanical and physical properties, and manufacturability as given in Table 3.1.

The objectives of this chapter are (1) the review of basic chemical and physical properties of the synthetic polymers, (2) the sterilization of the polymeric biomaterials, (3) the importance of the surface treatment for improving biocompatability, and (4) the application of the **chemogradient surface** for the study on cell to polymer interactions.

TABLE 3.1   Requirements for Biomedical Polymers

Properties	Description
Biocompatibility	Noncarcinogenesis, nonpyrogenicity, nontoxicity, and nonallergic response
Sterilizability	Autoclave, dry heating, ethylenoxide gas, and radiation
Physical property	Strength, elasticity, and durability
Manufacturability	Machining, molding, extruding, and fiber forming

Source: Modified from Ikada, Y. (Ed.) 1989. In: *High Technology Fiber*, Part B, Marcel Dekker, New York. With permission.

TABLE 3.2   Typical Condensation Polymers

Type	Interunit linkage
Polyester	$-\overset{\overset{\text{O}}{\|}}{\text{C}}-\text{O}-$
Polyamide	$-\overset{\overset{\text{O}}{\|}}{\text{C}}-\overset{\overset{\text{H}}{\|}}{\text{N}}-$
Polyurea	$-\overset{\overset{\text{H}}{\|}}{\text{N}}-\overset{\overset{\text{O}}{\|}}{\text{C}}-\overset{\overset{\text{H}}{\|}}{\text{N}}-$
Polyurethane	$-\text{O}-\overset{\overset{\text{O}}{\|}}{\text{C}}-\overset{\overset{\text{H}}{\|}}{\text{N}}-$
Polysiloxane	$-\overset{\overset{\text{R}}{\|}}{\underset{\underset{\text{R}}{\|}}{\text{Si}}}-\text{O}-$
Protein	$-\overset{\overset{\text{O}}{\|}}{\text{C}}-\overset{\overset{\text{H}}{\|}}{\text{N}}-$

3.2   Polymerization and Basic Structure

3.2.1   Polymerization

In order to link the small molecules one has to force them to lose their electrons by the chemical processes of condensation and addition. By controlling the reaction temperature, pressure, and time in the presence of catalyst(s), the degree to which **repeating units** are put together into chains can be manipulated.

3.2.1.1   Condensation or Step Reaction Polymerization

During **condensation polymerization** a small molecule such as water will be condensed out by the chemical reaction. For example

$$\underset{\text{(amine)}}{\text{R-NH}_2} + \underset{\text{(carboxylic acid)}}{\text{R}'\text{COOH}} \rightarrow \underset{\text{(amide)}}{\text{R}'\text{CONHR}} + \underset{\text{(condensed molecule)}}{\text{H}_2\text{O}} \tag{3.1}$$

This particular process is used to make polyamides (Nylons). Nylon was the first commercial polymer, made in the 1930s.

Some typical condensation polymers and their interunit linkages are given in Table 3.2. One major drawback of condensation polymerization is the tendency for the reaction to cease before the chains grow to a sufficient length. This is due to the decreased mobility of the chains and reactant chemical species as polymerization progresses. This results in short chains. However, in the case of Nylon the chains are polymerized to a sufficiently large extent before this occurs and the physical properties of the polymer are preserved.

Natural polymers, such as polysaccharides and proteins are also made by condensation polymerization. The condensing molecule is always water (H_2O).

3.2.1.2 Addition or Free Radical Polymerization

Addition polymerization can be achieved by rearranging the bonds within each monomer. Since each "mer" has to share at least two covalent electrons with other mers the monomer should have at least one double bond. For example, in the case of ethylene:

$$n \ \{ \overset{\overset{\text{H}}{|}}{\underset{\underset{\text{H}}{|}}{\text{C}}} = \overset{\overset{\text{H}}{|}}{\underset{\underset{\text{H}}{|}}{\text{C}} } \} \ \rightarrow \ -\overset{\overset{\text{H}}{|}}{\underset{\underset{\text{H}}{|}}{\text{C}}} + \overset{\overset{\text{H}}{|}}{\underset{\underset{\text{H}}{|}}{\text{C}}} - \overset{\overset{\text{H}}{|}}{\underset{\underset{\text{H}}{|}}{\text{C}}} + \overset{\overset{\text{H}}{|}}{\underset{\underset{\text{H}}{|}}{\text{C}}} - \tag{3.2}$$

The breaking of a double bond can be made with an **initiator**. This is usually a free radical such as benzoyl peroxide ($H_5C_6COO–OOCC_6H_5$). The initiation can be activated by heat, ultraviolet light, and other chemicals. The free radicals (initiators) can react with monomers and this free radical can react with another monomer and the process can continue on. This process is called propagation. The propagation process can be terminated by combining two free radicals, by transfer, or by disproportionate processes. Some of the free radical polymers are given in Table 3.3. There are three more types of initiating species for addition polymerization beside free-radicals; cations, anions, and coordination (stereospecific) catalysts. Some monomers can use two or more of the initiation processes but others can use only one process as given in Table 3.3.

TABLE 3.3 Monomers for Addition Polymerization and Suitable Process

Monomer names	Chemical structure	Polymerization mechanism			
		Radical	Cationic	Anionic	Coordination
Acrylonitrile	$CH_2 = CH$ $\quad\quad\ |$ $\quad\quad C{\equiv}N$	+	−	+	+
Ethylene	$CH_2 = CH_2$	+	+	−	−
Methacrylate	$CH_2 = CH$ $\quad\quad\ |$ $\quad\quad COOCH_3$	+	−	+	+
Methylmethacrylate	$CH_2 = CCH_3$ $\quad\quad\ |$ $\quad\quad COOCH_3$	+	−	+	+
Propylene	$CH_2 = CH$ $\quad\quad\ |$ $\quad\quad CH_3$	−	−	−	+
Styrene	$CH_2 = CH$ $\quad\quad\ |$ $\quad\quad C_6H_5$	+	+	+	+
Vinylchloride	$CH_2 = CH$ $\quad\quad\ |$ $\quad\quad Cl$	+	−	−	+
Vinylidenechloride	$CH_2 = C$ $\quad\quad\ |^{\ Cl}$ $\quad\quad Cl$	+	−	+	−

+: high polymer formed.−: no reaction or oligomers only.
Source: Modified form Billmeyer, F.W.Jr. 1984. *Text Book of Polymer Scence*, 3rd ed. John Wiley & Sons; New York. With permission.

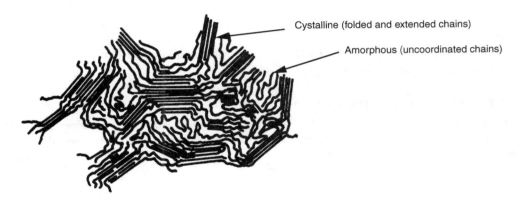

Cystalline (folded and extended chains)

Amorphous (uncoordinated chains)

FIGURE 3.1   Fringed-micelle model of a linear polymer with semi-crystalline structure.

3.2.2 Basic Structure

Polymers have very long chain molecules which are formed by **covalent bonding** along the backbone chain. The long chains are held together either by secondary bonding forces such as van der Waals and hydrogen bonds or primary covalent bonding forces through crosslinks between chains. The long chains are very flexible and can be tangled easily. In addition, each chain can have **side groups**, branches and copolymeric chains or blocks which can also interfere with the long-range ordering of chains. For example, paraffin wax has the same chemical formula as polyethylene (PE) $[(CH_2CH_2)_n]$, but will crystallize almost completely because of its much shorter chain lengths. However, when the chains become extremely long {from 40 to 50 repeating units $[-CH_2CH_2-]$ to several thousands as in linear PE} they cannot be crystallized completely (up to 80 to 90% crystallization is possible). Also, branched PE in which side chains are attached to the main backbone chain at positions normally occupied by a hydrogen atom, will not crystallize easily due to the **steric hindrance** of side chains resulting in a more noncrystalline structure. The partially crystallized structure is called semicrystalline which is the most commonly occurring structure for linear polymers. The semicrystalline structure is represented by disordered noncrystalline (amorphous) regions and ordered crystalline regions which may contain folded chains as shown in Figure 3.1.

The degree of polymerization (DP) is defined as an average number of mers, or repeating units, per molecule, that is, chain. Each chain may have a different number of mers depending on the condition of polymerization. Also, the length of each chain may be different. Therefore, it is assumed there is an average degree of polymerization or average molecular weight (MW). The relationship between molecular weight and degree of polymerization can be expressed as:

$$\text{MW of polymer} = \text{DP} \times \text{MW of mer (or repeating unit)} \tag{3.3}$$

The two average molecular weights most commonly used are defined in terms of the numbers of molecules, Ni, having molecular weight, Mi; or wi, the weight of species with molecular weights Mi as follows:

1. The number-average molecular weight, Mn, is defined by

$$\text{Mn} = \frac{\Sigma\,NiMi}{\Sigma\,NiMi} = \Sigma\,niMi = \frac{\Sigma\,Wi}{\Sigma\,(Wi/Mi)} = \frac{1}{\Sigma\,(wi\,Mi)} \tag{3.4}$$

2. The weight average molecular weight, Mw, is defined by

$$\text{Mw} = \frac{\Sigma\,WiMi}{\Sigma\,Wi} = \Sigma\,wiMi = \frac{\Sigma\,NiMi^2}{\Sigma\,NiMi} \tag{3.5}$$

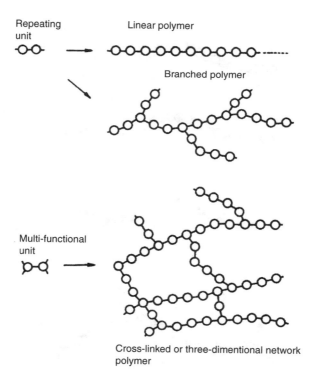

FIGURE 3.2 Arrangement of polymer chains into linear, branched, and network structure depending on the functionality of the repeating units.

An absolute method of measuring the molecular weight is one that depends on theoretical consider-ations, counting molecules and their weight directly. The relative methods require calibration based on an absolute method and include intrinsic viscosity and gel permeation chromatography (GPC). Absolute methods of determining the number-average molecular weight (Mn) include osmometry and other col-ligative methods, and end group analysis. Light-scattering yields an absolute weight-average molecular weight (Mw).

As the molecular chains become longer by the progress of polymerization, their relative mobility decreases. The chain mobility is also related to the physical properties of the final polymer. Generally, the higher the molecular weight, the less the mobility of chains which results in higher strength and greater thermal stability. The polymer chains can be arranged in three ways; linear, branched, and a cross-linked (or three-dimensional) network as shown in Figure 3.2. Linear polymers such as polyvinyls, polyamides, and polyesters are much easier to crystallize than the cross-linked or branched polymers. However, they cannot be crystallized 100% as with metals. Instead they become semicrystalline polymers. The arrangement of chains in crystalline regions is believed to be a combination of folded and extended chains. The chain folds, which are seemingly more difficult to form, are necessary to explain observed single crystal structures in which the crystal thickness is too small to accommodate the length of the chain as determined by electron and x-ray diffraction studies. The classical "fringed-micelle" model in which the amorphous and crystalline regions coexist has been modified to include chain folds in the crystalline regions. The cross-linked or three-dimensional network polymers such as polyphenolformaldehyde cannot be crystallized at all and they become noncrystalline, amorphous polymers.

Vinyl polymers have a repeating unit $-CH_2-CHX-$ where X is some monovalent side group. There are three possible arrangements of side groups (X) (1) atactic, (2) isotactic, and (3) syndiotactic. In atactic arrangements the side groups are randomly distributed while in syndiotactic and isotactic arrange-ments they are either in alternating positions or on one side of the main chain. If side groups are

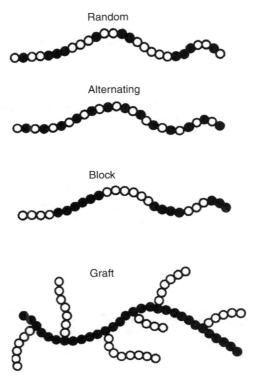

FIGURE 3.3 Possible arrangements of copolymers.

small like polyethylene (X = H) and the chains are linear, the polymer crystallizes easily. However, if the side groups are large as in polyvinyl chloride (X = Cl) and polystyrene (X = C_6H_5, benzene ring) and are randomly distributed along the chains (atactic), then a noncrystalline structure will be formed. The isotactic and syndiotactic polymers usually crystallize even when the side groups are large.

Copolymerization, in which two or more homopolymers (one type of repeating unit throughout its structure) are chemically combined, always disrupts the regularity of polymer chains thus promoting the formation of a noncrystalline structure. Possible arrangement of the different copolymerization is shown in Figure 3.3. The addition of **plasticizers** to prevent crystallization by keeping the chains separated from one another will result in more flexible polymers, a noncrystalline version of a polymer which normally crystallizes. An example is celluloid which is normally made of crystalline nitrocellulose plasticized with camphor. Plasticizers are also used to make rigid noncrystalline polymers like polyvinylchloride (PVC) into a more flexible solid (a good example is Tygon® tubing).

Elastomers, or rubbers, are polymers which exhibit large stretchability at room temperature and can snap back to their original dimensions when the load is released. The elastomers are non-crystalline polymers which have an intermediate structure consisting of long chain molecules in three-dimensional networks (see next section for more details). The chains also have "kinks" or "bends" in them which straighten when a load is applied. For example, the chains of *cis*-polyisoprene (natural rubber) are bent at the double bond due to the methyl group interfering with the neighboring hydrogen in the repeating unit [–CH_2–$C(CH_3)$=CH–CH_2–]. If the methyl group is on the opposite side of the hydrogen then it becomes *trans*-polyisoprene which will crystallize due to the absence of the steric hindrance present in the *cis* form. The resulting polymer is a very rigid solid called gutta percha which is not an elastomer. Below the **glass transition temperature** (T_g; second-order transition temperature between viscous liquid and solid) natural rubber loses its compliance and becomes a glass-like material. Therefore, to be flexible,

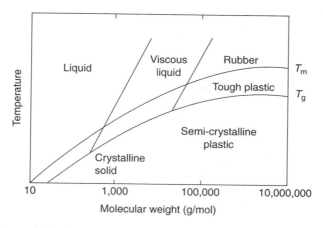

FIGURE 3.4 Approximate relations among molecular weight, T_g, T_m, and polymer properties.

all elastomers should have T_g well below room temperature. What makes the elastomers not behave like liquids above T_g is in fact due to the cross-links between chains which act as pinning points. Without cross-links the polymer would deform permanently. An example is latex which behaves as a viscous liquid. Latex can be cross-linked with sulfur (**vulcanization**) by breaking double bonds (C=C) and forming C–S–S–C bonds between the chains. The more cross-links are introduced the more rigid the structure becomes. If all the chains are cross-linked together, the material will become a three-dimensional rigid polymer.

3.2.3 Effect of Structural Modification on Properties

The physical properties of polymers can be affected in many ways. In particular, the chemical composition and arrangement of chains will have a great effect on the final properties. By such means the polymers can be tailored to meet the end use.

3.2.3.1 Effect of Molecular Weight and Composition

The molecular weight and its distribution have a great effect on the properties of a polymer since its rigidity is primarily due to the immobilization or entanglement of the chains. This is because the chains are arranged like cooked spaghetti strands in a bowl. By increasing the molecular weight the polymer chains become longer and less mobile and a more rigid material results as shown in Figure 3.4. Equally important is that all chains should be equal in length since if there are short chains they will act as plasticizers. Another obvious way of changing properties is to change the chemical composition of the backbone or side chains. Substituting the backbone carbon of a polyethylene with divalent oxygen or sulfur will decrease the melting and glass transition temperatures since the chain becomes more flexible due to the increased rotational freedom. On the other hand if the backbone chains can be made more rigid then a stiffer polymer will result.

3.2.3.2 Effect of Side Chain Substitution, Cross-Linking, and Branching

Increasing the size of side groups in linear polymers such as polyethylene will decrease the melting temperature due to the lesser perfection of molecular packing, that is, decreased crystallinity. This effect is seen until the side group itself becomes large enough to hinder the movement of the main chain as shown in Table 3.4. Very long side groups can be thought of as being branches.

Cross-linking of the main chains is in effect similar to the side-chain substitution with a small molecule, that is, it lowers the melting temperature. This is due to the interference of the cross-linking which causes

TABLE 3.4 Effect of Side
Chain Substitution on Melting
Temperature in Polyethylene

Side chain	T_m (°C)
$-H$	140
$-CH_3$	165
$-CH_2CH_3$	124
$-CH_2CH_2CH_3$	75
$-CH_2CH_2CH_2CH_3$	−55
$-CH_2CHCH_2CH_3$ CH_3	196
$-CH_2CCH_2CH_3$ (with CH_3 above and $-CH_3$ below)	350

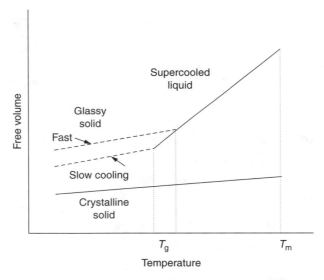

FIGURE 3.5 Change of volume vs. temperature of a solid. The glass transition temperature (T_g) depends on the rate of cooling and below (T_g) the material behaves as a solid like a window glass.

decreased mobility of the chains resulting in further retardation of the crystallization rate. In fact, a large degree of cross-linking can prevent crystallization completely. However, when the cross-linking density increases for a rubber, the material becomes harder and the glass transition temperature also increases.

3.2.3.3 Effect of Temperature on Properties

Amorphous polymers undergo a substantial change in their properties as a function of temperature. The glass transition temperature, T_g, is a boundary between the glassy region of behavior in which the polymer is relatively stiff and the rubbery region in which it is very compliant. T_g can also be defined as the temperature at which the slope of volume change versus temperature has a discontinuity in slope as shown in Figure 3.5. Since polymers are non-crystalline or at most semicrystalline, the value obtained in this measurement depends on how fast it is taken.

TABLE 3.5   Biomedical Application of Polymeric Biomaterials

Synthetic Polymers	Applications
Polyvinylchloride (PVC)	Blood and solution bag, surgical packaging, IV sets, dialysis devices, catheter bottles, connectors, and cannulae
Polyethylene (PE)	Pharmaceutical bottle, nonwoven fabric, catheter, pouch, flexible container, and orthopedic implants
Polypropylene (PP)	Disposable syringes, blood oxygenator membrane, suture, nonwoven fabric, and artificial vascular grafts
Polymethylmetacrylate (PMMA)	Blood pump and reservoirs, membrane for blood dialyzer, implantable ocular lens, and bone cement
Polystyrene (PS)	Tissue culture flasks, roller bottles, and filterwares
Polyethylenterephthalate (PET)	Implantable suture, mesh, artificial vascular grafts, and heart valve
Polytetrafluoroethylene (PTFE)	Catheter and artificial vascular grafts
Polyurethane (PU)	Film, tubing, and components
Polyamide (Nylon)	Packaging film, catheters, sutures, and mold parts

3.3   Polymers Used as Biomaterials

Although hundreds of polymers are easily synthesized and could be used as biomaterials only ten to twenty polymers are mainly used in medical device fabrications from disposable to long-term implants as given in Table 3.5. In this section, the general information of the characteristics, properties, and applications of the most commonly used polymers will be discussed [Billmeyer, 1984; Park, 1984; Leininger and Bigg, 1986; Shalaby, 1988; Brandrup and Immergut, 1989; Sharma and Szycher, 1991; Park and Lakes, 1992; Dumitriu, 1993; Lee and Lee, 1995; Ratner et al., 1996].

3.3.1   Polyvinylchloride (PVC)

The PVC is an amorphous, rigid polymer due to the large side group (Cl, chloride) with a T_g of 75 to 105°C. It has a high melt viscosity hence it is difficult to process. To prevent the thermal degradation of the polymer (HCl could be released), thermal stabilizers such as metallic soaps or salts are incorporated. Lubricants are formulated on PVC compounds to prevent adhesion to metal surfaces and facilitate the melt flow during processing. Plasticizers are used in the range of 10 to 100 parts per 100 parts of PVC resin to make it flexible. Di-2-ethylhexylphthalate (DEHP or DOP) is used in medical PVC formulation. However, the plasticizers of trioctyltrimellitate (TOTM), polyester, azelate, and phosphate ester are also used to prevent extraction by blood, aqueous solution, and hot water during autoclaving sterilization.

PVC sheets and films are used in blood and solution storage bags and surgical packaging. PVC tubing is commonly used in intravenous (IV) administration, dialysis devices, catheters, and cannulae.

3.3.2   Polyethylene (PE)

PE is available commercially in five major grades: (1) high density (HDPE), (2) low density (LDPE), (3) linear low density (LLDPE), (4) very low density (VLDPE), and (5) ultra high molecular weight (UHMWPE). HDPE is polymerized in a low temperature (60–80°C), and at a low pressure ($\sim$10 kg/cm^2) using metal catalysts. A highly crystalline, linear polymer with a density ranging from 0.94 to 0.965 g/cm^3 is obtained. LDPE is derived from a high temperature (150–300°C) and pressures (1000–3000 kg/cm^2) using free radical initiators. A highly branched polymer with lower crystallinity and densities ranging from 0.915 to 0.935 g/cm^3 is obtained. LLDPE (density: 0.91–0.94 g/cm^3) and VLDPE (density: 0.88–0.89 g/cm^3), which are linear polymers, are polymerized under low pressures and temperatures using metal catalysts with comonomers such as 1-butene, 1-hexene, or 1-octene to obtain the desired physical properties and density ranges.

HDPE is used in pharmaceutical bottles, nonwoven fabrics, and caps. LDPE is found in flexible container applications, nonwoven-disposable and laminated (or coextruded with paper) foil, and polymers for packaging. LLDPE is frequently employed in pouches and bags due to its excellent puncture resistance and VLDPE is used in extruded tubes. UHMWPE (MW $>2\times10^6$ g/mol) has been used for orthopedic implant fabrications, especially for load-bearing applications such as an acetabular cup of total hip and the tibial plateau and patellar surfaces of knee joints. Biocompatability tests for PE are given by ASTM standards in F981, F639, and F755.

3.3.3 Polypropylene (PP)

PP can be polymerized by a Ziegler-Natta stereospecific catalyst which controls the isotactic position of the methyl group. Thermal (T_g: $-12°C$, T_m: 125–167°C and density: 0.85–0.98 g/cm^3) and physical properties of PP are similar to PE. The average molecular weight of commercial PP ranges from 2.2 to 7.0×10^5 g/mol and has a wide molecular weight distribution (polydispersity) which is from 2.6 to 12. Additives for PP such as antioxidants, light stabilizer, nucleating agents, lubricants, mold release agents, antiblock, and slip agents are formulated to improve the physical properties and processability. PP has an exceptionally high flex life and excellent environment stress-cracking resistance, hence it had been tried for finger joint prostheses with an integrally molded hinge design [Park, 1984]. The gas and water vapor permeability of PP are in-between those of LDPE and HDPE. PP is used to make disposable hypothermic syringes, blood **oxygenator** membrane, packaging for devices, solutions, and drugs, **suture**, artificial vascular **grafts**, nonwoven fabrics, etc.

3.3.4 Polymethylmetacrylate (PMMA)

Commercial PMMA is an amorphous (T_g: 105°C and density: 1.15 to 1.195 g/cm^3) material with good resistance to dilute alkalis and other inorganic solutions. PMMA is best known for its exceptional light transparency (92% transmission), high **refractive index** (1.49), good weathering properties, and as one of the most biocompatible polymers. PMMA can be easily machined with conventional tools, molded, surface coated, and plasma etched with glow or corona discharge. PMMA is used broadly in medical applications such as a blood pump and reservoir, an IV system, membranes for blood dialyzer, and in *in vitro* diagnostics. It is also found in contact lenses and implantable ocular lenses due to excellent optical properties, dentures, and maxillofacial prostheses due to good physical and coloring properties, and **bone cement** for joint prostheses fixation (ASTM standard F451).

Another acrylic polymer such as polymethylacrylate (PMA), polyhydroxyethyl-methacrylate (PHEMA), and polyacrylamide (PAAm) are also used in medical applications. PHEMA and PAAm are **hydrogels**, lightly cross-linked by ethyleneglycoldimethylacrylate (EGDM) to increase their mechanical strength. The extended wear soft contact lenses are synthesized from PMMA and N-vinylpyrrolidone or PHEMA which have high water content (above 70%) and a high oxygen permeability.

3.3.5 Polystyrene (PS) and Its Copolymers

The PS is polymerized by free radical polymerization and is usually atactic. Three grades are available; unmodified general purpose PS (GPPS, T_g: 100°C), high impact PS (HIPS), and PS foam. GPPS has good transparency, lack of color, ease of fabrication, thermal stability, low specific gravity (1.04–1.12 g/cm^3), and relatively high modulus. HIPS contains a rubbery modifier which forms chemical bonding with the growing PS chains. Hence the ductility and impact strength are increased and the resistance to environmental stress-cracking is also improved. PS is mainly processed by injection molding at 180–250°C. To improve processability additives such as stabilizers, lubricants, and mold releasing agents are formulated. GPPS is commonly used in tissue culture flasks, roller bottles, vacuum canisters, and filterware.

Acrylonitrile–butadiene–styrene (ABS) **copolymers** are produced by three monomers; acrylonitrile, butadiene, and styrene. The desired physical and chemical properties of ABS polymers with a wide range of functional characteristics can be controlled by changing the ratio of these monomers. They are resistant to the common inorganic solutions, have good surface properties, and dimensional stability. ABS is used for IV sets, clamps, blood dialyzers, diagnostic test kits, and so on.

3.3.6 Polyesters

Polyesters such as polyethyleneterephthalate (PET) are frequently found in medical applications due to their unique chemical and physical properties. PET is so far the most important of this group of polymers in terms of biomedical applications such as artificial vascular graft, sutures, and meshes. It is highly crystalline with a high melting temperature (T_m: 265°C), hydrophobic and resistant to hydrolysis in dilute acids. In addition, PET can be converted by conventional techniques into molded articles such as luer filters, check valves, and catheter housings. Polycaprolactone is crystalline and has a low melting temperature (T_m: 64°C). Its use as a soft matrix or coating for conventional polyester fibers was proposed by recent investigation [Leininger and Bigg, 1986].

3.3.7 Polyamides (Nylons)

Polyamides are known as nylons and are designated by the number of carbon atoms in the repeating units. Nylons can be polymerized by step-reaction (or condensation) and ring-scission polymerization. They have excellent fiber-forming ability due to interchain **hydrogen bonding** and a high degree of crystallinity, which increases strength in the fiber direction.

The presence of –CONH– groups in polyamides attracts the chains strongly toward one another by hydrogen bonding. Since the hydrogen bond plays a major role in determining properties, the number and distribution of –CONH– groups are important factors. For example, T_g can be decreased by decreasing the number of –CONH– groups . On the other hand, an increase in the number of –CONH– groups improves physical properties such as strength as one can see that Nylon 66 is stronger than Nylon 610 and Nylon 6 is stronger than Nylon 11.

In addition to the higher Nylons (610 and 11) there are aromatic polyamides named aramids. One of them is poly (*p*-phenylene terephthalate) commonly known as **Kevlar®**, made by DuPont. This material can be made into fibers. The specific strength of such fibers is five times that of steel, therefore, it is most suitable for making composites.

Nylons are hygroscopic and lose their strength *in vivo* when implanted. The water molecules serve as plasticizers which attack the amorphous region. Proteolytic enzymes also aid in hydrolyzing by attacking the amide group. This is probably due to the fact that the proteins also contain the amide group along their molecular chains which the proteolytic enzymes could attack.

3.3.8 Fluorocarbon Polymers

The best known fluorocarbon polymer is polytetrafluoroethylene (PTFE), commonly known as **Teflon®** (DuPont). Other polymers containing fluorine are polytrifluorochloroethylene (PTFCE), polyvinylfluoride (PVF), and fluorinated ethylene propylene (FEP). Only PTFE will be discussed here since the others have rather inferior chemical and physical properties and are rarely used for implant fabrication.

PTFE is made from tetrafluoroethylene under pressure with a peroxide catalyst in the presence of excess water for removal of heat. The polymer is highly crystalline (over 94% crystallinity) with an average molecular weight of 0.5–5 $\times 10^6$ g/mol. This polymer has a very high density (2.15–2.2 g/cm^3), low modulus of elasticity (0.5 GPa) and tensile strength (14 MPa). It also has a very low surface tension (18.5 erg/cm^2) and friction coefficient (0.1).

Standard specifications for the implantable PTFE are given by ASTM F754. PTFE also has an unusual property of being able to expand on a microscopic scale into a microporous material which is an excellent thermal insulator. PTFE cannot be injection molded or melt extruded because of its very high melt viscosity and it cannot be plasticized. Usually the powders are sintered to above 327°C under pressure to produce implants.

3.3.9 Rubbers

Silicone, natural, and synthetic rubbers have been used for the fabrication of implants. Natural rubber is made mostly from the latex of the Hevea brasiliensis tree and the chemical formula is the same as that of *cis*-1,4 polyisoprene. Natural rubber was found to be compatible with blood in its pure form. Also, cross-linking by x-ray and organic peroxides produces rubber with superior blood compatibility compared with rubbers made by the conventional sulfur vulcanization.

Synthetic rubbers were developed to substitute for natural rubber. The Ziegler-Natta types of stereospecific polymerization techniques have made this variety possible. The synthetic rubbers have rarely been used to make implants. The physical properties vary widely due to the wide variations in preparation recipes of these rubbers.

Silicone rubber, developed by Dow Corning company, is one of the few polymers developed for medical use. The repeating unit is dimethyl siloxane which is polymerized by a condensation polymerization. Low molecular weight polymers have low viscosity and can be cross-linked to make a higher molecular weight, rubber-like material. Medical grade silicone rubbers contain stannous octate as a catalyst and can be mixed with a base polymer at the time of implant fabrication.

3.3.10 Polyurethanes

Polyurethanes are usually thermosetting polymers: they are widely used to coat implants. Polyurethane rubbers are produced by reacting a prepared prepolymer chain with an aromatic di-isocyanate to make very long chains possessing active isocyanate groups for cross-linking. The polyurethane rubber is quite strong and has good resistance to oil and chemicals.

3.3.11 Polyacetal, Polysulfone, and Polycarbonate

These polymers have excellent mechanical, thermal, and chemical properties due to their stiffened main backbone chains. Polyacetals and polysulfones are being tested as implant materials, while polycarbonates have found their applications in the heart/lung assist devices, food packaging, etc.

Polyacetals are produced by reacting formaldehyde. These are also sometimes called polyoxymethylene (POM) and known widely as **Delrin®** (DuPont). These polymers have a reasonably high molecular weight ($>2 \times 10^4$ g/mol) and have excellent mechanical properties. More importantly, they display an excellent resistance to most chemicals and to water over wide temperature ranges.

Polysulfones were developed by Union Carbide in the 1960s. These polymers have a high thermal stability due to the bulky side groups (therefore, they are amorphous) and rigid main backbone chains. They are also highly stable to most chemicals but are not so stable in the presence of polar organic solvents such as ketones and chlorinated hydrocarbons.

Polycarbonates are tough, amorphous, and transparent polymers made by reacting bisphenol A and diphenyl carbonate. It is noted for its excellent mechanical and thermal properties (high T_g: 150°C), hydrophobicity, and antioxidative properties.

3.3.12 Biodegradable Polymers

Recently, several biodegradable polymers such as polylactide (PLA), polyglycolide (PGA), poly(glycolide-*co*-lactide) (PLGA), poly(dioxanone), poly(trimethylene carbonate), poly(carbonate), and so on are extensively used or tested on a wide range of medical applications due to their good biocompatibility, controllable biodegradability, and relatively good processability [Khang et al., 1997]. PLA, PGA, and PLGA are bioresorbable polyesters belonging to the group of poly α-hydroxy acids. These polymers degrade by nonspecific hydrolytic scission of their ester bonds. The hydrolysis of PLA yields lactic acid which is a normal byproduct of anaerobic metabolism in the human body and is incorporated in the tricarboxylic acid (TCA) cycle to be finally excreted by the body as carbon dioxide and water. PGA biodegrades by a combination of hydrolytic scission and enzymatic (esterase) action producing glycolic acid which can either enter the TCA cycle or is excreted in urine and can be eliminated as carbon dioxide and water. The degradation time of PLGA can be controlled from weeks to over a year by varying the ratio of monomers and the processing conditions. It might be a suitable biomaterial for use in tissue engineered repair systems in which cells are implanted within PLGA films or scaffolds and in drug delivery systems in which drugs are loaded within PLGA microspheres. PGA (T_m: 225–230°C, T_g: 35–40°C) can be melt spun into fibers which can be converted into bioresorbable sutures, meshes, and surgical products. PLA (T_m: 173–178°C, T_g: 60–65°C) exhibit high tensile strength and low elongation resulting in a high modulus suitable for load-bearing applications such as in bone fracture fixation. Poly-p-dioxanone (T_m: 107–112°C, T_g: ~10°C) is a bioabsorbable polymer which can be fabricated into flexible monofilament surgical sutures.

3.4   Sterilization

Sterilizability of biomedical polymers is an important aspect of the properties because polymers have lower thermal and chemical stability than other materials such as ceramics and metals, consequently, they are also more difficult to sterilize using conventional techniques. Commonly used sterilization techniques are dry heat, autoclaving, radiation, and ethylene oxide gas [Block, 1977].

In dry heat sterilization, the temperature varies between 160 and 190°C. This is above the melting and softening temperatures of many linear polymers like polyethylene and PMMA. In the case of polyamide (Nylon), oxidation will occur at the dry sterilization temperature although this is below its melting temperature. The only polymers which can safely be dry sterilized are PTFE and silicone rubber.

Steam sterilization (autoclaving) is performed under high steam pressure at relatively low temperature (125–130°C). However, if the polymer is subjected to attack by water vapor, this method cannot be employed. PVC, polyacetals, PE (low-density variety), and polyamides belong to this category.

Chemical agents such as ethylene and propylene oxide gases [Glaser, 1979], and phenolic and hypochloride solutions are widely used for sterilizing polymers since they can be used at low temperatures. Chemical agents sometimes cause polymer deterioration even when sterilization takes place at room temperature. However, the time of exposure is relatively short (overnight), and most polymeric implants can be sterilized with this method.

Radiation sterilization [Sato, 1983] using the isotopic ^{60}Co can also deteriorate polymers since at high dosage the polymer chains can be dissociated or cross-linked according to the characteristics of the chemical structures, as shown in Table 3.6. In the case of PE, at high dosage (above 10^6 Gy) it becomes a brittle and hard material. This is due to a combination of random chain scission cross-linking. PP articles will often discolor during irradiation giving the product an undesirable color tint but the more severe problem is the embrittlement resulting in flange breakage, luer cracking, and tip breakage. The physical properties continue to deteriorate with time, following irradiation. These problems of coloration and changing physical properties are best resolved by avoiding the use of any additives which discolor at the sterilizing dose of radiation [Khang, 1996c].

3.5   Surface Modifications for Improving Biocompatability

Prevention of **thrombus** formation is important in clinical applications where blood is in contact such as hemodialysis membranes and tubes, artificial heart and heart–lung machines, prosthetic valves, and artificial vascular grafts. In spite of the use of anticoagulants, considerable platelet deposition and thrombus formation take place on the artificial surfaces [Branger, 1990].

Heparin, one of the complex carbohydrates known as mucopolysaccharides or glycosaminoglycan is currently used to prevent formation of clots. In general, heparin is well tolerated and devoid of serious consequences. However, it allows platelet adhesion to foreign surfaces and may cause hemorrhagic complications such as subdural hematoma, retroperitoneal hematoma, gastrointestinal bleeding, hemorrage

TABLE 3.6   Effect of Gamma Irradiation on
Polymers Which Could be Cross-Linked or
Degraded

Cross-linking polymers	Degradable polymers
Polyethylene	Polyisobutylene
Polypropylene	Poly-α-methylstyrene
Polystyrene	Polymethylmetacrylate
Polyarylates	Polymethacrylamide
Polyacrylamide	Polyvinylidenechloride
Polyvinylchloride	Cellulose and derivatives
Polyamides	Polytetrafluoroethylene
Polyesters	Polytrifluorochloroethylene
Polyvinylpyrrolidone	
Polymethacrylamide	
Rubbers	
Polysiloxanes	
Polyvinylalcohol	
Polyacroleine	

into joints, ocular and retinal bleeding, and bleeding at surgical sites [Lazarus, 1980]. These difficulties give rise to an interest in developing new methods of hemocompatible materials.

Many different groups have studied immobilization of heparin [Kim and Feijen, 1985; Park et al., 1988] on the polymeric surfaces, heparin analogues and heparin-prostaglandin or heparin-fibrinolytic enzyme conjugates [Jozefowicz and Jozefowicz, 1985]. The major drawback of these surfaces is that they are not stable in the blood environment. It has not been firmly established that a slow leakage of heparin is needed for it to be effective as an immobilized antithrombogenic agent, if not its effectiveness could be hindered by being "coated over" with an adsorbed layer of more common proteins such as albumin and **fibrinogen**. Fibrinolytic enzymes, urokinase, and various prostaglandins have also been immobilized by themselves in order to take advantage of their unique fibrin dissolution or antiplatelet aggregation actions [Ohshiro, 1983].

Albumin-coated surfaces have been studied because surfaces that resisted platelet adhesion *in vitro* were noted to adsorb albumin preferentially [Keogh et al., 1992]. Fibronectin coatings have been used in *in vitro* endothelial cell seeding to prepare a surface similar to the natural blood vessel lumen [Lee et al., 1989]. Also, algin-coated surfaces have been studied due to their good biocompatibility and biodegradability [Lee et al., 1990b; 1997b].

Recently, plasma gas discharge [Khang et al., 1997a] and corona treatment [Khang et al., 1996d] with reactive groups introduced on the polymeric surfaces have emerged as other ways to modify biomaterial surfaces [Lee et al., 1991; 1992].

Hydrophobic coatings composed of silicon- and fluorine-containing polymeric materials as well as polyurethanes have been studied because of the relatively good clinical performances of Silastic®, Teflon®, and polyurethane polymers in cardiovascular implants and devices. Polymeric fluorocarbon coatings deposited from a tetrafluoroethylene gas discharge have been found to greatly enhance resistance to both acute thrombotic occlusion and embolization in small diameter **Dacron®** grafts.

Hydrophilic coatings have also been popular because of their low interfacial tension in biological environments [Hoffman, 1981]. Hydrogels as well as various combinations of hydrophilic and hydrophobic monomers have been studied on the premise that there will be an optimum polar-dispersion force ratio which could be matched on the surfaces of the most passivating proteins. The passive surface may induce less clot formation. Polyethylene oxide coated surfaces have been found to resist protein adsorption and cell adhesion and have therefore been proposed as potential "blood compatible" coatings [Lee et al., 1990a]. General physical and chemical methods to modify the surfaces of polymeric biomaterials are listed in Table 3.7 [Ratner et al., 1996].

TABLE 3.7 Physical and Chemical Surface Modification Methods for Polymeric Biomaterials

To modify blood compatibility	Octadecyl group attachment to surface
	Silicon containing block copolymer additive
	Plasma fluoropolymer deposition
	Plasma siloxane polymer deposition
	Radiation-grafted hydrogels
	Chemically modified polystyrene for heparin-like activity
To influence cell adhesion and growth	Oxidized polystyrene surface
	Ammonia plasma-treated surface
	Plasma-deposited acetone or methanol film
	Plasma fluoropolymer deposition
To control protein adsorption	Surface with immobilized polyethyelenglycol
	Treated ELISA dish surface
	Affinity chromatography particulates
	Surface cross-linked contact lens
To improve lubricity	Plasma treatment
	Radiation-grafted hydrogels
	Interpenetrating polymeric networks
To improve wear resistance and corrosion resistance	Ion implantation
	Diamond deposition
	Anodization
To alter transport properties	Plasma deposition (methane, fluoropolymer, siloxane)
To modify electrical characteristics	Plasma deposition
	Solvent coatings
	Parylene coatings

Source: Ratner, B.D. et al. 1996. Academic Press, NY, p. 106.

Another way of making antithrombogenic surfaces is the saline perfusion method, which is designed to prevent direct contacts between blood and the surface of biomaterials by means of perfusing saline solution through the porous wall which is in contact with blood [Park and Kim, 1993; Khang et al., 1996a, b]. It has been demonstrated that the adhesion of the blood cells could be prevented by the saline perfusion through PE, alumina, sulfonated/nonsulfonated PS/SBR, ePTFE (expanded polytetrafluoroethylene), and polysulfone porous tubes.

3.6 Chemogradient Surfaces for Cell and Protein Interaction

The behavior of the adsorption and desorption of blood proteins or adhesion and proliferation of different types of mammalian cells on polymeric materials depends on the surface characteristics such as wettability, hydrophilicity/hydrophobicity ratio, bulk chemistry, surface charge and charge distribution, surface roughness, and rigidity.

Many research groups have studied the effect of the surface wettability on the interactions of biological species with polymeric materials. Some have studied the interactions of different types of cultured cells or blood proteins with various polymers with different wettabilities to correlate the surface wettability and blood- or tissue-compatibility [Baier et al., 1984]. One problem encountered from the study using different kinds of polymers is that the surfaces are heterogeneous, both chemically and physically (different surface chemistry, roughness, rigidity, crystallinity, etc.), which caused widely varying results. Some others have studied the interactions of different types of cells or proteins with a range of methacrylate copolymers with different wettabilities and have the same kind of chemistry but are still physically heterogeneous [van Wachem et al., 1987]. Another methodological problem is that such studies are often tedious, laborious, and time-consuming because a large number of samples must be prepared to characterize the complete range of the desired surface properties.

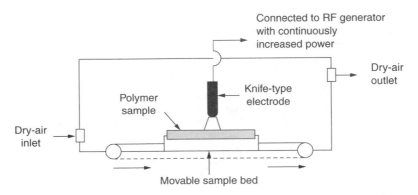

FIGURE 3.6 Schematic diagram showing corona discharge apparatus for the preparation of wettability chemogradient surfaces.

Many studies have been focused on the preparation of surfaces whose properties are changed gradually along the material length. Such chemogradient surfaces are of particular interest in basic studies of the interactions between biological species and synthetic materials surfaces since the affect of a selected property can be examined in a single experiment on one surface preparation. A chemogradient of methyl groups was formed by diffusion of dimethyldichlorosilane through xylene on flat hydrophilic silicone-dioxide surfaces [Elwing et al., 1989]. The wettability chemogradient surfaces were made to investigate hydrophilicity-induced changes of adsorbed proteins.

Recently, a method for preparing wettability chemogradients on various polymer surfaces was developed [Lee et al., 1989, 1990; Khang et al., 1997b]. The wettability chemogradients were produced via radio frequency (RF) and plasma discharge treatment by exposing the polymer sheets continuously to the plasma [Lee et al., 1991]. The polymer surfaces oxidized gradually along the sample length with increasing plasma exposure time and thus the wettability chemogradient was created. Another method for preparing a wettability chemogradient on polymer surfaces using corona discharge treatment has been developed as shown in Figure 3.6 [Lee et al., 1992]. The wettability chemogradient was produced by treating the polymer sheets with corona from a knife-type electrode whose power was gradually changed along the sample length. The polymer surface gradually oxidized with the increasing power and the wettability chemogradient was created. Chemogradient surfaces with different functional gruops such as $-COOH$, $-CH_2OH$, $-CONH_2$, and $-CH_2NH_2$ were produced on PE surfaces by the above corona treatment followed by vinyl monomer grafting and substitution reactions [Kim et al., 1993; Lee et al., 1994a, b]. We have also prepared chargeable functional groups [Lee et al., 1997c, d, 1998a], comb-like polyethyleneoxide (PEO) [Jeong et al., 1996; Lee et al., 1997a] and phospholipid polymer chemogradient surfaces [Iwasaki et al., 1997] by the corona discharge treatment, followed by the graft copolymerization with subsequent substitution reaction of functional vinyl monomers as acrylic acid, sodium p-sulfonic styrene and N, N-dimethyl aminopropyl acrylamide, poly(ethyleneglycol) mono-methacrylate, and ω-methacryloyloxyalkyl phosphorylcholine (MAPC), respectively.

The water contact angles of the corona-treated PE surfaces gradually decrease along the sample length with increasing corona power (from about $95°$ to about $45°$) as shown in Figure 3.7. The decrease in contact angles, that is, the increase in wettability along the sample length was due to the oxygen-based polar functionalities incorporated on the surface by the corona treatment. It was also confirmed also by fourier-transform infrared spectroscopy in the attenuated total reflectance mode and electron spectroscopy for chemical analysis (ESCA).

In order to investigate the interaction of different types of cells in terms of the surface hydrophilicity/hydrophobicity of polymeric materials, Chinese hamster ovaries (CHO), fibroblasts, and bovine aortic endothelial cells (EC) were cultured for 1 and 2 days on the PE wettability chemogradient surfaces. The maximum adhesion and growth of the cells appeared around a water contact angle of 50 to 55°

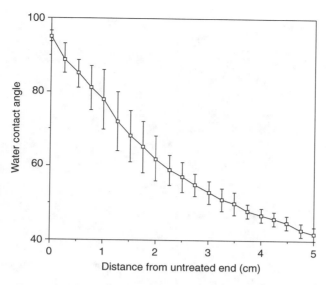

FIGURE 3.7 Changes in water contact angle of corona-treated PE surface along the sample length. Sample numbers, $n = 3$.

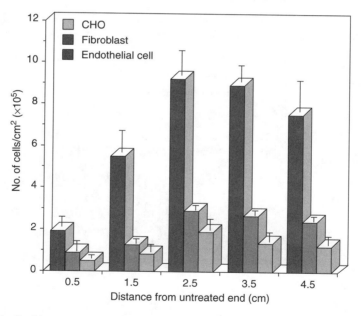

FIGURE 3.8 CHO, fibroblast, and endothelial cell growth on wettability chemogradient PE surfaces after 2 days culture (number of seeded cells, $4 \times 104/cm^2$). $n = 3$.

as shown in Figure 3.8. The observation of scanning electron microscopy (SEM) also verified that the cells are more adhered, spread, and grown onto the sections with moderate hydrophilicity as shown in Figure 3.9.

To determine the cell proliferation rates, the migration of fibroblasts on PE wettability chemogradient surfaces were observed [Khang et al., 1998b]. After the change of culture media at 24 h, cell growth morphology was recorded for 1 or 2 h intervals at the position of 0.5, 1.5, 2.5, and 4.5 cm for the counting

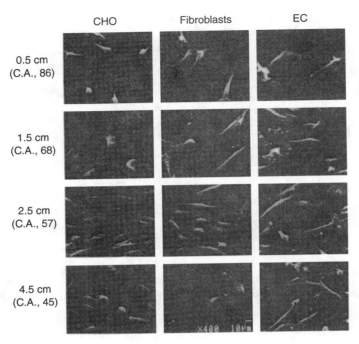

FIGURE 3.9 SEM microphotographs of CHO, fibroblast, and endothelial cells grown on PE wettability chemogradient surface along the sample length after 2 days culture (original magnification; ×400).

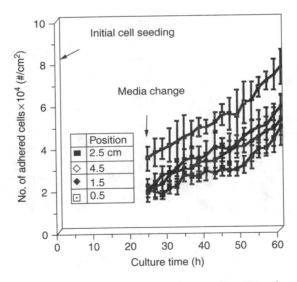

FIGURE 3.10 Fibroblast cell proliferation rates on wettability chemogradient PE surfaces (24 to 60 h culture).

of grown cells and the observation of cell morphology with a video tape recorder. The proliferation rates of fibroblast cells were calculated from the slopes of Figure 3.10 as given in Table 3.8. The proliferation rates on the PE surfaces with wettability chemogradient showed that as the surface wettability increased, it increased and then decreased. The maximum proliferation rate of the cells as 1111 cells/h · cm^2 appeared at around the position 2.5 cm.

TABLE 3.8 Proliferation Rates of Fibroblast Cells on Wettability Gradient PE Surfaces

Positions (cm)	Contact angle (°)	Cell proliferation rate (#cell/h cm^2)
2.5	55	1111
4.5	45	924
1.5	67	838
0.5	85	734

Note: 24 to 60 h culture.

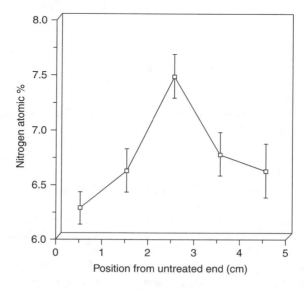

FIGURE 3.11 Serum protein adsorption on PE wettability chemogradient surface (1 h adsorption). $n = 3$.

To observe the effect of serum proteins on the cell adhesion and growth behaviors, fetal bovine serum (FBS), which contains more than 200 kinds of different proteins, was adsorbed onto the wettability gradient PE surfaces for 1 h at 37°C. Figure 3.11 shows the relative adsorbed amount of serum proteins on the wettability gradient surfaces determined by ESCA. The maximum adsorption of the proteins appeared at around the 2.5 cm position, which is the same trend as the cell adhesion, growth, and migration behaviors. It can be explained that preferential adsorption of some serum proteins, like fibronectin and vitronectin from culture medium, onto the moderately wettable surfaces may be a reason for better cell adhesion, spreading, and growth. Proteins like fibronectin and vitronectin are well known as cell-adhesive proteins. Cells attached on surfaces are spread only when they are compatible on the surfaces. It seems that surface wettability plays an important role for cell adhesion, spreading, and migration.

Also investigated were (1) platelet adhesion on wettability chemogradient [Lee and Lee, 1998b], (2) cell interaction on microgrooved PE surfaces (groove depth, 0.5 μm; groove width, 0.45 μm; and pitch, 0.9 μm) with wettability chemogradient [Khang et al., 1997c], (3) detachment of human endothelial under flow from wettability gradient surface with different functional groups [Ruardy et al., 1997], (4) cell interaction on microporous polycarbonate membrane with wettability chemogradient [Lee et al., 1998c], and (5) cell interaction on poly(lactide-*co*-glycolide) surface with wettability chemogradient [Khang et al., 1998a].

During the last several years, "chemogradient surfaces" have evolved into easier and more popular tools for the study of protein adsorption and platelet or cell interactions continuously which relate to the surface properties such as wettability, chemistry and charge, or dynamics of polymeric materials. In many studies, different kinds of polymeric materials with widely varying surface chemistries are used and the explanation of the results is often in controversy due to the surface heterogeneity. In addition, these studies are tedious, laborious, and time-consuming, and biological variations are more likely to occur. The application of chemogradient surfaces for these studies can reduce these discomforts and problems, and eventually save time and money. Also, chemogradient surfaces are valuable in investigating the basic mechanisms by which complicated systems such as proteins or cells interact with surfaces, since a continuum of selected and controlled physical–chemical properties can be studied in one experiment on the polymeric surface.

The possible applications of chemogradient surfaces in the near future are (1) separation devices of cells and/or biological species by different surface properties, (2) column packing materials for separation, (3) biosensoring, etc.

Defining Terms

Acetabulum: The socket portion of the hip joint.

Addition (or free radical) polymerization: Polymerization in which monomers are added to the growing chains, initiated by free radical agents.

Biocompatibility: Acceptance of an artificial implant by the surrounding tissues and as a whole. The implant should be compatible with tissues in terms of mechanical, chemical, surface, and pharmacological properties.

Biomaterials: Synthetic materials used to replace part of a living system or to function in intimate contact with living tissue.

Bone cement: Mixture of polymethylmethacrylate powder and methylmethacrylate monomer liquid to be used as a grouting material for the fixation of orthopedic joint implants.

Branching: Chains grown from the sides of the main backbone chains.

Chemogradient surface: The surface whose properties such as wettability, surface charge, and hydrophilicity/hydrophobicity ratio are changed gradually along the material length.

Condensation (step reaction) polymerization: Polymerization in which two or more chemicals are reacted to form a polymer by condensing out small molecules such as water and alcohol.

Copolymers: Polymers made from two or more monomers which can be obtained by grafting, block, alternating, or random attachment of the other polymer segment.

Covalent bonding: Bonding of atoms or molecules by sharing valence electrons.

Dacron®: Polyethyleneterephthalate polyester that is made into fiber. If the same polymer is made into a film, it is called Mylar®.

Delrin®: Polyacetal made by Union Carbide.

Elastomers: Rubbery materials. The restoring force comes from uncoiling or unkinking of coiled or kinked molecular chains. They can be highly stretched.

Embolus: Any foreign matter, as a blood clot or air bubble, carried in the blood stream.

Fibrinogen: A plasma protein of high molecular weight that is converted to fibrin through the action of thrombin. This material is used to make (absorbable) tissue adhesives.

Filler: Materials added as a powder to a rubber to improve its mechanical properties.

Free volume: The difference in volume occupied by the crystalline state (minimum) and non-crystalline state of a material for a given temperature and a pressure.

Glass transition temperature: Temperature at which solidification without crystallization takes place from viscous liquid.

Grafts: A transplant.

Heparin: A substance found in various body tissues, especially in the liver, that prevents the clotting of blood.

Hydrogel: Polymer which can absorb 30% or more of its weight in water.

Hydrogen bonding: A secondary bonding through dipole interactions in which the hydrogen ion is one of the dipoles.

Hydroquinone: Chemical inhibitor added to the bone cement liquid monomer to prevent accidental polymerization during storage.

Initiator: Chemical used to initiate the addition polymerization by becoming a free radical which in turn reacts with a monomer.

Ionic bonding: Bonding of atoms or molecules through electrostatic interaction of positive and negative ions.

Kevlar®:   Aromatic polyamides made by DuPont.

Lexan®:   Polycarbonate made by General Electric.

Oxygenator: An apparatus by which oxygen is introduced into blood during circulation outside the body, as during open-heart surgery.

Plasticizer: Substance made of small molecules, mixed with (amorphous) polymers to make the chains slide more easily past each other, making the polymer less rigid.

Refractive index: Ratio of speed of light in vacuum to speed of light in a material. It is a measure of the ability of a material to refract (bend) a beam of light.

Repeating unit: Basic molecular unit which can represent a polymer backbone chain. The average number of repeating units is called the degree of polymerization.

Repeating unit: The smallest unit representing a polymer molecular chain.

Semi-crystalline solid: Solid which contains both crystalline and noncrystalline regions and usually occurs in polymers due to their long chain molecules.

Side group: Chemical group attached to the main backbone chain. It is usually shorter than the branches and exists before polymerization.

Steric hindrance: Geometrical interference which restrains movements of molecular groups such as side chains and main chains of a polymer.

Suture: Material used in closing a wound with stitches.

Tacticity: Arrangement of asymmetrical side groups along the backbone chain of polymers. Groups could be distributed at random (atactic), one side (isotactic), or alternating (syndiotactic).

Teflon®:   Polytetrafluoroethylene made by DuPont.

Thrombus: The fibrinous clot attached at the site of thrombosis.

Udel®:   Polysulfone made by General Electric.

Valence electrons: The outermost (shell) electrons of an atom.

van der Waals bonding: A secondary bonding arising through the fluctuating dipole-dipole interactions.

Vinyl polymers: Thermoplastic linear polymers synthesized by free radical polymerization of vinyl monomers having a common structure of $CH_2=CHR$.

Vulcanization: Cross-linking of a (natural) rubber by adding sulfur.

Ziegler–Natta catalyst: Organometallic compounds which have the remarkable capacity of polymerizing a wide variety of monomers to linear and stereoregular polymers.

Acknowledgments

This work was supported by grants from the Korea Ministry of Health and Welfare (grant Nos. HMP-95-G-2-33 and HMP-97-E-0016) and the Korea Ministry of Science and Technology (grant No. 97-N1-02-05-A-02).

References

Baier, R.E., Meyer, A.E., Natiella, J.R., Natiella, R.R., and Carter, J.M. 1984. Surface properties determine bioadhesive outcomes; methods and results, *J. Biomed. Mater. Res.*, 18: 337–355.

Billmeyer, F.W. Jr. 1984. *Textbook of Polymer Science*, 3rd ed. John Wiley & Sons, NY.

Block, S.S. (Ed.) 1977. *Disinfection, Sterilization, and Preservation*, 2nd ed. Rea and Febiger, Philadelphia, PA.

Bloch, B. and Hastings, G.W. 1972. *Plastic Materials in Surgery*, 2nd ed. C.C. Thomas, Springfield, IL.

Brandrup, J. and Immergut, E.H., Ed. 1989. Polymer Handbook, 3rd ed. Wiley-Interscience Pub., NY.

Branger, B., Garreau, M., Baudin, G., and Gris, J.C. 1990. Biocompatibility of blood tubings, *Int. J. Artif. Organs*, 13: 697–703.

Dumitriu, S. (Ed.) 1993. *Polymeric Biomaterials*, Marcell Dekker, Inc., NY.

Elwing, E., Askendal, A., and Lundstorm, I. 1989. Desorption of fibrinogen and γ-globulin from solid surfaces induced by a nonionic detergent, *J. Colloid Interface Sci.*, 128: 296–300.

Glaser, Z.R. 1979. Ethylene oxide: toxicology review and field study results of hospital use, *J. Environ. Pathol. Toxicol.*, 2: 173–208.

Hoffman, A.S. 1981. Radiation processing in biomaterials: A review, *Radiat. Phys. Chem.*, 18: 323–340.

Ikada, Y. (Ed.) 1989. Bioresorbable fibers for medical use. In: *High Technology Fiber*, Part B., Marcel Dekker, NY.

Iwasaki, Y., Ishihara, K., Nakabayashi, N., Khang, G., Jeon, J.H., Lee, J.W., and Lee, H.B. 1997. Preparation of gradient surfaces grafted with phospholipid polymers and evaluation of their blood compatibility. In: *Advances in Biomaterials Science*, Vol. 1, T. Akaike, T. Okano, M. Akashi, M. Terano, and N. Yui, Eds. pp. 91–100, CMC Co., LTD., Tokyo.

Jeong, B.J., Lee, J.H., and Lee, H.B. 1996. Preparation and characterization of comb-like PEO gradient surfaces. *J. Colloid Interface Sci.*, 178: 757–763.

Jozefowicz, M. and Jozefowicz, J. 1985. New approaches to anticoagulation: heparin-like biomaterials. *J. Am. Soc. Art. Intern. Org.* 8: 218–222.

Keogh, J.R., Valender, F.F., and Eaton, J.W. 1992. Albumin-binding surfaces for implantable devices, *J. Biomed. Mater. Res.*, 26: 357–372.

Khang, G., Park J.B., and Lee, H.B. 1996a. Prevention of platelet adhesion on the polysulfone porous catheter by saline perfusion, I. *In vitro* investigation, *Bio-Med. Mater. Eng.*, 6: 47–66.

Khang, G., Park, J.B., and Lee, H.B. 1996b. Prevention of platelet adhesion on the polysulfone porous catheter by saline perfusion, II. *Ex vivo* and *in vivo* investigation, *Bio-Med. Mater. Eng.*, 6: 123–134.

Khang, G., Lee, H.B., and Park, J.B. 1996c. Radiation effects on polypropylene for sterilization, *Bio-Med. Mater. Eng.*, 6: 323–334.

Khang, G., Kang, Y.H., Park, J.B., and Lee, H.B, 1996d. Improved bonding strength of poly-ethylene/polymethylmetacrylate bone cement — a preliminary study, *Bio-Med. Mater. Eng.*, 6: 335–344.

Khang, G., Jeon, J.H., Lee, J.W., Cho, S.C., and Lee, H.B. 1997a. Cell and platelet adhesion on plasma glow discharge-treated poly(lactide-co-glycolide), *Bio-Med. Mater. Eng.*, 7: 357–368.

Khang, G., Lee, J.H., and Lee, H.B. 1997b. Cell and platelet adhesion on gradient surfaces, In: *Advances in Biomaterials Science*. Vol. 1, T. Akaike, T. Okano, M. Akashi, Terano, and N. Yui, Eds. pp. 63–70, CMC Co., LTD., Tokyo.

Khang, G., Lee, J.W., Jeon, J.H., Lee, J.H., and Lee, H.B., 1997c. Interaction of fibroblasts on microgrooved polyethylene surfaces with wettabililty gradient, *Biomat. Res.*, 1: 1–6.

Khang, G., Cho, S.Y., Lee, J.H., Rhee, J.M. and Lee, H.B., 1998a. Interactions of fibroblast, osteo-blast, hepatoma, and endothelial cells on poly(lactide-co-glycolide) surface with chemogradient (to appear).

Khang, G., Jeon, J.H., and Lee, H.B., 1998b. Fibroblast cell migration on polyethylene wettability chemogradient surfaces, *to appear*.

Kim, H.G., Lee, J.H., Lee, H.B., and Jhon, M.S. 1993. Dissociation behavior of surface-grafted poly(acrylic acid): Effects of surface density and counterion size, *J. Colloid Interface Sci.*, 157: 82–87.

Kim, S.W. and Feijen, J. 1985. Surface modification of polymers for improved blood biocompatibility, *CRC Crit. Rev. Biocompat.*, 1: 229–260.

Lazarus, J.M. 1980. Complications in hemodialysis: An overview, *Kidney Int.*, 18: 783–796.

Lee, H.B. 1989. Application of synthetic polymers in implants. In: *Frontiers of Macromolecular Science*, T. Seagusa, T., Higashimura, and A. Abe, Eds. pp. 579–584, Blackwell Scientific Publications, Oxford.

Lee, H.B. and Lee, J.H. 1995. Biocompatibility of solid substrates based on surface wettability. In: *Encyclopedic Handbook of Biomaterials and Bioengineering: Materials*, Vol. 1., D.L. Wise, D.J. Trantolo, D.E. Altobelli, M.J. Yasemski, J.D. Gresser, and E.R. Schwartz, pp. 371–398, Marcel Dekker, New York.

Lee, J.H., Khang, G., Park, K.H., Lee, H.B., and Andrade, J.D. 1989. Polymer surfaces for cell adhesion: I. Surface modification of polymers and ESCA analysis. *J. Korea Soc. Med. Biol. Eng.*, 10: 43–51.

Lee, J.H., Khang, G., Park, J. W., and Lee, H. B. 1990a. Plasma protein adsorption on polyethyleneoxide gradient surfaces, *33rd IUPAC International Symposium on Macromolecules*, July 8–13, Montreal, Canada.

Lee, J.H., Shin, B.C., Khang, G., and Lee, H.B. 1990b. Algin impregnated vascular graft: I. *In vitro* investigation. *J. Korea Soc. Med. Biol. Eng.*, 11: 97–104.

Lee, J.H., Park, J.W., and Lee, H.B. 1991. Cell adhesion and growth on polymer surfaces with hydroxyl groups prepared by water vapor plasma treatment, *Biomaterials*, 12: 443–448.

Lee, J.H., Kim, H.G., Khang, G., Lee, H.B., and Jhon, M.S. 1992. Characterization of wettability gradient surfaces prepared by corona discharge treatment, *J. Colloid Interface Sci.*, 151: 563–570.

Lee, J.H., Kim, H.W., Pak, P.K., and Lee, H.B. 1994a. Preparation and characterization of functional group gradient surfaces, *J. Polym. Sci., Part A, Polym. Chem.*, 32: 1569–1579.

Lee, J.H., Jung, H.W., Kang, I.K., and Lee, H.B. 1994b. Cell behavior on polymer surfaces with different functional groups, *Biomaterials*, 15: 705–711.

Lee, J.H., Jeong, B.J., and Lee, H.B. 1997a. Plasma protein adsorption and platelet adhesion onto comb-like PEO gradient surface, *J. Biomed. Mater. Res.*, 34: 105–114.

Lee, J.H., Kim, W.G., Kim, S.S., Lee, J.H., and Lee, H.B., 1997b. Development and characterization of an alginate-impregnated polyester vascular graft, *J. Biomed. Mater. Res.*, 36: 200–208.

Lee, J.H., Khang, G., Lee, J.H., and Lee, H.B. 1997c. Interactions of protein and cells on functional group gradient surfaces, *Macromol. Symp.*, 118: 571–576.

Lee, J.H., Khang, G., Lee, J.H., and Lee, H.B. 1997d. Interactions of cells on chargeable functional group gradient surfaces, *Biomaterials*, 18: 351–358.

Lee, J.H., Khang, G., Lee, J.H., and Lee, H.B. 1998a. Platelet adhesion onto chargeable functional group gradient surfaces, *J. Biomed. Mater. Res.*, 40: 180–186.

Lee, J.H. and Lee, H.B. 1998b. Platelet adhesion onto wettability gradient surfaces in the absence and presence of plasma protein, *J. Biomed. Mater. Res.*, 41: 304–311.

Lee, J.H., Lee, S.J., Khang, G., and Lee, H.B. 1998c. Interactions of cells onto microporous polycarbonate membrane with wettability gradient surfaces, *J. Biomat. Sci., Polm. Edn.* (in press).

Leininger, R.I. and Bigg, D.M. 1986. Polymers. In: *Handbook of Biomaterials Evaluation*, pp. 24–37, Macmillian Publishing Co., NY.

Oshiro, T. 1983. Thrombosis, antithrombogenic characteristics of immobilized urokinase on synthetic polymers, In: *Biocompatible Polymers, Metals, and Composites*, M. Szycher, Ed. pp. 275–299. Technomic, Lancaster, PA.

Park, J.B. 1984. *Biomaterials Science and Engineering*, Plenum Publication, NY.

Park, J.B. and Lakes, R. 1992. *Biomaterials: An Introduction*, 2nd ed. pp. 141–168, Plenum Press, NY.

Park, J.B. and Kim S.S. 1993. Prevention of mural thrombus in porous inner tube of double-layered tube by saline perfusion. *Bio-Med. Mater. Eng.*, 3: 101–116.

Park, K.D., Okano, T., Nojiri, C., and Kim S.W. 1988. Heparin immobilized onto segmented polyurethane effect of hydrophillic spacers. *J. Biomed. Mater. Res.*, 22: 977–992.

Ratner, B.D., Hoffman, A.S., Schoen, F.J., and Lemons, J.E. 1996. *Biomaterials Science: An Introduction to Materials in Medicine*, Academic Press, NY.

Raurdy, T.G., Moorlag, H.E., Schkenraad, J.M., van der Mei, H.C., and Busscher, H.J. 1997. Detachment of human endothelial under flow from wettability gradient surface with different functional groups, *Cell Mat.*, 7: 123–133.

Sato, K. 1983. Radiation sterilization of medical products. *Radioisotopes*, 32: 431–439.

Shalaby, W.S. 1988. Polymeric materials. In: *Encyclopedia of Med. Dev. Instr.*, J.G. Webster, Ed. pp. 2324–2335, Wiley-Interscience Pub., NY.

Sharma, C.P. and Szycher, M. Eds., *Blood Compatible Materials and Devices: Perspective Toward the 21st Century*, Technomic Publishing Co. Inc., Lancaster, PA.

van Wachem, P.B., Beugeling. T., Feijen, J., Bantjes, A., Detmers, J.P., and van Aken, W.G. 1985. Interaction of cultured human endothelial cells with polymeric surfaces of different wettabilities, *Biomaterials*, 6: 403–408.

4

Composite Biomaterials

Roderic S. Lakes
University of Wisconsin-Madison

Composite materials are solids which contain two or more distinct constituent materials or phases, on a scale larger than the atomic. The term "composite" is usually reserved for those materials in which the distinct phases are separated on a scale larger than the atomic, and in which properties such as the elastic modulus are significantly altered in comparison with those of a homogeneous material. Accordingly, reinforced plastics such as fiberglass as well as natural materials such as bone are viewed as composite materials, but alloys such as brass are not. A foam is a composite in which one phase is empty space. Natural biological materials tend to be composites. Natural composites include bone, wood, dentin, cartilage, and skin. Natural foams include lung, cancellous bone, and wood. Natural composites often exhibit hierarchical structures in which particulate, porous, and fibrous structural features are seen on different micro-scales [Katz, 1980; Lakes, 1993]. In this segment, composite material fundamentals and applications in biomaterials [Park and Lakes, 1988] are explored. Composite materials offer a variety of advantages in comparison with homogeneous materials. These include the ability for the scientist or engineer to exercise considerable control over material properties. There is the potential for stiff, strong, lightweight materials as well as for highly resilient and compliant materials. In biomaterials, it is important that each constituent of the composite be biocompatible. Moreover, the interface between constituents should not be degraded by the body environment. Some applications of composites in biomaterial applications are (1) dental filling composites, (2) reinforced methyl methacrylate bone cement and ultra-high molecular weight polyethylene, and (3) orthopedic implants with porous surfaces.

4.1 Structure

The properties of composite materials depend very much upon *structure*. Composites differ from homogeneous materials in that considerable control can be exerted over the larger scale structure, and hence over the desired properties. In particular, the properties of a composite material depend upon the *shape* of the heterogeneities, upon the *volume fraction* occupied by them, and upon the *interface* among the constituents. The shape of the heterogeneities in a composite material is classified as follows. The principal inclusion shape categories are (1) the particle, with no long dimension, (2) the fiber, with one long dimension, and (3) the platelet or lamina, with two long dimensions, as shown in Figure 4.1. The inclusions may vary in size and shape within a category. For example, particulate inclusions may be spherical, ellipsoidal, polyhedral, or irregular. If one phase consists of voids, filled with air or liquid, the material is known as a cellular solid. If the cells are polygonal, the material is a honeycomb; if the cells are polyhedral, it is a foam. It is necessary in the context of biomaterials to distinguish the above structural cells from biological cells, which occur only in living organisms. In each composite structure, we may moreover make the distinction between random orientation and preferred orientation.

4.2 Bounds on Properties

Mechanical properties in many composite materials depend on structure in a complex way, however for some structures, the prediction of properties is relatively simple. The simplest composite structures are the idealized Voigt and Reuss models, shown in Figure 4.2. The dark and light areas in these diagrams represent the two constituent materials in the composite. In contrast to most composite structures, it is easy to calculate the stiffness of materials with the Voigt and Reuss structures, since in the Voigt structure the strain is the same in both constituents; in the Reuss structure the stress is the same. The Young's modulus, E, of the Voigt composite is:

$$E = E_i V_i + E_m [1 - V_i] \tag{4.1}$$

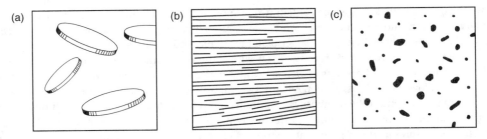

FIGURE 4.1 Morphology of basic composite inclusions. (a) Particle, (b) fiber, and (c) platelet.

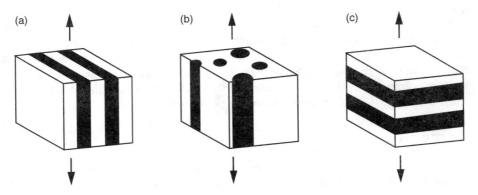

FIGURE 4.2 Voigt (a, laminar; b, fibrous) and Reuss (c) composite models, subjected to tension force indicated by arrows.

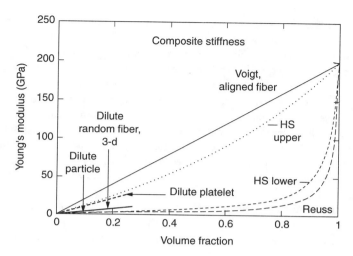

FIGURE 4.3 Stiffness vs. volume fraction for Voigt and Reuss models, as well as for dilute isotropic suspensions of platelets, fibers, and spherical particles embedded in a matrix. Phase moduli are 200 and 3 GPa.

in which E_i is the Young's modulus of the inclusions, and V_i is the volume fraction of inclusions, and E_m is the Young's modulus of the matrix. The Voigt relation for the stiffness is referred to as the rule of mixtures.

The Reuss stiffness E,

$$E = \left[\frac{V_i}{E_i} + \frac{1 - V_i}{E_m} \right]^{-1} \tag{4.2}$$

is less than that of the Voigt model. The Voigt and Reuss models provide upper and lower bounds respectively upon the stiffness of a composite of arbitrary phase geometry [Paul, 1960]. The bounds are far apart if, as is commonplace, the phase moduli differ a great deal, as shown in Figure 4.3. For composite materials which are isotropic, the more complex relations of Hashin and Shtrikman [1963] provide tighter bounds upon the moduli (Figure 4.3); both the Young's and shear moduli must be known for each constituent to calculate these bounds.

4.3 Anisotropy of Composites

Observe that the Reuss laminate is identical to the Voigt laminate, except for a rotation with respect to the direction of load. Therefore, the stiffness of the laminate is *anisotropic*, that is, dependent on direction [Lekhnitskii, 1963; Nye, 1976; Agarwal and Broutman, 1980]. Anisotropy is characteristic of composite materials. The relationship between stress σ_{ij} and strain ε_{kl} in anisotropic materials is given by the tensorial form of Hooke's law as follows:

$$\sigma_{ij} = \sum_{k=1}^{3} \sum_{l=1}^{3} C_{ijkl} \varepsilon_{kl} \tag{4.3}$$

Here C_{ijkl} is the elastic modulus tensor. It has $3^4 = 81$ elements, however since the stress and strain are represented by symmetric matrices with six independent elements each, the number of independent modulus tensor elements is reduced to 36. An additional reduction to 21 is achieved by considering elastic materials for which a strain energy function exists. Physically, C_{2323} represents a shear modulus since it couples a shear stress with a shear strain. C_{1111} couples axial stress and strain in the 1 or x direction, but it is not the same as Young's modulus. The reason is that Young's modulus is measured with the lateral strains free to occur via the Poisson effect, while C_{1111} is the ratio of axial stress to strain when there is only one nonzero strain value; there is no lateral strain. A modulus tensor with 21 independent elements describes a *triclinic* crystal, which is the least symmetric crystal form. The unit cell has three different oblique angles and three different side lengths. A triclinic composite could be made with groups of fibers of three different spacings, oriented in three different oblique directions. Triclinic modulus elements such as C_{2311}, known as cross-coupling constants, have the effect of producing a shear stress in response to a uniaxial strain; this is undesirable in many applications. An *orthorhombic* crystal or an *orthotropic* composite has a unit cell with orthogonal angles. There are nine independent elastic moduli. The associated engineering constants are three Young's moduli, three Poisson's ratios, and three shear moduli; the cross-coupling constants are zero when stresses are aligned to the symmetry directions. An example of such a composite is a unidirectional fibrous material with a rectangular pattern of fibers in the cross-section. Bovine bone, which has a laminated structure, exhibits orthotropic symmetry, as does wood. In a material with *hexagonal* symmetry, out of the nine C elements, there are five independent elastic constants. For directions in the transverse plane the elastic constants are the same, hence the alternate name transverse isotropy. A unidirectional fiber composite with a hexagonal or random fiber pattern has this symmetry, as does human Haversian bone. In *cubic* symmetry, there are three independent elastic constants, a Young's modulus, E, a shear modulus, G, and an independent Poisson's ratio, ν. Cross-weave fabrics have cubic symmetry. Finally, an *isotropic* material has the same material properties in any direction. There are only two independent elastic constants, hence E, G, ν, and also the bulk modulus B are related in an isotropic material. Isotropic materials include amorphous solids, polycrystalline metals in which the grains are randomly oriented, and composite materials in which the constituents are randomly oriented.

Anisotropic composites offer superior strength and stiffness in comparison with isotropic ones. Material properties in one direction are gained at the expense of properties in other directions. It is sensible, therefore, to use anisotropic composite materials only if the direction of application of the stress is known in advance.

4.4 Particulate Composites

It is often convenient to stiffen or harden a material, commonly a polymer, by the incorporation of particulate inclusions. The shape of the particles is important [see Christensen, 1979]. In isotropic systems, stiff platelet (or flake) inclusions are the most effective in creating a stiff composite, followed by fibers; and the least effective geometry for stiff inclusions is the spherical particle, as shown in Figure 4.3. A dilute concentration of spherical particulate inclusions of stiffness E_i and volume fraction V_i, in a matrix (with Poisson's ratio assumed to be 0.5) denoted by the subscript m, gives rise to a composite with a stiffness E:

$$E = \frac{5(E_i - E_m)V_i}{3 + 2(E_i/E_m)} + E_m \tag{4.4}$$

The stiffness of such a composite is close to the Hashin–Shtrikman lower bound for isotropic composites. Even if the spherical particles are perfectly rigid compared with the matrix, their stiffening effect at low concentrations is modest. Conversely, when the inclusions are more compliant than the matrix, spherical ones reduce the stiffness the least and platelet ones reduce it the most. Indeed, soft platelets are suggestive of crack-like defects. Soft platelets, therefore result not only in a compliant composite, but also a weak one. Soft spherical inclusions are used intentionally as crack stoppers to enhance the toughness of polymers such as polystyrene (high impact polystyrene), with a small sacrifice in stiffness.

Particle reinforcement has been used to improve the properties of bone cement. For example, inclusion of bone particles in PMMA cement somewhat improves the stiffness and improves the fatigue life considerably [Park et al., 1986]. Moreover, the bone particles at the interface with the patient's bone are

FIGURE 4.4 Microstructure of a dental composite. Miradapt® [Johnson & Johnson] 50% by volume filler: barium glass and colloidal silica [Park and Lakes, 1992].

ultimately resorbed and are replaced by ingrown new bone tissue. This approach is in the experimental stages.

Rubber used in catheters, rubber gloves, etc. is usually reinforced with very fine particles of silica (SiO_2) to make the rubber stronger and tougher.

Teeth with decayed regions have traditionally been restored with metals such as silver amalgam. Metallic restorations are not considered desirable for anterior teeth for cosmetic reasons. Acrylic resins and silicate cements had been used for anterior teeth, but their poor material properties led to short service life and clinical failures. Dental composite resins have virtually replaced these materials and are very commonly used to restore posterior teeth as well as anterior teeth [Cannon, 1988].

The dental composite resins consist of a polymer matrix and stiff inorganic inclusions [Craig, 1981]. A representative structure is shown in Figure 4.4. The particles are very angular in shape. The inorganic inclusions confer a relatively high stiffness and high wear resistance on the material. Moreover, since they are translucent and their index of refraction is similar to that of dental enamel, they are cosmetically acceptable. Available dental composite resins use quartz, barium glass, and colloidal silica as fillers. Fillers have particle size from 0.04 to 13 μm, and concentrations from 33 to 78% by weight. In view of the greater density of the inorganic filler phase, a 77% weight percent of filler corresponds to a volume percent of about 55%. The matrix consists of a polymer, typically BIS-GMA. In restoring a cavity, the dentist mixes several constituents, then places them in the prepared cavity to polymerize. For this procedure to be successful the viscosity of the mixed paste must be sufficiently low and the polymerization must be controllable. Low viscosity liquids such as triethylene glycol dimethacrylate are used to lower the viscosity and inhibitors such as BHT (butylated trioxytoluene) are used to prevent premature polymerization. Polymerization can be initiated by a thermochemical initiator such as benzoyl peroxide, or by a photochemical initiator (benzoin alkyl ether) which generates free radicals when subjected to ultraviolet light from a lamp used by the dentist.

Dental composites have a Young's modulus in the range 10 to 16 GPa, and the compressive strength from 170 to 260 MPa [Cannon, 1988]. As shown in Table 4.1, these composites are still considerably less stiff than dental enamel, which contains about 99% mineral. Similar high concentrations of mineral particles in synthetic composites cannot easily be achieved, in part because the particles do not pack densely. Moreover, an excessive concentration of particles raises the viscosity of the unpolymerized paste. An excessively high viscosity is problematical since it prevents the dentist from adequately packing the paste into the prepared cavity; the material will then fill in crevices less effectively.

The thermal expansion of dental composites, as with other dental materials, exceeds that of tooth structure. Moreover, there is a contraction during polymerization of 1.2 to 1.6%. These effects are thought

TABLE 4.1   Properties of Bone, Teeth, and Biomaterials

Material	Young's modulus E(GPa)	Density ρ (g/cm^3)	Strength (MPa)	References
Hard Tissue				
Tooth, bone, human compact bone, longitudinal direction	17	1.8	130 (tension)	Craig and Peyton, 1958; Reilly and Burstein, 1975: Peters et al., 1984; Park and Lakes, 1992
Tooth dentin	18	2.1	138 (compression)	
Tooth enamel	50	2.9		
Polymers				Park and Lakes, 1992
Polyethylene (UHMW)	1	0.94	30 (tension)	
Polymethyl methacrylate, PMMA	3	1.1	65 (tension)	
PMMA bone cement	2	1.18	30 (tension)	
Metals				Park and Lakes, 1992
316L Stainless steel (wrought)	200	7.9	1000 (tension)	
Co-Cr-Mo (cast)	230	8.3	660 (tension)	
Co Ni Cr Mo (wrought)	230	9.2	1800 (tension)	
Ti6A14V	110	4.5	900 (tension)	
Composites				
Graphite-epoxy (unidirectional fibrous, high modulus)	215	1.63	1240 (tension)	Schwartz, 1997
Graphite-epoxy (quasi-isotropic fibrous)	46	1.55	579 (tension)	Schwartz, 1997
Dental composite resins (particulate)	10–16		170–260 (compression)	Cannon, 1988
Foams				Gibson and Ashby, 1988
Polymer foams	10^{-4}–1	0.002–0.8	0.01–1 (tension)	

to contribute to leakage of saliva, bacteria, etc., at the interface margins. Such leakage in some cases can cause further decay of the tooth.

Use of colloidal silica in the so-called "microfilled" composites allows these resins to be polished, so that less wear occurs and less plaque accumulates. It is more difficult, however, to make these with a high fraction of filler. All the dental composites exhibit creep. The stiffness changes by a factor of 2.5 to 4 (depending on the particular material) over a time period from 10 sec to 3 h under steady load [Papadogianis et al., 1985]. This creep may result in indentation of the restoration, but wear seems to be a greater problem.

Dental composite resins have become established as restorative materials for both anterior and posterior teeth. The use of these materials is likely to increase as improved compositions are developed and in response to concern over long term toxicity of silver-mercury amalgam fillings.

4.5   Fibrous Composites

Fibers incorporated in a polymer matrix increase the stiffness, strength, fatigue life, and other properties [Agarwal and Broutman, 1980; Schwartz, 1992]. Fibers are mechanically more effective in achieving a stiff, strong composite than are particles. Materials can be prepared in fiber form with very few defects which concentrate stress. Fibers such as graphite are stiff (Young's modulus is 200–800 GPa) and strong (the tensile strength is 2.7–5.5 GPa). Composites made from them can be as strong as steel but much lighter, as shown in Table 4.1. The stiffness of a composite with aligned fibers, if it is loaded along the fibers, is equivalent to the Voigt upper bound, Equation 4.1. Unidirectional fibrous composites, when

FIGURE 4.5   Knee prostheses with polyethylene tibial components reinforced with carbon fiber.

loaded along the fibers, can have strengths and stiffnesses comparable to that of steel, but with much less weight (Table 4.1). However if it is loaded transversly to the fibers, such a composite will be compliant, with a stiffness not much greater than that of the matrix alone. While unidirectional fiber composites can be made very strong in the longitudinal direction, they are weaker than the matrix alone when loaded transversely, as a result of stress concentration around the fibers. If stiffness and strength are needed in all directions, the fibers may be oriented randomly. For such a three-dimensional isotropic composite, for a low concentration of fibers,

$$E = \frac{E_i V_i}{6} + E_m \qquad (4.5)$$

so the stiffness is reduced by about a factor of six in comparison with an aligned composite as illustrated in Figure 4.3. However if the fibers are aligned randomly in a plane, the reduction in stiffness is only a factor of three. The degree of anisotropy in fibrous composites can be very well controlled by forming laminates consisting of layers of fibers embedded in a matrix. Each layer can have fibers oriented in a different direction. One can achieve quasi-isotropic behavior in the laminate plane; such a laminate is not as strong or as stiff as a unidirectional one, as illustrated in Table 4.1. Strength of composites depends on such particulars as the brittleness or ductility of the inclusions and the matrix. In fibrous composites failure may occur by (1) fiber breakage, buckling, or pullout, (2) matrix cracking, or (3) debonding of fiber from matrix.

Short fiber composites are used in many applications. They are not as stiff or as strong as composites with continuous fibers, but they can be formed economically by injection molding or by *in situ* polymerization. Choice of an optimal fiber length can result in improved toughness, due to the predominance of fiber pull-out as a fracture mechanism.

Carbon fibers have been incorporated in the high density polyethylene used in total knee replacements (Figure 4.5). The standard ultra high molecular weight polyethylene (UHMWPE) used in these implants

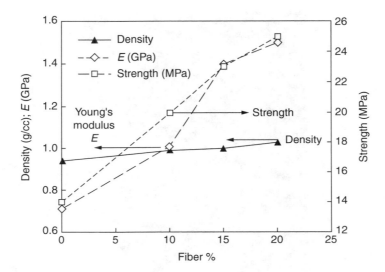

FIGURE 4.6 Properties of carbon fiber reinforced ultra high molecular weight polyethylene. (Replotted from Sclippa, E. and Piekarski, K. 1973. *J. Biomed. Mater. Res.*, **7**, 59–70. With permission.)

is considered adequate for most purposes for implantation in older patients. A longer wear-free implant lifetime is desirable for use in younger patients. It is considered desirable to improve the resistance to creep of the polymeric component, since excessive creep results in an indentation of that component after long term use. Representative properties of carbon reinforced ultra high molecular weight polyethylene are shown in Figure 4.6 [Sclippa and Piekarski, 1973]. Enhancements of various properties by a factor of two are feasible.

Polymethyl methacrylate (PMMA) used in bone cement is compliant and weak in comparison with bone. Therefore several reinforcement methods have been attempted. Metal wires have been used clinically as macroscopic "fibers" to reinforce PMMA cement used in spinal stabilization surgery [Fishbane and Pond, 1977]. The wires are made of a biocompatible alloy such as cobalt–chromium alloy or stainless steel. Such wires are not currently used in joint replacements owing to the limited space available. Graphite fibers have been incorporated in bone cement [Knoell et al., 1975] on an experimental basis. Significant improvements in the mechanical properties have been achieved. Moreover, the fibers have an added beneficial effect of reducing the rise in temperature which occurs during the polymerization of the PMMA in the body. Such high temperature can cause problems such as necrosis of a portion of the bone into which it is implanted. Thin, short titanium fibers have been embedded in PMMA cement [Topoleski et al., 1992]; a toughness increase of 51% was observed with a 5% volumetric fiber content. Fiber reinforcement of PMMA cement has not found much acceptance since the fibers also increase the viscosity of the unpolymerized material. It is consequently difficult for the surgeon to form and shape the polymerizing cement during the surgical procedure.

Metals are currently used in bone plates for immobilizing fractures and in the femoral component of total hip replacements. A problem with currently used implant metals is that they are much stiffer than bone, so they shield the nearby bone from mechanical stress. Stress-shielding results in a kind of disuse atrophy: the bone resorbs [Engh and Bobyn, 1988]. Therefore composite materials have been investigated as alternatives [Bradley et al., 1980; Skinner, 1988]. Fibrous composites can deform to higher strains (to about 0.01) than metals (0.001 for a mild steel) without damage. This resilience is an attractive characteristic for more flexible bone plates and femoral stems. Flexible composite bone plates are effective in promoting healing [Jockish, 1992]. Composite hip replacement prostheses have been made with carbon fibers in a matrix of polysulfone and polyetherether ketone (PEEK). These prostheses experience heavy load with a static component. Structural metals such as stainless steel and cobalt chromium alloys do not

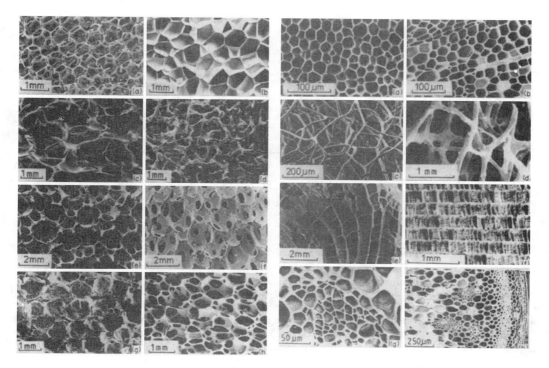

FIGURE 4.7 Cellular solids structures, after Gibson and Ashby [1988]. Left: synthetic cellular solids: (a) open-cell polyurethane, (b) closed-cell polyethylene, (c) foamed nickel, (d) foamed copper, (e) foamed zirconia, (f) foamed mullite, (g) foamed glass, (h) polyester foam with both open and closed cells. Right: natural cellular solids: (a) cork, (b) balsa wood, (c) sponge, (d) cancellous bone, (e) coral, (f) cuttlefish bone, (g) iris leaf, (h) plant stalk.

creep significantly at room or body temperature. In composites which contain a polymer constituent, creep behavior is a matter of concern. The carbon fibers exhibit negligible creep, but polymer constituents tend to creep. Prototype composite femoral components were found to exhibit fiber dominated creep of small magnitude and are not expected to limit the life of the implant [Maharaj and Jamison, 1993].

Fibrous composites have also been used in external medical devices such as knee braces [Yeaple, 1989], in which biocompatibility is not a concern but light weight is crucial.

4.6 Porous Materials

The presence of voids in porous or cellular solids will reduce the stiffness of the material. For some purposes, that is both acceptable and desirable. Porous solids are used for many purposes: flexible structures such as (1) seat cushions, (2) thermal insulation, (3) filters, (4) cores for stiff and lightweight sandwich panels, (5) flotation devices, and (6) to protect objects from mechanical shock and vibration; and in biomaterials, as coatings to encourage tissue ingrowth. Representative cellular solid structures are shown in Figure 4.7.

The stiffness of an open-cell foam is given by [Gibson and Ashby, 1988]

$$E = E_s[V_s]^2 \tag{4.6}$$

in which E_s is the Young's modulus and V_s is the volume fraction of the solid phase of the foam; V_s is also called the relative density.

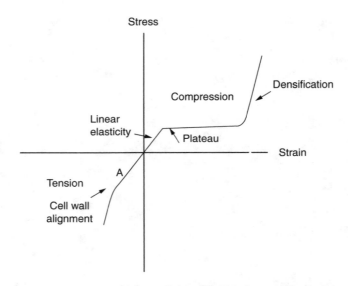

FIGURE 4.8   Representative stress-strain curve for a cellular solid. The plateau region for compression in the case of elastomeric foam (a rubbery polymer) represents elastic buckling; for an elastic-plastic foam (such as metallic foam), it represents plastic yield, and for an elastic-brittle foam (such as ceramic) it represents crushing. On the tension side, point 'A' represents the transition between cell wall bending and cell wall alignment. In elastomeric foam, the alignment occurs elastically, in elastic plastic foam it occurs plastically, and an elastic-brittle foam fractures at A.

The strength for crushing of a brittle foam and the elastic collapse of an elastomeric foam is given, respectively, by

$$\sigma_{\text{crush}} = 0.65 \, \sigma_{\text{f,s}}[V_{\text{s}}]^{3/2} \tag{4.7}$$

$$\sigma_{\text{coll}} = 0.05 \, E_{\text{s}}[V_{\text{s}}]^{2} \tag{4.8}$$

Here $\sigma_{\text{f,s}}$ is the fracture strength of the solid phase. These strength relations are valid for relatively small density. Their derivation is based on the concept of *bending* of the cell ribs and is presented by Gibson and Ashby [1988]. Most man-made closed cell foams tend to have a concentration of material at the cell edges, so that they behave mechanically as open cell foams. The salient point in the relations for the mechanical properties of cellular solids is that the *relative density* dramatically influences the stiffness and the strength. As for the relationship between stress and strain, a representative stress strain curve is shown in Figure 4.8. The physical mechanism for the deformation mode beyond the elastic limit depends on the material from which the foam is made. Trabecular bone, for example, is a natural cellular solid, which tends to fail in compression by crushing. Many kinds of trabecular bone appear to behave mechanically as an open cell foam. For trabecular bone of unspecified orientation, the stiffness is proportional to the cube of the density and the strength as the square of the density [Gibson and Ashby, 1988], which indicates behavior dominated by bending of the trabeculae. For bone with oriented trabeculae, both stiffness and strength in the trabecular direction are proportional to the density, a fact which indicates behavior dominated by axial deformation of the trabeculae.

Porous materials have a high ratio of surface area to volume. When porous materials are used in biomaterial applications, the demands upon the inertness and biocompatibility are likely to be greater than for a homogeneous material.

Porous materials, when used in implants, allow tissue ingrowth [Spector et al., 1988a,b]. The ingrowth is considered desirable in many contexts, since it allows a relatively permanent anchorage of the implant to the surrounding tissues. There are actually two composites to be considered in porous implants (1) the implant prior to ingrowth, in which the pores are filled with tissue fluid which is ordinarily of no

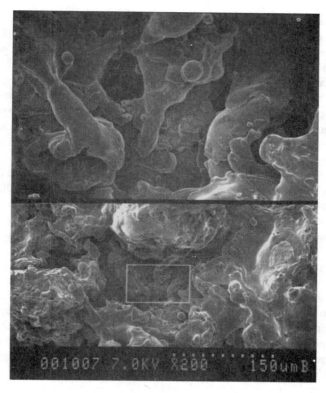

FIGURE 4.9 Irregular pore structure of porous coating in Ti5Al4V alloy for bony ingrowth. The top scanning electron microscopic picture is a 5× magnification of the rectangular region of the bottom picture (200×). (From Park, J.B. and Lakes, R.S. 1992. *Biomaterials*, Plenum, New York.)

mechanical consequence; and (2) the implant filled with tissue. In the case of the implant prior to ingrowth, it must be recognized that the stiffness and strength of the porous solid are much less than in the case of the solid from which it is derived.

Porous layers are used on bone compatible implants to encourage bony ingrowth [Galante et al., 1971; Ducheyne, 1984]. The pore size of a cellular solid has no influence on its stiffness or strength (though it does influence the toughness), however pore size can be of considerable biological importance. Specifically, in orthopedic implants with pores larger than about 150 μm, bony ingrowth into the pores occurs and this is useful to anchor the implant. This minimum pore size is on the order of the diameter of osteons in normal Haversian bone. It was found experimentally that pores <75 μm in size did not permit the ingrowth of bone tissue. Moreover, it was difficult to maintain fully viable osteons within pores in the 75 to 150 μm size range. Representative structure of such a porous surface layer is shown in Figure 4.9. Porous coatings are also under study for application in anchoring the artificial roots of dental implants to the underlying jawbone. Porous hydroxyapatite has been studied for use in repairing large defects in bone [Meffert et al., 1985; Holmes et al., 1986]. Hydroxyapatite is the mineral constituent of bone, and it has the nominal composition $Ca_{10}(PO_4)_6(OH)_2$. Implanted hydroxyapatite is slowly resorbed by the body over several years and replaced by bone. Tricalcium phosphate is resorbed more quickly and has been considered as an implant constituent to speed healing.

When a porous material is implanted in bone, the pores become filled first with blood which clots, then with osteoprogenitor mesenchymal cells, then, after about 4 weeks, bony trabeculae. The ingrown bone then becomes remodeled in response to mechanical stress. The bony ingrowth process depends on a degree of mechanical stability in the early stages of healing. If too much motion occurs, the ingrown tissue will be collagenous scar tissue, not bone.

Porous materials used in soft tissue applications include polyurethane, polyamide, and polyester velours used in percutaneous devices. Porous reconstituted collagen has been used in artificial skin, and braided polypropylene has been used in artificial ligaments. As in the case of bone implants, the porosity encourages tissue ingrowth which anchors the device.

Blood vessel replacements are made with porous materials which encourage soft tissue to grow in, eventually forming a new lining, or neointima. The new lining consists of the patient's own cells. It is a natural nonthrombogenic surface resembling the lining of the original blood vessel. This is a further example of the biological role of porous materials as contrasted with the mechanical role.

Ingrowth of tissue into implant pores is not always desirable. For example, sponge (polyvinyl alcohol) implants used in early mammary augmentation surgery underwent ingrowth of fibrous tissue, and contracture and calcification of that tissue, resulting in hardened, calcified breasts. Current mammary implants make use of a balloon-like nonporous silicone rubber layer enclosing silicone oil or gel, or perhaps a saline solution in water. A porous layer of polyester felt or velour attached to the balloon is provided at the back surface of the implant so that limited tissue ingrowth will anchor it to the chest wall and prevent it from migrating.

Foams are also used externally to protect the human body from injury. Examples include knee pads, elbow pads, wrestling mats, and wheelchair cushions. Since these foams are only in contact with skin rather than any internal organs, they are not subject to rigorous biocompatibility requirements. They are therefore designed based on mechanical considerations. Foam used in sports equipment must have the correct compliance to limit impact force without bottoming out. Foam used in wheelchair cushions is intended to prevent pressure sores in people who suffer limited mobility. The properties of cushions are crucial in reducing illness and suffering in people who are confined to wheelchairs or hospital beds for long periods. Prolonged pressure on body parts can obstruct circulation in the capillaries. If this lasts too long it may cause a sore or ulcer called a pressure sore, also called a bed sore. In its most severe manifestation, a pressure sore can form a deep crater-like ulcer in which underlying muscle or bone is exposed [Dinsdale, 1974]. A variety of flexible cushion materials have been tried to minimize the incidence and severity of pressure sores [Garber, 1985]. Viscoelastic foam allows the cushion to progressively conform to the body shape. However, progressive densification of the foam due to creep results in a stiffer cushion which must be periodically replaced.

Porous materials are produced in a variety of ways. For example, in the case of bone compatible surfaces they are formed by sintering of beads or wires. Vascular and soft tissue implants are produced by weaving or braiding fibers as well as by nonwoven "felting" methods. Protective foams for use outside the body are usually produced by use of a "blowing agent" which is a chemical which evolves gas during the polymerization of the foam. An interesting approach to producing micro-porous materials is the replication of structures found in biological materials: the *replamineform* process [White et al., 1976]. The rationale is that the unique structure of communicating pores is thought to offer advantages in the induction of tissue ingrowth. The skeletal structure of coral or echinoderms (such as sea urchins) is replicated by a casting process in metals and polymers; these have been tried in vascular and tracheal prostheses as well as in bone substitutes.

4.7 Biocompatibility

Carbon itself has been successfully used as a biomaterial. Carbon based fibers used in composites are known to be inert in aqueous (even seawater) environments, however they do not have a track record in the biomaterials setting. *In vitro* studies by Kovacs [1993] disclose substantial electrochemical activity of carbon fiber composites in an aqueous environment. If such composites are placed near a metallic implant, galvanic corrosion is a possibility. Composite materials with a polymer matrix absorb water when placed in a hydrated environment such as the body. Moisture acts as a plasticizer of the matrix and shifts the glass transition temperature towards lower values [DeIasi and Whiteside, 1978], hence a reduction in stiffness and an increase in mechanical damping. Water immersion of a graphite epoxy cross-ply composite [Gopalan et al., 1989] for 20 days reduced the strength by 13% and the stiffness by 9%. Moisture absorption by polymer constituents also causes swelling. Such swelling can be beneficial in dental composites since it offsets some of the shrinkage due to polymerization.

Flexible composite bone plates are effective in promoting healing [Jockish, 1992], but particulate debris from composite bone plates gives rise to a foreign body reaction similar to that caused by ultra high molecular weight polyethylene.

4.8 Summary

Composite materials are a relatively recent addition to the class of materials used in structural applications. In the biomaterials field, the ingress of composites has been even more recent. In view of their potential for high performance, composite materials are likely to find increasing use as biomaterials.

References

Agarwal, A.G. and Broutman, L.J. 1980. *Analysis and Performance of Fiber Composites*, John Wiley & Sons, New York.

Bradley, J.S., Hastings, G.W., and Johnson-Hurse, C. 1980. Carbon fiber reinforced epoxy as a high strength, low modulus material for internal fixation plates, *Biomaterials* 1, 38–40.

Cannon, M.L. 1988. Composite resins, in: *Encyclopedia of Medical Devices and Instrumentation*, J.G. Webster, Ed., John Wiley & Sons, New York.

Christensen, R.M. 1979. *Mechanics of Composite Materials*, John Wiley & Sons, New York.

Craig, R. 1981. Chemistry, composition, and properties of composite resins, In: *Dental Clinics of North America*, H. Horn, Ed., W.B. Saunders, Philadelphia, PA.

Craig, R.G. and Peyton, F.A. 1958. Elastic and mechanical properties of human dentin, *J. Dental Res.*, 37, 710–718.

DeIasi, R. and Whiteside, J.B. 1978. Effect of moisture on epoxy resins and composites: advanced composite materials — environmental effects, J.R. Vinson, Ed. *ASTM Publication STP 658*, Philadelphia, PA.

Dinsdale, S.M. 1974. Decubitus ulcers: role of pressure and friction in causation, *Arch. Phys. Med. Rehabil.*, 55, 147–152.

Ducheyne, P. 1984. Biological fixation of implants, in: *Functional Behavior of Orthopaedic Biomaterials*, G.W. Hastings and P. Ducheyne, Eds. CRC Press, Boca Raton, FL.

Engh, C.A. and Bobyn, J.D. 1988. Results of porous coated hip replacement using the AML prosthesis, In: *Non-Cemented Total Hip Arthroplasty*, Raven Press, New York.

Fishbane, B.M. and Pond, R.B. 1977. Stainless steel fiber reinforcement of polymethylmethacrylate, *Clin. Orthop.*, 128, 490–498.

Galante, J., Rostoker, W., Lueck, R., and Ray, R.D. 1971. Sintered fiber metal composites as a basis for attachment of implants to bone, *J. Bone Joint Surg.*, 53A, 101–114.

Garber, S.L. 1985. Wheelchair cushions: a historical review, *Am. J. Occup. Ther.*, 39, 453–459.

Gibson, L.J. and Ashby, M.F. 1988. *Cellular Solids*, Cambridge, England.

Gopalan, R., Somashekar, B.R., and Dattaguru, B. 1989. Environmental effects on fiber-polymer composites, *Polym. Degradat. Stabil.*, 24, 361–371.

Hashin, Z. and Shtrikman, S. 1963. A variational approach to the theory of the elastic behavior of multiphase materials, *J. Mech. Phys. Solids*, 11, 127–140.

Holmes, D.E., Bucholz, R.W., and Mooney, V. 1986. Porous hydroxyapatite as a bone graft substitute in metaphyseal defects, *J. Bone Jnt. Surg.*, 68, 904–911.

Katz, J.L. 1980. Anisotropy of Young's modulus of bone. *Nature*, 283, 106–107.

Kovacs, P. 1993. *In vitro* studies of the electrochemical behavior of carbon-fiber composites, in: *Composite Materials for Implant Applications in the Human Body: Characterization and Testing, ASTM STP 1178*. R.D. Jamison and L.N. Gilbertson, Eds. ASTM, Philadelphia, PA, pp. 41–52.

Jockish, K.A., Brown, S.A., Bauer, T.W., and Merritt, K. 1992. Biological response to chopped carbon reinforced PEEK, *J. Biomed. Mater. Res.*, 26, 133–146.

Knoell, A., Maxwell, H., and Bechtol, C. 1975. Graphite fiber reinforced bone cement, *Ann. Biomed. Eng.*, 3, 225–229.

Lakes, R.S. 1993. Materials with structural hierarchy, *Nature*, 361, 511–515.

Lekhniitski, 1963. Elasticity of an anisotropic elastic body, Holden Day.

Maharaj, G.R. and Jamison, R.D. 1993. Creep testing of a composite material human hip prosthesis, In: *Composite Materials for Implant Applications in the Human Body: Characterization and Testing*, ASTM STP 1178. R.D. Jamison and L.N. Gilbertson, Eds. Am. Soc. Testing, Materials, Philadelphia, PA, 86–97.

Meffert, R.M., Thomas, J.R., Hamilton, K.M., and Brownstein, C.N. 1985. Hydroxylapatite as allopathic graft in the treatment of periodontal osseous defects, *J. Periodontol.*, 56, 63–73.

Park, H.C., Liu, Y.K., and Lakes, R.S. 1986. The material properties of bone-particle impregnated PMMA, *J. Biomech. Eng.*, 108, 141–148.

Nye, J.F. 1976. *Physical Properties of Crystals*, Oxford University Press, Oxford.

Papadogianis, Y., Boyer, D.B., and Lakes, R.S. 1985. Creep of posterior dental composites, *J. Biomed. Mat. Res.*, 19, 85–95.

Park, J.B. and Lakes, R.S. 1992. *Biomaterials*, Plenum, New York.

Paul, B. 1960. Prediction of elastic constants of multiphase materials, *Trans. AIME*, 218, 36–41.

Peters, M.C., Poort, H.W., Farah, J.W., and Graig, R.G. 1983. Stress analysis of a tooth restored with a post and a core, *J. Dental Res.*, 62, 760–763.

Reilly, D.T. and Burstein, A.H. 1975. The elastic and ultimate properties of compact bone tissue, *J. Biomech.*, 8, 393–405.

Schwartz, M.M. 1992. *Composite Materials Handbook*, 2nd ed. McGraw-Hill, New York.

Sclippa, E. and Piekarski, K. 1973. Carbon fiber reinforced polyethylene for possible orthopaedic usage, *J. Biomed. Mater. Res.*, 7, 59–70.

Skinner, H.B. 1988. Composite technology for total hip arthroplasty. *Clin. Orthop. Rel. Res.*, 235, 224–236.

Spector, M., Miller, M., and Beals, N. 1988a. Porous materials, In: *Encyclopedia of Medical Devices and Instrumentation*, J.G. Webster, Ed. John Wiley & Sons, New York.

Spector, M., Heyligers, I., and Robertson, J.R. 1988b. Porous polymers for biological fixation, *Clin. Orthop. Rel. Res.*, 235, 207–219.

Topoleski, L.D.T., Ducheyne, P., and Cackler, J.M. 1992. The fracture toughness of titanium fiber reinforced bone cement, *J. Biomed. Mater. Res.*, 26, 1599–1617.

White, R.A., Weber, J.N., and White, E.W. 1976. Replamineform: a new process for preparing porous ceramic, metal, and polymer prosthetic materials, *Science*, 176, 922.

Yeaple, F. 1989. Composite knee brace returns stability to joint, *Design News*, 46, 116.

5

Biodegradable Hydrogels: Tailoring Properties and Function through Chemistry and Structure

Andrew T. Metters
Chien-Chi Lin
Clemson University

5.1   Introduction

Hydrogels are water-swollen polymeric networks. They remain insoluble when placed in aqueous environments due to chemical or physical cross-linking of individual polymer chains. Chemical cross-links may be ionic or covalent, while physical cross-links may be entanglements, crystallites, or weak associations, such as hydrogen bonding or van der Waals interactions [Lowman and Peppas, 1999].

Hydrogels have played an important role in biomedical applications since the late 1950s and the development of poly(2-hydroxyethyl methacrylate) (PHEMA) as a soft contact lens material [Wichterle and Lim, 1960]. Since then hydrogels have found application in drug delivery. In the last decade hydrogels have played an ever-increasing role in the revolutionary field of tissue engineering where they are used as scaffolds to guide the growth of new tissues. Their widespread acceptance in these fields is primarily due to the structural similarity that they exhibit compared to macromolecular-based components in the body and their ability to replicate the properties of natural tissue better than any synthetic material. Hydrogels such as those produced from PHEMA have high water contents at equilibrium, exhibit rubbery behavior, and show low surface coefficients of friction. Biocompatible hydrogels are currently used in numerous biomedical applications including opthalmologic devices, biosensors, biomembranes, and carriers for controlled delivery of drugs and proteins [Wichterle and Lim, 1960; Andrade, 1976; Lowman and Peppas, 1999; Peppas et al., 2000].

More recently, the design and application of biodegradable hydrogels has dramatically increased the potential impact of hydrogel materials in the biomedical field and enabled the development of exciting advances in controlled drug delivery and tissue engineering. While all polymers will eventually degrade under extreme environmental conditions (i.e., high temperatures or low pH solutions), biodegradable hydrogels of interest to this discussion degrade over clinically relevant timescales under relatively mild conditions (i.e., aqueous solutions, physiological temperature and pH). This degradation capability eliminates the need for long-term *in vivo* biocompatibility or surgical removal of the gels. Biodegradable hydrogels, if correctly designed, will break down into lower molecular weight, water-soluble fragments *in vivo* that can then be resorbed or excreted by the body once the desired function of the gel is accomplished. In addition to minimizing surgical invasiveness, the use of these erodible gels facilitates a wide variety of new applications and delivery strategies, such as degradation-controlled drug delivery, *in situ* scaffold formation and tissue regeneration, and controlled release via intravenous or pulmonary administration of degradable polymeric microspheres.

This chapter provides an overview of the chemistry, design, fabrication, and application of biodegradable hydrogels for drug delivery and tissue engineering. The first section briefly describes the various structural classes of degradable hydrogels that exist. Following an overview of the most commonly used gel types, the next section focuses on the fabrication and characterization of bulk-degrading, covalently cross-linked hydrogels. The chemical and structural parameters that quantify the physico-chemical properties of these gels before and during gel degradation are identified and detailed from an experimental and theoretical perspective. The inherent factors that are known to significantly impact degradation rates and macroscopic degradation behavior are also described in an effort to demonstrate how efforts to intelligently engineer the chemistry and structure of these gels on the molecular level directly correlate with improved device performance in tissue engineering and drug delivery applications.

The final section of this chapter delivers an overview of degradable hydrogel chemistries with proven or potential applications in tissue engineering and drug delivery. Gels are divided into two categories according to the natural or synthetic origin of their predominant polymer chemistry. Hydrogels from natural polymers have already gained widespread use in the biomedical field. However, gels obtained from natural polymers exhibit distinct limitations that have motivated approaches to modify these naturally occurring polymers as well as to develop synthetic derivatives and entirely novel synthetic chemistries. An emphasis of this concluding section is to highlight recent efforts to develop hybrid hydrogel systems that display the most advantageous properties derived from both natural and synthetic materials.

5.2 Hydrogel Classifications

5.2.1 Degradation vs. Erosion

For the discussions in this chapter, *degradation* refers to bond cleavage or cross-link dissolution within a network, while *erosion* refers to the subsequent mass loss from the network that occurs as a result of gel degradation. Degradation can occur via dissolution of physical cross-links. It can also occur in covalently cross-linked systems through the cleavage of hydrolytically labile bonds such as anhydride or ester groups, or enzymatically cleavable peptide or protein linkages. These labile bonds can be present in the cross-link segments (predominant in synthetic polymer networks) or along the backbone chains (predominant in naturally derived polymer networks).

The *degradation behavior* of a hydrogel pertains to the time-dependent evolution of the chemical, physical, and structural properties of the cross-linked network that occur as labile bonds on the surface or within the bulk of the gel are cleaved. Important phenomena that occur to varying extents during gel degradation include changes in hydrogel swelling ratios or equilibrium water contents, network mechanics, and solute diffusivities within the swollen matrices. The rate and profile of mass loss from the hydrogel is also important. One or more of these listed properties will play a role in determining the successful function of a degradable gel for a particular drug delivery or tissue engineering application. In addition, the molecular weight, chemistry, and local concentration of the degradation products produced by an eroding gel must always be considered to ensure complete biocompatibility.

In addition to the macroscopic gel properties themselves, the rate of hydrogel degradation must be carefully controlled both in drug delivery and tissue engineering applications. In drug-delivery systems, hydrogel degradation and erosion rates help determine drug availability and pharmacokinetic effects on surrounding cells and tissues. For tissue engineering applications, it is usually desirable to have biodegradable scaffolding to promote cell infiltration and tissue growth. It is commonly believed that the degradation rates of tissue scaffolds must be matched to the rates of various cellular processes in order to optimize tissue regeneration [Hubbell, 1999; Lee et al., 2001]. Therefore, the degradation behavior of all biodegradable hydrogels should be well defined, reproducible, and tunable via hydrogel chemistry or structure.

For physically cross-linked gels, the degradation kinetics of dissolving cross-links are hard to define and control. The degradation kinetics of chemically cross-linked gels are, however, more easily defined and much work has recently been done to link the kinetics of labile bond degradation to the overall gel degradation behavior in hopes of engineering biodegradable hydrogels with precise properties.

5.2.2 Bulk vs. Surface-Degradation

In general, degradation of cross-linked networks occurs in one of two forms: nonuniform, surface-degradation or uniform, bulk degradation [Kohn and Langer, 1996]. Surface-degrading networks maintain their cross-linking density and structural integrity throughout the degradation process, because degradation is limited to the surface of the material. For example, surface-degradation often results when the rate of bond hydrolysis is much faster than the rate of water transport into a polymeric device. Surface-degrading networks are advantageous for drug-delivery applications because zero-order release of entrapped species at a desired rate can be obtained by choosing the appropriate device geometry or altering the kinetics of degradation [Davis and Anseth, 2002].

In bulk-degrading networks, infiltration and transport of species critical to the particular degradation mechanism employed by the hydrogel chemistry are faster than the inherent degradation kinetics. By definition, all hydrolytically degradable hydrogels will exhibit bulk-degrading characteristics owing to the presence of a relatively high concentration of water molecules throughout the gel architecture. The same water that swells these gels will also homogeneously degrade the labile bonds present throughout the network. However, hydrogels that are degraded by species other than water may degrade by a bulk, surface, or combined mechanism depending on the permeability of the degrading species within the gel. For example, peptide-cross-linked hydrogels that degrade through the actions of a particular enzyme may exhibit extremely high water content in excess of 90% by volume. However, because the gel chemistry or limited mesh size prevents the uptake of the degradative, macromolecular enzyme within the bulk of the gel, only the labile peptide bonds at the surface will be exposed to the enzyme. The limited enzyme permeation will produce the observation of surface-mediated gel degradation and erosion.

While the cross-linking density and physical properties of surface-degrading gels remain constant during the biodegradation process, the properties of bulk-degrading gels are altered in a systematic fashion. The evolution of microscopic and macroscopic properties in bulk-degrading hydrogels is inherently tied to their polymer chemistry, network structure, and degradation kinetics. For example, highly swollen hydrogels formed from dimethacrylated poly(lactic acid)-b-poly(ethylene glycol)-b-poly(lactic acid) (PLA-PEG-PLA) undergo bulk degradation. The modulus, solute permeability, water content, and many other gel properties depend on the cross-linking density of the PEG-based hydrogel [Sawhney et al., 1993; Metters et al., 2000a]. As the lactic acid bonds are cleaved uniformly throughout the gel via hydrolysis, the cross-linking density of the still insoluble network systematically decreases, increasing swelling and network mesh size while lowering gel modulus. These property changes then, in turn, lead to macroscopically observable changes in the water content, permeability, and elasticity of the hydrogel as it degrades. The quantitative relationships between these dynamic properties during hydrogel degradation are detailed later in this chapter.

5.2.3 Homogeneous vs. Heterogeneous Networks

Because the microscopic structure of a degradable hydrogel network plays such an important role in its degradation behavior, degradable hydrogels can be classified according to their network structure. Hydrogel network morphologies can be described as being homogeneous or heterogeneous. Homogeneous gels exhibit a random distribution of relatively mobile chains and pores within the cross-linked network. Examples of homogeneous gels include networks derived from synthetic polymers such as PEG, poly(vinyl alcohol) (PVA), or poly(acrylamide). Homogeneous gels may be amorphous or semi-crystalline. They may be cross-linked using ionic, covalent, or noncovalent methods. They can also be either neutral or ionic depending on the ionization of their pendant groups [Peppas, 1986].

Heterogeneous hydrogels, on the other hand, exhibit an anisotropic network structure characterized by a high degree of interpolymer interaction. Examples of heterogeneous hydrogels include many insoluble networks derived from naturally occurring polymers such as calcium alginate, agarose, and κ-carrageenan [Muhr and Blanshard, 1982]. Additionally, supermolecular fibrils and fiber bundles of size-scales much greater than individual polymer chains can be formed in these networks via complex, thermodynamically driven self-assembly processes [Stupp, 2005]. While such structures exhibit high porosities between immobile, large-scale fiber bundles, their overall degradation more closely resembles the characteristics of surface-eroding systems due to the inability of water or macromolecular enzymes to penetrate within the small-scale, self-assembled architectures [Ehrbar et al., 2005]. Fibrous, anisotropic networks are found in gels made of natural or synthetic macromolecules such as collagen, fibrin, and synthetic polypeptides [Voet and Voet, 1995; Lutolf and Hubbell, 2005].

Both homogeneous and heterogeneous networks may be cast in the form of macroporous, microporous, or nonporous gels [Peppas, 1986]. Both types of morphologies may also display anisotropies in macroscopic properties such as swelling, elasticity, and porosity. Furthermore the degradability of both gel types is based on a limited number of biodegradable bonds. Therefore, their degradation during biomedical application occurs because of identical mechanisms — hydrolytic or enzymatic cleavage of covalent cross-links or dissolution of physical cross-links [Shalaby et al., 1991; Sawhney et al., 1993; West and Hubbell, 1999; Jeong et al., 2002].

However, the effect of labile bond cleavage on the evolution of macroscopic gel properties differs significantly between homogeneous and heterogeneous gels due to differences in cross-linking, supermolecular polymer-chain organization, and overall network structure. Thermodynamic relationships developed for nondegradable, homogeneous gels have been applied to help understand the dynamic degradation behavior of homogeneous gels. These relationships are detailed in the next section of this chapter. However, the anisotropic network structure present in heterogeneous gels greatly limits our current ability to correlate the extent of gel degradation with predicted or experimentally measurable changes in gel properties. While advances in the design and application of heterogeneous hydrogels for biomedical applications have been made and will be discussed in this review, the overview of structural characteristics presented in the next section focuses on identifying the key parameters that determine the microscopic and macroscopic behavior of bulk-degrading, homogeneous hydrogels.

5.3 Bulk-Degrading, Covalently Cross-Linked Hydrogels

As detailed in the final section of this chapter, a large array of techniques exists for creating biodegradable hydrogels from both synthetic and naturally derived polymer chemistries. In recent years, covalent cross-linking has emerged as a preferred method due to its wide compatibility with a number of polymer chemistries and its ability to fine tune hydrogel properties [Hennink and van Nostrum, 2002]. Therefore, in this section the fabrication and characterization of covalently cross-linked, degradable hydrogels are detailed. The quantitative analyses provided for correlating degradation behavior to labile bond cleavage kinetics apply to a specific class of covalently cross-linked, bulk-degrading hydrogels with hydrolytically degradable cross-links. However, the observed degradation behavior is not necessarily unique to this class of hydrogel and many of the structure–function relationships developed to describe this system can readily be extended to describe the degradation behavior of other types of degradable hydrogels.

5.3.1 Fabrication and Network Structure

The degradation behavior of a biodegradable hydrogel depends significantly on its method of fabrication. Various mechanisms for forming these materials have been investigated, including ionic cross-linking, thermally induced physical cross-linking, and enzymatic or pH-induced gelation. Unfortunately, most of these methodologies yield limited control over the gelation kinetics, material properties, and degradation behavior. In contrast, covalent cross-linking methods remove the need for interpolymer interactions and lead to the formation of homogeneous networks with uniform and precise cross-linking densities. This high degree of engineerability permits fine-tuning of polymer diffusivity and permeability, degradation rate, equilibrium water content, elasticity, and modulus.

Three main polymerization mechanisms are used to form covalently cross-linked, degradable hydrogels including step-growth, chain-growth, and mixed-mode chain and step-growth mechanisms. Figure 5.1 illustrates the methods of gel fabrication and potential degradation for each of these polymerization techniques [Rydholm et al., 2005]. It should be noted that the distinct site of bond cleavage within each of these biodegradable networks depend on the method of polymerization as well as on the chemistry and functionality of the chosen macromers and monomers. Modifications to the monomer chemistry or reaction conditions directly impacts the density and degradability of the network cross-links and allows the degradation behavior as well as the moduli, elasticity, permeability, and gel water content to be tailored in each system.

Network formation via the step-growth mechanism is based on the reaction of gel precursors exhibiting a stoichiometric ratio of at least two mutually reactive chemical groups (Figure 5.1a). These traditional A-B type polymerizations lead to the formation of an insoluble network if the average precursor functionality is greater than or equal to two as first dictated by Flory [1953]. The simplest form of the step-growth mechanism related to biodegradable hydrogel fabrication is the straightforward cross-linking of highly multifunctional natural polymers using small, bifunctional cross-linking agents. For example, hyaluronic acid (HA) or other polysaccharides can be cross-linked with glutaraldehyde. For these networks, degradation will occur via cleavage of chemical bonds along the backbone of the naturally derived polymer chain. Alternatively, biodegradable hydrogels can also be formed via the step-wise cross-linking of nondegradable, synthetic polymers using degradable cross-linkers such as peptides, proteins, or even cells.

As the size and number of reactive functionalities per monomer molecule decrease, the step-growth mechanism assumes the character of a true polymerization rather than a simple cross-linking reaction. Degradable networks that result from step-growth polymerizations of small, multifunctional monomers have been developed. For example, as illustrated in Figure 5.1a, Hubbell and coworkers developed degradable networks using Michael-type addition reactions between thiol- and acrylate-functionalized monomers [West and Hubbell, 1999; Elbert et al., 2001; Lutolf et al., 2003; Lutolf and Hubbell, 2003;

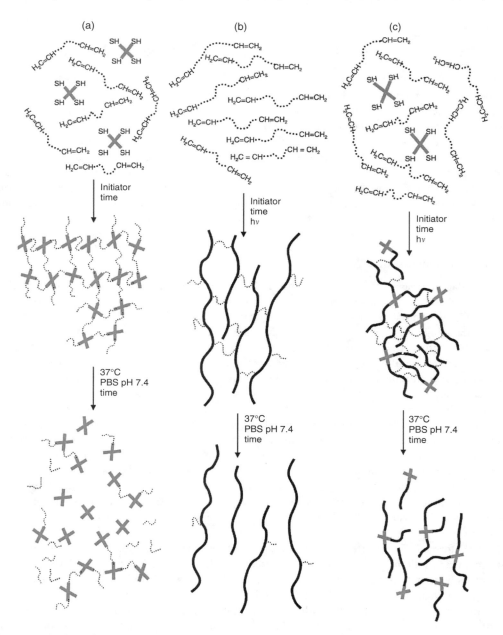

FIGURE 5.1 Pictorial representation of the initial monomer molecules, cross-linked polymer networks, and degradation products for materials formed from (a) step-growth polymerization mechanism, (b) chain-growth polymerization mechanism, and (c) mixed-mode chain and step-growth mechanism. (From Rydholm, A.E., Bowman, C.N., et al. 2005. *Biomaterials* 26: 4495–4506. With permission.)

Pratt et al., 2004; Seliktar et al., 2004]. Since gel degradation occurs by hydrolytic cleavage of a thio-ether ester bond formed during thiol–acrylate coupling, low molecular weight, nondegradable monomers can be used to form degradable hydrogels. Elbert et al. [2001] fabricated hydrolytically degradable hydrogels from multiarmed PEG acrylates and linear PEG di-thiols. Additionally, enzymatic degradability was imparted to these gels through the use of protease-sensitive di-thiols made from short oligopeptide sequences [Lutolf and Hubbell, 2003; Lutolf et al., 2003].

As shown by Metters and Hubbell [2005], the degradation rates of networks formed via the step-growth mechanism depend on the molecular weight, hydrophilicity, and degree of functionality of the starting monomers. Step-growth polymerizations are known to produce few structural defects during network formation, which permits precise control of the cross-linking density and degradation behavior [Dusek and Duskov-Smrckova, 2000]. Additionally, using relatively low molecular weight monomers to form the degradable, Michael-type gels eliminates the subsequent production of high molecular weight degradation products that commonly result from using polymeric gel precursors [Metters et al., 2000; Lovestead et al., 2002].

Contrary to the high concentration of reactive intermediates present during step-growth network formation, a low concentration of active centers is generated during formation of a typical chain-growth network (Figure 5.1b). These active centers are typically radicals and are generated by a variety of methods including thermal energy, redox reactions, and cleavage of a photoinitiator molecule when irradiated with ultraviolet (UV) or visible light [Odian, 1991]. They rapidly propagate through monomers containing multiple carbon–carbon double bonds to form high molecular weight, kinetic chains that are covalently cross-linked. The highly stable carbon–carbon bonds that result from chain-growth polymerization are generally nondegradable under biological conditions. Rather, degradation is incorporated into the networks through specially designed multi-vinyl macromers with hydrolytically or enzymatically cleavable segments [Sawhney et al., 1993; Davis et al., 2003; Lutolf and Hubbell, 2005]. Upon network formation, these linkages are present in the network cross-links. The degradation products from such networks are comprised of the degraded segments from the cross-linking molecules as well as the higher molecular weight kinetic chains generated during polymerization [Anseth et al., 2002].

An illustration of degradable, chain-polymerized networks comes from the pioneering work of Sawhney et al. where linear PEGs of various molecular weights were used as initiators for the ring-opening polymerization of a-hydroxy acids (lactic, glycolic), followed by reaction with acid halides to produce vinyl terminated macromonomers containing degradable ester linkages (Figure 5.2)[Sawhney et al., 1993]. Chain-growth polymerization of the PLA-PEG-PLA tri-block copolymer macromers was accomplished using mild photopolymerization conditions that permitted *in situ* network formation under physiological conditions.

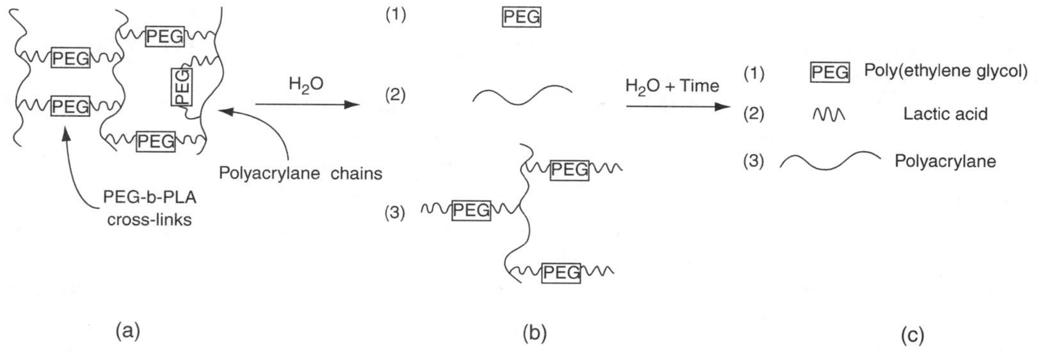

FIGURE 5.2 Illustration of three different stages during the bulk degradation of a PLA-b-PEG-b-PLA hydrogel network: (a) initial, nondegraded PLA-b-PEG-b-PLA network, (b) primary erosion products that are released during degradation, and (c) final degradation products after complete hydrolysis. (From Metters, A.T., Bowman, C.N., et al. 2000. *J. Phys. Chem. B* 104: 7043–7049. With permission.)

The degradation behavior of chain-polymerized networks with hydrolytically or enzymatically labile cross-links can be tailored through a variety of parameters. Although physically cross-linked networks can also be formed using degradable ABA block copolymers such as nonacrylated PLA-PEG-PLA, Sawhney's work was the first to incorporate polymerizable moieties into the macromer design. Covalent cross-linking of the degradable macromers provides dramatically improved control over the resulting network structure and subsequent degradation rate. For example, the degradation rate of covalently cross-linked (PLA-PEG-PLA) hydrogels can be tailored by varying the molecular weights of the PEG and PLA copolymer blocks within the cross-linker, the chemistry and degree of vinyl group functionalization, or the type and amount of comonomers added to the system [Metters et al., 2000a,b, Metters et al., 2001].

Finally, as their name suggests, networks formed from mixed-mode polymerizations exhibit characteristics between chain and step-growth polymerizations (Figure 5.1c). One relatively new type of degradable hydrogel based on the mixed-mode polymerization of acrylated PLA-PEG-PLA monomers and multifunctional thiols has been developed by Bowman and coworkers [Cramer et al., 2004; Reddy et al., 2004; Lu et al., 2005; Okay and Bowman, 2005; Okay et al., 2005; Reddy et al., 2005]. The network structure that results from this mixed-mode polymerization mechanism is unique from networks formed by chain and step-growth polymerizations and is directly impacted by reactive group ratios. As the ratio of thiol to acrylate groups increases in the system, the networks transition from being chain-like to more step-like. In addition, the erosion profile and swelling changes that occur during degradation are controlled by variations in thiol-acrylate ratios that impact network structure. Additionally, changing the thiol mole fraction in the network provides control of the degradation products' molecular weight distributions [Reddy et al., 2005].

5.3.2 Function and Degradation

In addition to the obvious need for biocompatibility during the lifetime of the degradable hydrogel, three material properties critical to the successful biomedical application of any hydrogel are water content, mechanical stiffness or elasticity, and permeability. By definition hydrogels must exhibit high water contents. While no exact water content value is required to describe a hydrophilic, cross-linked material as a hydrogel, most hydrogels currently used to encapsulate living cells, for example, swell to greater than 90% water by weight when placed in suitable physiological fluids. The highly solvated gel environment is critical to maintaining cell viability and also minimizes nonspecific adsorption of proteins and other macromolecules present in a biological environment that would otherwise lead to harmful inflammatory responses [Bryant and Anseth, 2002].

The mechanical properties of hydrogels are also particularly important in tissue engineering applications where the gel must create and maintain a space for cell infiltration and tissue development. In addition, results from a number of investigations have demonstrated that the adhesion, structure, metabolism, and gene expression of encapsulated cells are strongly influenced by the mechanical properties of the polymer scaffold [Huang and Ingber, 1999].

Finally, gel permeability is important for successful gel function. Controlled gel permeability permits sustained drug release over long periods as well as prevents the infiltration of harmful species such as enzymes or inflammatory/immune cells that would affect the stability of encapsulated proteins or transplanted cells, respectively [Langer, 1990, 1991]. In addition, correct gel permeability is also important for the transportation of nutrients and metabolic wastes to support growth of gel-encapsulated cells and tissues.

The swelling characteristics, mechanical properties, and permeability of hydrogels depend on several factors including the supermolecular structure of the original polymer chains; the type of cross-linking molecules and the cross-linking density; and the hydrophilic/hydrophobic balance of the cross-links and backbone polymer chains within the cross-linked network [Lee and Mooney, 2001]. As previously mentioned, the high degree of interpolymer interactions and structural anisotropies that occur in heterogeneous gels can be extremely difficult to characterize. However, to describe the structure, chemistry, and resultant material properties of homogeneous gels, the multitude of system design variables can generally be condensed to a few critical parameters (1) the polymer volume fraction in the swollen state, $v_{2,s}$; (2) the number average molecular weight between cross-links, $\bar{M}_c$; (3) the degree of polymer–solvent interaction, χ_{12}; and (4) the network mesh size, ξ [Peppas, 1986].

5.3.3 Swelling

By definition, the equilibrium polymer volume fraction in a degradable or nondegradable hydrogel, $v_{2,s}$, is the ratio of the volume of polymer, V_p, to the volume of the swollen gel, V_{gel}, and the reciprocal of the volume swelling ratio Q:

$$v_{2,s} = \frac{V_p}{V_{gel}} = Q^{-1}$$

The polymer volume fraction can be determined by equilibrium swelling measurements before or during degradation [Peppas, 1986; Metters et al., 2000]. The degree of swelling is also commonly reported as a mass swelling ratio, Q_m or q, which can also be related to $v_{2,s}$ as follows [Kong et al., 2002]:

$$v_{2,s} = Q^{-1} = \frac{1/\rho_2}{[(Q_m/\rho_2) + (1/\rho_2)]}$$

where

$$Q_m = \frac{M_{gel}}{M_p}$$

Here, M_{gel} and M_p are the masses of the swollen gel and dried polymer, respectively. ρ_1 and ρ_2 are the densities of the solvent and polymer, respectively.

For further hydrogel characterization, the number-average molecular weight between cross-links, $\bar{M}_c$, is the most common parameter used to represent the level of cross-linking within the network. Assuming an ideal network structure is formed, $\bar{M}_c$ can be theoretically calculated based on the size, chemistry, and functionality of the gel precursors. However, defects in network structure that increase its value occur during the fabrication of almost every hydrogel [Dusek and Duskov-Smrckova, 2000; Elliott et al., 2003, 2004; DuBose et al., 2005]. Therefore, $\bar{M}_c$ is best-determined using theories that correlate its true value to experimental measurements of gel swelling and mechanical strength.

The swelling behavior of hydrogels in biological fluids can be reasonably described by a variety of nonideal thermodynamic models. Due to the highly complex behavior of polymer networks in electrolyte solutions, no theory can predict exact swelling behavior. However, the Flory–Rehner analysis and its various modifications, continues to be used with reasonable success [Flory and Rehner, 1943]. This theoretical framework describes gels as neutral, cross-linked networks with a Gaussian distribution of polymer chains. When placed in aqueous solution, this model assumes that swelling equilibrium will occur at the point where the swelling force due to the thermodynamic compatibility of the polymer and water balances the retractive force induced by the stretching of the network cross-links. This analysis leads to the Flory–Rehner expression for the true $\bar{M}_c$ of a nonionized hydrogel:

$$\frac{1}{\bar{M}_c} = \frac{2}{\bar{M}_n} - \frac{(\bar{v}/V_1)\lfloor\ln(1 - v_{2,s}) + v_{2,s} + \chi_{12}v_{2,s}^2\rfloor}{(v_{2,s}^{1/3} - (2/\phi)v_{2,s})}$$

Here, χ_{12} is the polymer–water interaction parameter, V_1 is the molar volume of water, $\bar{v}$ is the specific volume of the polymer, $\bar{M}_n$ is the average molecular weight of linear polymer chains prepared at the same conditions without cross-linking, and ϕ is the functionality of the cross-linker (e.g., $\phi = 4$ for a chain-polymerized, divinyl cross-linker).

Under the common conditions of high network swelling where $Q > 10$ ($v_{2,s} < 10\%$) the original Flory–Rehner equation can be simplified to show a more direct relationship between Q and $\bar{M}_c$:

$$Q = \left[\frac{\bar{v}((1/2) - 2\chi_{12})\bar{M}_c}{V_1} \right]^{3/5} = \beta(\bar{M}_c)^{3/5}$$

Here, β is a constant. This simplification assumes that chain-end effects can be neglected ($\bar{M}_c \ll \bar{M}_n$) and that all physical parameters remain constant [Flory, 1953].

Although $\bar{M}_c$ cannot be directly measured, the power-law relationship between Q and $\bar{M}_c$ outlined above is indirectly evident in the experimentally observed swelling behavior of degrading hydrogels. In a bulk-degrading hydrogel, degradable linkages present along network cross-links or backbone polymer chains will be cleaved homogeneously throughout the entire gel at a rate controlled by the reaction kinetics of labile bond cleavage (e.g., hydrolysis of PLA ester bonds within PLA-PEG-PLA cross-links). This ongoing bond cleavage systematically decreases the cross-linking density of the overall network and increases $\bar{M}_c$. As predicted by the simplified equation given above, the hydrogel swelling ratio will increase as degradation proceeds and $\bar{M}_c$ increases. This behavior is observed during the bulk degradation of a wide variety of hydrogels where Q and Q_m are seen to increase with degradation time (Figure 5.3) [Lee et al., 2000; Metters et al., 2000; Elbert et al., 2001; Lutolf et al., 2003; Metters and Hubbell, 2005]. The exact

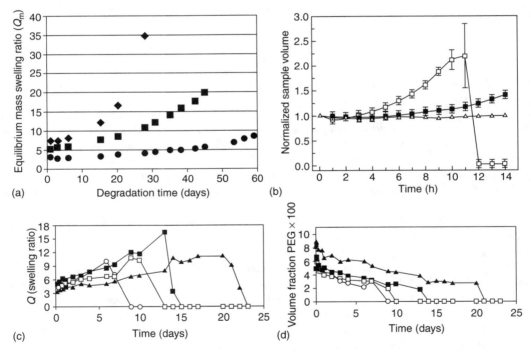

FIGURE 5.3 Examples of systematically increasing network swelling ratio with degradation time for different bulk-degrading hydrogels. (a) As the molecular weight of an octa-acrylate PEG precursor is increased, both the initially observed swelling ratio (at $t = 0$) and the apparent rate of degradation (slope of swelling curve) increase. (From Metters, A. and Hubbell, J. 2005. *Biomacromolecules* 6: 290–301.) (b) Swelling measurements of enzymatically degradable hydrogels with three different peptide cross-linkers exhibiting varying susceptibility to exogenously added proteases. (From Lutolf, M.P., Lauer-fields, J.L., et al. 2003. *Proc. Natl Acad. Sci.* USA 100: 5413–5418.) (c) Gel swelling and (d) calculated polymer volume fraction during degradation of gels made from PEG-dithiol and PEG-multiacrylates. (From Elbert, D.L., Pratt, A.B., et al. 2001. *J. Control. Release* 76: 11–25.)

function describing the rate of increase in gel swelling with degradation time will depend on the kinetics of individual bond cleavage as well as the gel structure [Metters et al., 2000; Metters et al., 2001; DuBose et al., 2005; Metters and Hubbell, 2005].

In some covalently cross-linked, degradable hydrogels the swelling behavior as a function of degradation and the dependence of $\bar{M}_c$ on the labile-bond cleavage kinetics can be described more quantitatively. For example, dimethacrylated PLA-b-PEG-b-PLA macromers, once polymerized, form hydrolytically degradable cross-links within swollen hydrogel structures as shown in Figure 5.1b and Figure 5.2 [Metters et al., 2000; Rydholm et al., 2005]. Assuming pseudo-first-order hydrolysis kinetics of the individual PLA ester bonds leads to a first-order decrease in the gel cross-linking density, since the cleavage of any ester bond in the macromer will lead to cross-link cleavage [Metters et al., 2000]. Combining the ester-bond hydrolysis kinetics with knowledge of the tri-block cross-link structure yields the following exponential relationship for $\bar{M}_c$ as a function of degradation time (t):

$$\bar{M}_c(t) = \bar{M}_c|_{t=0}\,e^{2jk_E' t}$$

where $t = 0$ represents the initial time prior to any network degradation, j is the degree of polymerization of the two PLA blocks in the PLA-PEG-PLA macromer (equivalent to the number of ester bonds per block), and k_E' is the pseudo first-order kinetic rate constant for hydrolysis of those ester bonds. Combining this time-dependent expression for $\bar{M}_c$ with the simplified form of the Flory–Rehner equation provided above yields an equation predicting a similar exponential increase in gel swelling with degradation time:

$$Q(t) = Q|_{t=0}\,e^{\frac{6}{5}jk_E' t}$$

Thus, for a system where mass-transfer limitations are not significant and the system is reaction controlled, the swelling ratio of the hydrogel at any point during degradation can be predicted based on knowledge of the hydrolysis kinetics of the individual bonds as well as the composition of the cross-links and overall network structure of the degradable gel. As shown in Figure 5.4, the typical swelling behavior of a degrading PLA-PEG-PLA hydrogel exhibits this predicted exponential increase in gel swelling ratio with degradation time [Metters et al., 2000; Anseth et al., 2002].

The time or degradation-dependent swelling behavior of highly permeable, degradable hydrogels formed via step-growth polymerizations can also be predicted with adequate knowledge of bond cleavage kinetics and network structure. DuBose et al. [2005] showed how the bond hydrolysis kinetics and network structure could be varied independently to affect the dynamic swelling profiles of PEG-based

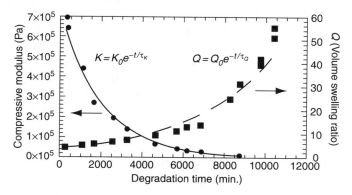

FIGURE 5.4 Typical *in vitro* degradation behavior of a PLA-b-PEG-b-PLA hydrogel: compressive modulus (●) and volumetric swelling ratio (■). The solid and dashed lines are exponential curves fit to each property with time constants of $\tau_Q = 4200$ min and $\tau_K = 2000$ min. (From Anseth, K.S., Metters, A.T., et al. 2002. *J. Control. Release* 78: 199–209. With permission.)

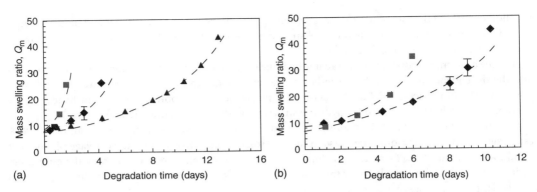

FIGURE 5.5 Swelling of degradable PEG-acrylate/dithiol gels formed via step-growth polymerization. (a) Gels fabricated from 30 wt % eight-armed PEG-acrylate/DTT precursor solutions and degraded at varying temperatures: 37°C (▲), 46°C (♦), and 57°C (■). (b) Gels fabricated with either four-arm/10 kDa (■) or eight-arm/20 kDa (♦) PEG were measured and compared with model predictions (– – –). (From DuBose, J.W., Cutshall, C., et al. 2005. *J. Biomed. Mater. Res. A* 74A: 104–116. With permission.)

hydrogels formed using Michael-type addition reactions with small-molecule di-thiols [DuBose et al., 2005]. Figure 5.5 demonstrates how the rate of swelling can be increased in a predictable fashion by increasing temperature during degradation (i.e., increasing the rate constant for bond hydrolysis) or by decreasing the number of acrylate groups per PEG cross-link in the gel network.

5.3.4 Mechanics

The mechanical behavior of hydrogels is best described using the theories of rubber elasticity and viscoelasticity [Flory, 1953; Treloar, 1975; Aklonis and MacKnight, 1983]. These theories are based on the time-independent and time-dependent recovery of the chain orientation and structure, respectively. Rubbers are materials that respond to stresses with nearly instantaneous and fully reversible deformation [Bueche, 1962]. Normal rubbers are lightly cross-linked networks with a rather large free volume that allows them to respond to external stresses with a rapid rearrangement of the polymer segments. In their swollen state, most hydrogels satisfy these criteria. When a hydrogel is in the region of rubber-like behavior, the mechanical behavior of the gel is dependent mainly on the architecture of the polymer network. Only at very low temperatures will these gels lose their rubber elastic properties and exhibit viscoelastic behavior. General characteristics of rubber elastic behavior include high extensibility generated by low mechanical stress, complete recovery after removal of the deformation, and high extensibility and recovery that are driven by entropic rather than enthalpic changes.

By using rubber elasticity theory to describe the mechanical behaviors of biomedical hydrogels, it is possible to analyze the polymer structure and determine the effective molecular weight between cross-links ($\bar{M}_c$) as well as elucidate information about the number of elastically active chains and deviations from ideal network structure (defects) that occur under a variety of reaction conditions. Using this theory, the shear modulus of an unswollen, dry gel (G_d) depends on the average molecular weight between cross-links as follows [Anseth et al., 1996]:

$$G_d = \frac{\rho RT}{\bar{M}_c}\left(1 - \frac{2\bar{M}_c}{\bar{M}_n}\right)$$

Here, ρ is the polymer density, R is the gas constant (8.314 J mol^{-1} K^{-1}), and T is temperature (K). The effects of chain ends have been included. This equation is easily modified to describe the mechanical behavior of swollen hydrogels. The shear modulus of a swollen gel (G_s) is dependent on its network structure and degree of swelling as given by:

$$G_s = \frac{\rho RT}{\bar{M}_c}\left(1 - \frac{2\bar{M}_c}{\bar{M}_n}\right)(v_{2,s})^{1/3} = \frac{\rho RT}{\bar{M}_c}\left(1 - \frac{2\bar{M}_c}{\bar{M}_n}\right)\frac{1}{Q^{1/3}}$$

Therefore, for a given hydrogel at a fixed temperature, a higher degree of swelling results in a reduction of the shear modulus. Alternatively, if the degree of cross-linking is increased (i.e., $\bar{M}_c$ is decreased), the modulus is increased. This interpretation is adequate for low strains.

Assuming analysis of a bulk-degrading hydrogel exhibiting a high degree of swelling ($Q > 10$), as well as neglecting the influence of chain ends, the relationship between Q and $\bar{M}_c$ obtained from using the simplified Flory–Rehner equation can be inserted to yield G_s as a function of $\bar{M}_c$:

$$G_d = \frac{\rho RT}{\bar{M}_c}\frac{1}{[\beta(\bar{M}_c)^{3/5}]^{1/3}} = \frac{\gamma}{\bar{M}_c^{6/5}}$$

Here, β and γ are constants. This equation explains why the modulus of a bulk-degrading PLA-PEG-PLA hydrogel decreases exponentially with time as shown in Figure 5.4. As $\bar{M}_c$ increases exponentially with time for the PLA-PEG-PLA hydrogel, G_s decreases in the following manner:

$$G_s(t) = \frac{G_{s,o}}{e^{-12/5jk'_E t}} = G_{s,o}e^{-12/5jk'_E t}$$

Here, $G_{s,o}$ is the initial shear modulus of the swollen hydrogel network prior to degradation.

As shown by the equation above, the elasticity of a bulk-degrading network is dependent upon bond cleavage kinetics and gel microstructure. These dependencies are similar to what were previously calculated for gel swelling. However, comparison of the exponential rate constants for the modulus and swelling degradation curves reported in Figure 5.4 indicate that the rate of modulus decay is approximately twice as fast as the rate at which swelling increases for degrading PLA-PEG-PLA gels. These experimental results are supported through first-principle predictions based on the simplified thermodynamic $\propto$ relationships given above which relate both gel swelling and modulus to $\bar{M}_c$ and predict [Metters et al., 2001].

$$G_s = \frac{\gamma}{\bar{M}_c^{6/5}}\frac{\gamma}{[(Q/\beta)^{5/3}]^{6/5}} \propto \frac{1}{Q^2}$$

This scaling argument shows that for degradable and nondegradable hydrogels, the shear modulus of the gel (G_s) is approximately twice as sensitive to changes in $\bar{M}_c$ as the volumetric swelling (Q).

It is important to note that while the results presented in Figure 5.4 illustrate good agreement between experimental observations of hydrogel degradation and the predictive thermodynamic equations, many assumptions were made to reach the scaling argument provided above. These assumptions include a high degree of swelling, constant physical and thermodynamic parameters, and an ideal network structure founded on degradable cross-links. The presence of nonidealities in the network structure, changes in hydrogel chemistry during degradation (e.g., production of numerous carboxylic acid end groups during PLA ester bond cleavage), mass transfer limitations, or autocatalytic effects will produce characteristics in the gel degradation behavior that cannot be appropriately described by the simplified analysis presented in this section [Metters et al., 2001]. For example, Amsden et al. [2004] have developed an alternative strategy for modeling the degradation of hydroxyethylmethacrylate-grafted dextran hydrogels, where bond cleavage occurs along the backbone chains of the network and changes in polymer chemistry during degradation are significant.

5.3.5 Diffusivity

The ability of a hydrogel to restrict the diffusive movement of a solute plays a key role in applications as diverse as cell encapsulation, chromatography, biosensors, and drug delivery [Langer and Peppas, 1981]. Diffusion-controlled drug-delivery systems based on hydrogel materials, for example, can be matrix or reservoir systems [Lowman and Peppas, 1999; Mallapragada and Narasimhan, 1999]. In the reservoir system, the active agent is located in a core and a polymer membrane surrounds it. In matrix systems, the drug or protein is homogeneously distributed throughout the membrane and is slowly released from it. In either system the hydrogel membrane can be biodegradable [Heller, 1980].

It is important to have an understanding of the mechanisms and underlying parameters governing solute diffusion within hydrogels. For this reason, a number of mathematical relationships have been developed in an effort to model solute diffusion in hydrogels. Drug or protein diffusion within swollen hydrogel networks is best described by Fick's equation or by the Stefan–Maxwell equations that correlate the flux of a particular solute with its chemical–potential gradient in the system [Peppas et al., 2000]. For porous gels with pore sizes much larger than the molecular dimensions of the solute, the diffusion coefficient is related to the porosity and the tortuosity of the porous structure [Peppas, 1986]. For porous hydrogels with pore sizes comparable to the solute molecular size and for nonporous hydrogels, various expressions have been proposed for the diffusion coefficients [Peppas, 1986; Amsden, 1998; Masaro and Zhu, 1999]. The polymer chains within these cross-linked networks have been proposed to retard solute movement by reducing the average free volume per molecule available to the solute, by increasing the hydrodynamic drag experienced by the solute, or by acting as physical obstructions that increase the solute path length [Mackie and Meares, 1955; Cohen and Turnbull, 1959; Bird et al., 1960]. Model complexity increases as these mechanisms are combined as well as when nonspherical solutes (e.g., linear polymer chains of high molecular weight) and polymer–solute interactions are considered [Muhr and Blanshard, 1982]. An excellent review of the most prevalent models along with a quantitative assessment of their predictive abilities has been presented by Amsden [1998].

Solute transport within hydrogels is assumed to occur primarily within the water-filled regions delineated by the polymer chains. The average size of these spaces can be quantified through the correlation length, ξ, also known as the network mesh size [Peppas, 1986; Lustig and Peppas, 1988]. The mesh size of a hydrogel network can be determined as described by Canal and Peppas [1989]:

$$\xi = v_{2,s}^{-1/3}(\bar{r}_0^2)^{1/2} = Q^{1/3}(\bar{r}_0^2)^{1/2}$$

where $(\bar{r}_0^2)^{1/2}$ is the root-mean-squared end-to-end distance of network chains in the unperturbed state and is directly proportional to $\bar{M}_c$. Any factor that reduces the relative size of these solvent-filled spaces compared to the size of the solute will further retard solute diffusion. In general, the diffusivity of a solute through a covalently cross-linked hydrogel decreases as cross-linking density increases ($\bar{M}_c$ decreases), as the size of the solute (r_s) increases, and as the volume fraction of polymer within the gel ($v_{2,s}$) increases [Yasuda et al., 1968; Peppas, 1986; Johansson et al., 1991].

An exponential increase in both Q and $\bar{M}_c$ with degradation time has already been predicted for bulk-degrading gels with cross-links that degrade according to pseudo-first order kinetics. Therefore the network mesh size for these systems will also increase in an exponential manner with degradation time as shown below [Mason et al., 2001]:

$$\xi = Q^{1/3}(\bar{r}_0^2)^{1/2} \sim \bar{M}_c^{7/10} = \left[\bar{M}_c|_{t=0}e^{2jk_E't}\right]^{7/10} = \eta e^{7/5jk_E't}$$

where η is a constant.

Theoretical models for predicting solute diffusion coefficients take the general form:

$$\frac{D_g}{D_0} = f(r_s, v_{2,s}, \xi)$$

where D_g is the solute diffusion coefficient in the swollen hydrogel network, D_0 is the diffusion coefficient of solute in pure solvent, and r_s is the size of the solute molecules. Thus, the structure and pore size of the gel, the polymer composition, the water content, and the nature and size of the solutes are all taken into account by the diffusion coefficient of the solute [Peppas et al., 2000]. This general framework can be used to predict the rate of diffusion of all species within the gel network including the influx of degrading moieties (e.g., water or enzymes) as well as the release rate of encapsulated species such as drugs and other therapeutic agents. For a degradable hydrogel the previously discussed increase in mesh size and decrease in polymer volume fraction with network degradation will influence D_g and cause it to change with the extent of degradation as well.

Numerous variations of the general framework given above have been developed by several research groups to more clearly describe the relationship between solute diffusivity and network structure for nondegrading gels. As an example, one such model is given by:

$$\frac{D_g}{D_0} = \left(1 - \frac{r_s}{\xi}\right) \exp\left(-Y\left(\frac{v_{2,s}}{1 - v_{2,s}}\right)\right)$$

Here, Y is physically defined as the ratio of the critical volume required for a successful translational movement of the solute molecule and the average free volume per molecule of solvent. For most purposes a good approximation for Y is unity. This model was developed by Lustig and Peppas [1988] using a free-volume approach and has proven useful for predicting solute diffusivities within PEG networks.

For highly swollen, degradable PLA-PEG-PLA gels the diffusivity correlation provided by Lustig and Peppas can be simplified to

$$\frac{D_g}{D_0} = \left(1 - \frac{r_s}{\xi}\right) = \left(1 - \frac{r_s}{\eta e^{-7/5jk_E' t}}\right)$$

using the time-dependent expression for the mesh size given above. From this modified expression it can be clearly observed that as degradation proceeds, solute diffusivity within the gel (D_g) will increase in a systematic, predictable fashion and approach D_0. Like the shear modulus of the degrading gel (G_s), the rate of decrease in solute diffusivity depends on network structure and bond cleavage kinetics. This type of analysis has been used to explain the diffusivities and release profiles of various drugs from bulk-degrading networks compared to nondegrading matrices [Lu and Anseth, 2000; Mason et al., 2001]. It can also be used with alternative free-volume approaches, hydrodynamic scaling models, and obstruction theories to predict the time-dependent solute diffusivities within other degradable gels.

5.3.6 Characteristic Erosion

Mass loss or erosion from bulk-degrading, covalently cross-linked hydrogels is a complex process that depends upon the network structure and degradation kinetics [Heller, 1980; Metters et al., 2000a,b]. In this respect, mass loss from degrading hydrogels is similar to the swelling or mechanical properties. However, unlike most other macroscopic gel properties that are only related to the time-dependent cross-linking density or average molecular weight between cross-links ($\bar{M}_c$), mass loss also relies upon additional structural parameters. For example, models developed to predict mass loss from photo-cross-linked PLA-PEG-PLA hydrogels show that the erosion profiles for these systems are linked to network parameters such as the number of cross-links per backbone chain and the mass fraction of the network contained in the backbone chains relative to the cross-links [Metters et al., 2000, 2001; Martens et al., 2001, 2004]. Predictions of mass loss vs. degradation time also depend upon kinetic parameters such as the order and rate constant of the degradation reaction.

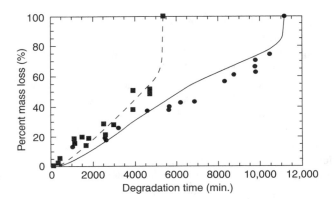

FIGURE 5.6 Experimental mass loss data as a function of degradation time for two identical PLA-PEG-PLA macromers polymerized at different concentrations to form gels with unique degradation behavior: (■) 25 wt% and (●) 50 wt%. The solid and dashed lines represent the percent mass loss predicted by a statistical model: (dashed) 25 wt% and (solid) 50 wt%. (From Metters, A.T., Bowman, C.N., et al. 2000. *J. Phys. Chem.* B 104: 7043–7049. With permission.)

In Figure 5.6, experimental and predicted mass loss profiles are plotted vs. degradation time for two PLA-PEG-PLA hydrogels polymerized in solution at different macromer concentrations [Metters et al., 2000]. These degrading hydrogels have the same chemical composition, yet different initial cross-linking densities and microstructures that result from the behavior of the chain-growth polymerization used in their fabrication. Increasing the initial macromer concentration during polymerization increases the average length and functionality of the backbone polymer chains and decreases the number of structural defects within the network. These architectural changes are reflected in an increased time until complete network dissolution as well as a lower apparent degradation rate (slope of linear portion of curve) due to the additional cross-linking and lower swelling ratios of these networks.

A characteristic of bulk-degrading gels that is evident in Figure 5.6 is the occurrence of gel dissolution prior to 100% bond degradation. Gel dissolution, or *reverse gelation* as it has been termed, is an abrupt solid to liquid transition that leads to almost instantaneous erosion of a substantial fraction of the cross-linked gel mass [Metters et al., 2000]. It is analogous to the liquid to solid transition that occurs at the gel point during hydrogel formation [Flory, 1953]. The exact fraction of the total network mass lost during reverse gelation increases as the number of cross-links per backbone chain decreases as shown in Figure 5.6.

The burst of mass loss that occurs at the onset of reverse gelation can have broad implications on the successful application of the degradable hydrogel. A large fraction of network mass lost during reverse gelation will result in a large, localized concentration of potentially cytotoxic degradation products. For example, there is evidence that the macrophage response to implanted PLA-PEG-PLA gels appears strongest immediately following gel dissolution. This is most likely a response to the high concentration of soluble, acidic degradation products that are commonly seen with PLA-based materials and can lead to adverse cellular responses *in vivo*. These acidic degradation products can also decrease the stability of encapsulated proteins when these hydrogels are used for the controlled release of pharmaceutics. Therefore, to ensure biocompatibility and proper application of any degradable hydrogel it is important to understand the factors that determine when gel dissolution will occur.

Finally, the erosion profiles of bulk degrading hydrogels can also be engineered for specific drug-delivery applications. The same kinetic and structural factors that impact mass loss from a cross-linked gel can also be used to control the delivery of therapeutic biomolecules covalently attached to the gel network [Heller, 1980; Zhao and Harris, 1997, 1998; Seliktar et al., 2004; DuBose et al., 2005]. For example, fluorescently labeled probe molecules resembling clinically relevant peptide drugs were covalently incorporated within the three-dimensional network structure of PEG-based hydrogels formed via step-growth polymerizations. As shown in Figure 5.7, hydrolytic cleavage of the covalent bonds within the cross-linked

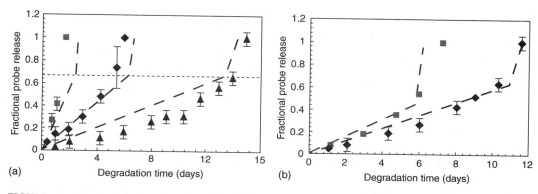

FIGURE 5.7 Fractional probe release from degradable PEG-acrylate/dithiol gels formed via step-growth polymerization (a) Gels fabricated from 30 wt % eight-armed PEG-acrylate/DTT precursor solutions and degraded at varying temperatures: 37°C (▲), 46°C (♦), and 57°C (■). (b) Gels fabricated with either four-arm/10 kDa (■) or eight-arm/20 kDa (♦) PEG were measured and compared with model predictions (– – –). (From DuBose, J.W., Cutshall, C., et al. 2005. *J. Biomed. Mater. Res. A* 74A: 104–116. With permission.)

PEG network, including those used to immobilize the probe molecule, resulted in a biphasic release profile consisting of a relatively slow, constant release rate of probe prior to gel dissolution and an almost instantaneous release following the onset of gel dissolution [DuBose et al., 2005]. The various release profiles shown in Figure 5.7 demonstrate how the release rate as well as the fraction of probe released prior to network dissolution can be controlled through intelligent choice of cross-linker functionality (tetra-functional PEG vs. an octa-functional PEG in Figure 5.7a) or degradation kinetics (varying temperature, pH, or chemistry of the degradable thio-ether ester bond in the network in Figure 5.7b). Such characteristic erosion profiles are seen in a variety of bulk-degrading hydrogel systems.

5.4 Degradable Hydrogel Chemistries

Degradable hydrogels can be prepared from naturally derived biopolymers, synthetic polymers, or the combination of the two. The major advantages of using naturally derived biopolymers such as chitosan, alginate, fibrin, collagen, gelatin, and HA derivatives include their inherent biodegradability and biocompatibility. However, natural biopolymers may not provide adequate mechanical strength as well as precise functionalities compared to synthetic polymers. Moreover, precautions against pathogenic contamination must be taken when many natural biopolymers are used in clinical applications.

On the other hand, many synthetic polymers, although not inherently degradable themselves, can be modified or copolymerized with labile groups to create degradable hydrogels. The good mechanical properties and well-defined chemistries enable hydrogels made from synthetic polymers such as PEG, PVA, poly(*N*-isopropylacrylamide) (PNIPAAM), and poly(methyl methacrylate) (PMMA) to be used for many biomedical applications.

Recently, many fundamental studies as well as clinical applications using degradable hydrogels derived from natural or synthetic polymers have been conducted. In this final section of the chapter, the chemistries as well as applications that comprise the most common degradable hydrogels are described.

5.5 Degradable Hydrogels Derived from Natural Biomaterials

5.5.1 Chitosan-Based Hydrogels

Chitosan, with a subunit of β-(1,4)-2-amido-2-deoxy-D-glucopyranose, is a form of deacetylated chitin that has been shown to be biocompatible. Chitosan can be dissolved in weak acidic solution and is positively charged in natural or basic environment due to its amino groups. It has been demonstrated that chitosan can accelerate wound healing processes and therefore is a well accepted material for various biomedical applications including tissue engineering, controlled delivery of proteins, and gene delivery. Several review papers have been published addressing the chemistry and applications of chitosan hydrogels. For example, Berger et al. [2004] have reviewed the chemistry of chitosan hydrogels formed via electrostatic assembly. Kumar et al. [2004] extensively reviewed chitosan chemistry as well as its pharmaceutical applications.

Chitosan hydrogels can be fabricated via either physical or chemical cross-linking. In previously used methods to induce physical cross-linking, positively charged chitosan was mixed with negatively charged polymers including alginate [Murata et al., 2002; Lee et al., 2004; Lin et al., 2005] and PVA [Koyano et al., 1998; Shin et al., 2002; Wang et al., 2004] to form a hydrogel network via electrostatic interaction. By adjusting the ratio of the positive and negative charged components during the fabrication process, a tunable gel swelling behavior can be readily achieved. Although physical cross-linking may be a convenient way to prepare chitosan hydrogels, disadvantages also exist such as poor mechanical strength and uncontrollable dissolution.

Chemical or covalent cross-linking is favorable for tissue engineering applications when adequate mechanical strength is required. In order to form covalent cross-links, additional functionalities must be introduced onto the chitosan backbone. Modifications have been made to the hydroxyl or amino groups on chitosan to facilitate covalent cross-linking. For example, several chemistries have been proposed to graft synthetic polymers or reactive groups onto chitosan including PEG [Gupta and Kumar, 2001; Park et al., 2001; Hu et al., 2005], PNIPAAM [Cho et al., 2004; Lee et al., 2004], polyurethane [Gong et al., 1998; Silva et al., 2003], and glycidyl methacrylate (GMA) [Navarro and Tatsumi, 2001; Flores-Ramirez et al., 2005]. Figure 5.8 illustrates the chemical structure of chitosan and summarizes chitosan-based hydrogels fabricated via both electrostatic and covalent cross-linking.

Chitosan hydrogels can be degraded via its β-1,4-glycosidic linkage by various enzymes including chitosanase and lysozyme. The degradability of chitosan-based hydrogels can be controlled by the degree of substitution of grafted side chains such as PEG. For example, the degradation rate of chitosan-g-PEG hydrogels by lysozyme has been shown to decrease with increasing degree of substitution of PEG on

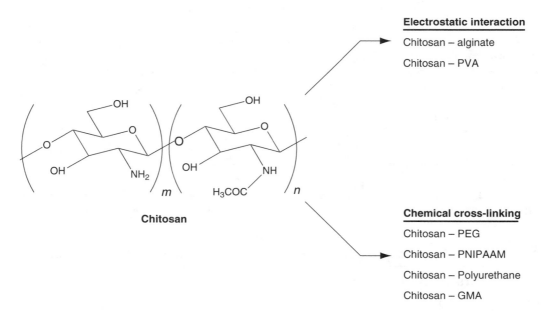

FIGURE 5.8   Structure of chitosan and its derivatives for hydrogel synthesis.

chitosan backbone [Hu et al., 2005]. This behavior is attributed to the fact that the grafted PEG increases the hydrostability of chitosan, which disrupts the accessibility of lysozyme to chitosan and thus decreases its enzymatic degradability.

5.5.2 Alginate-Based Hydrogels

Alginate is a linear polysaccharide composed of 1,4-linked poly(α-L-guluronic acid) and poly(β-D-mannuronic acid). Alginate can be physically cross-linked through its poly(guluronic acid) residues by adding calcium ions. Ionically cross-linked alginate does not form a stable hydrogel because the gel can be disintegrated or dissolved once calcium ions diffusive away or are stripped off by chelating agents. This disadvantage can be overcome by introducing cationic chitosan or polylysine to form a polyelectrolyte reinforced composite, or by grafting covalent cross-linking functionalities to alginate chains. The use of alginate hydrogels has been primarily in the field of controlled release of growth factors such as vascular growth factor (VEGF) [Lee et al., 2004], bone morphogenetic protein-2 (BMP-2) and TGF-β 3 [Simmons et al., 2004] as well as the encapsulation and transplantation of pancreatic islet cells for diabetes treatments [Sun, 1988].

Differing from chitosan, alginate is negatively charged at physiological pH because of the carboxylic acid groups on the backbone. Because of this negative charge, alginate can be fabricated electrostatically via self-assembly. Multilayers of hydrogel microspheres have been successfully produced by noncovalent conjugation of cationic polymers such as chitosan and polylysine to the anionic alginate.

In addition to ionic cross-linking, alginate hydrogels can also be formed via covalent cross-linking. As shown in Figure 5.9, Mooney and coworkers developed a chemical cross-linking method for alginate-based hydrogels where they oxidized poly(guluronate), the cross-linking portion of alginate, and used adipic dihydrazide as the cross-linking agent to form poly(aldehyde guluronate) hydrogels [Bouhadir et al., 1999; Lee et al., 2000]. The resulting alginate-derived hydrogels possess a wide range of mechanical properties and are suitable for tissue engineering application when cell-adhesive peptides were incorporated onto the otherwise nonadhesive alginate backbone.

While physical and chemical cross-linking of alginate hydrogels have been explored extensively by several research groups, few reports have been made regarding the biological cross-linking of alginate gels. In this particular area, Mooney's group developed a novel cross-linking method based on cell–ligand

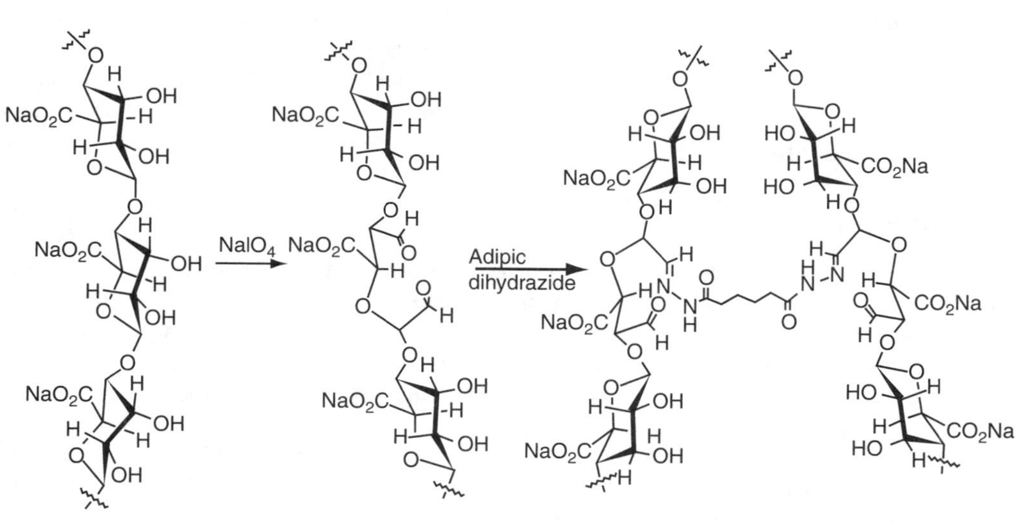

FIGURE 5.9 Scheme for the synthesis and cross-linking of poly(aldehyde guluronate). (From Bouhadir, K.H., Hausman, D.S., et al. 1999. *Polymer* 40: 3575–3584. With permission.)

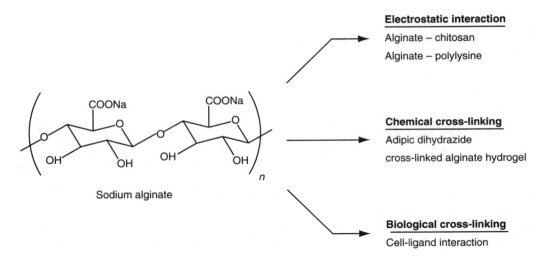

FIGURE 5.10   Structure of sodium alginate and methods for hydrogel fabrication.

interaction [Lee et al., 2003]. They first immobilized RGD peptide sequences on an alginate backbone and then utilized the integrin receptors on cell surfaces to cross-link the alginate into a three-dimensional hydrogel structure. There are several advantages of using this novel method for gel cross-linking. First of all, the immobilization of the target cells can be achieved simultaneously during gel formation. Secondly, this biological cross-linking system can be used to study cell–ligand interactions. Figure 5.10 shows the structure of sodium alginate and the methods for hydrogel fabrication including electrostatic, chemical, and cellular cross-linking.

Unlike chitosan hydrogels, which degrade by bond cleavage along the chitosan backbone, the degradation of ionically cross-linked alginate hydrogels occurs when calcium ions, the cross-linker, are removed from the hydrogel network. If alginate is oxidized to form a covalent cross-linked network, the hydrolytic degradation takes place on the acyl hydrazone bonds grafted onto alginate [Bouhadir et al., 1999]. Mooney and colleagues have extensively studied the degradation behavior [Kong, Alsberg, Kaigler et al., 2004; Kong et al., 2004] as well as biomedical applications of alginate hydrogels [Simmons et al., 2004]. They showed that the mechanical rigidity and the degradation rate of alginate hydrogels can be controlled by adjusting the molecular weight distribution of alginate hydrogels assembled from ionic or covalent cross-linking [Kong, Alsberg, Kaigler et al., 2004; Kong et al., 2004]. By controlling the molecular weights of alginate hydrogels in conjunction with delivery of multiple growth factors (TGF-$\beta3$ and BMP-2), they were able to enhance bone regeneration *in vivo* [Simmons et al., 2004].

5.5.3   Fibrin-Based Hydrogels

Fibrin, another source of polymer for fabricating degradable hydrogels, has undergone extensive investigation. Fibrin, derived from fibrinogen, is found in the blood and polymerized by factor XIIIa to form a clot in response to injuries. Fibrin clot degradation is associated with a series of cellular enzymatic activities during wound healing. The most important fibrin-degrading enzyme is plasmin. Clinically, fibrin is applied to wound sites as a glue to stop bleeding after surgeries or dental procedures. Fibrin glue is obtained from mixing fibrinogen and thrombin solutions to form a fibrin clot [Thompson et al., 1988]. Although not approved in the United States because of the potential for blood-borne transmission of diseases, fibrin glue is commonly used in Europe for controlling blood loss.

Due to the delicate design of fibrin activation and degradation by nature, researchers are able to fabricate degradable hydrogels based on fibrin chemistry. Hubbell and coworkers utilized modified fibrin hydrogels as matrices for various biomedical applications such as the controlled delivery of VEGF

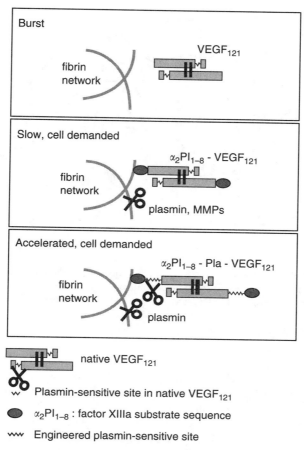

FIGURE 5.11 Model schemes: differential control of VEGF release from fibrin gel matrices by differential suscept-ibility to local cell-associated proteolytic activities. Top box, native VEGF$_{121}$ is freely diffusible in the aqueous milieu of the fibrin matrix. Middle box, VEGF variant, a$_2$PI$_{1-8}$-VEGF$_{121}$, is protected from diffusion, and its liberation is dependent on the cleavage of the fiber network by cell-associated fibrinolytic enzymes (slow, cell-demanded release). Bottom box, a new VEGF variant, a$_2$PI$_{1-8}$-Pla-VEGF$_{121}$, was designed to couple to fibrin networks via a plasmin-sensitive anchor. Cleavage of this plasmin-sensitive site by low and local plasmin could occur independent of fiber network degradation and enhance VEGF release rate (accelerated, cell-demanded release). (From Ehrbar, M., Metters, A., et al. 2005. *J. Control. Release* 101: 93–109. With permission.)

[Zisch et al., 2001, 2003; Hubbell et al., 2003; Ehrbar et al., 2004; Ehrbar et al., 2005; Urech et al., 2005]. There are several novelties of their fibrin-based materials. First, VEGF was covalently incorpor-ated into fibrin hydrogels utilizing the transglutaminase activity of factor XIIIa. Second, the delivery of VEGF was mediated by cellular activity. As shown in Figure 5.11, the liberation of growth factors was achieved either nonspecifically by the degradation of fibrin network by plasmin [Zisch et al., 2001; Hubbell et al., 2003; Zisch et al., 2003; Ehrbar et al., 2004] or specifically by the incorporation of substrate peptides for enzymes such as matrix metalloproteinases (MMPs) secreted by cells [Ehrbar et al., 2005; Urech et al., 2005]. Not only can this release strategy largely preserve the bioactivity of the growth factors, the cell-demanded release profile can also match to the rate of cell infiltration or tissue regeneration.

Sakiyama-Elbert and Hubbell [2000a,b] have developed fibrin hydrogels containing linker peptides and heparins for affinity-based drug delivery specifically used in promoting nerve regeneration. Theoretical model predictions indicate that the release rate of growth factors from these gels can be modulated through

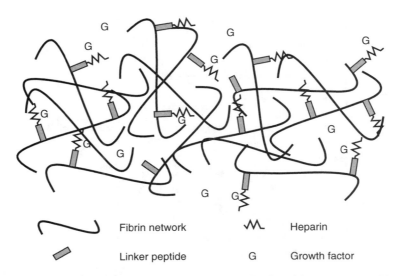

FIGURE 5.12 Diagram showing the components of the heparin-binding delivery system. a_2PI_{1-7}–$ATIII_{121-134}$ peptide is cross-linked into the fibrin gel via the transglutaminase activity of Factor XIIIa; heparin can bind to the peptide by electrostatic interactions. NT-3 can bind to the bound heparin, creating a gel-bound ternary complex that is not diffusible. NT-3 can also exist in the diffusible state, alone, or in a complex with free heparin. (From Taylor, S.J., McDonald, J.W., et al. 2004. *J. Control. Release* 98: 281–294. With permission.)

the affinity between the peptides and heparin and the concentration of heparin presented in the fibrin gels [Taylor et al., 2004]. Figure 5.12 shows the schematic diagram of the heparin-containing fibrin hydrogels for growth factor delivery [Taylor et al., 2004].

In addition to controlled growth-factor delivery, fibrin hydrogels have also been used as tissue engineering scaffolds. For example, fibrin hydrogels have been used as three-dimensional cell-culture matrices for cardiovascular tissue engineering. It was shown that fibrin gels can support homogenous cell growth, confluent collagen production, and tissue development [Ye et al., 2000; Jockenhoevel et al., 2001].

Fibrin hydrogels have also been shown to be biocompatible and can be formed *in situ* during cartilage repair [Homminga et al., 1993]. However, a potential shortcoming of fibrin hydrogels is that they do not possess significant mechanical strength when applied on load-bearing tissue such as cartilage. Composite scaffolds made from fibrin-polyglycolic acid [Ameer et al., 2002] and fibrin-polyurethane [Lee et al., 2005] have been developed to increase the mechanical properties of fibrin-based gels for articular cartilage tissue engineering.

5.5.4 Collagen and Gelatin-Based Hydrogels

Collagen, a triple-helix protein, is the major component of connective tissues. Due to its physiological abundance, collagen is considered biocompatible. Collagen can be biodegraded by enzymes such as collagenase. The application of collagen hydrogels, however, is limited due to its laborious, batch-production procedures as well as the inconsistency of its biological and mechanical properties between batches.

Modified collagen gels are still favored for many tissue-engineering applications. Composite scaffolds such as collagen-alginate [Bohl et al., 1998] or collagen-hyaluronan [Segura, Anderson et al., 2005; Segura, Chung et al., 2005] have been fabricated and used for several tissue engineering and DNA delivery applications. Matrigel, a type IV collagen-based and a commercially available hydrogel, mimics the ECM environment and is commonly used in *in vivo* or *in vitro* studies including cell growth and differentiation [Taub et al., 1990], angiogenesis, and tumor augmentation [Benelli and Albini, 1999; Auerbach et al.,

2003]. Collagen hydrogels have also been used to immobilize human neuroblastoma cells for cell-based biosensing [Mao and Kisaalita, 2004].

An important consideration of using collagen hydrogels in tissue engineering is that the gels significantly shrink after cell-seeding [Bell et al., 1979; Nakagawa et al., 1989]. Several methods have been developed to suppress the contraction of the collagen hydrogels such as increased cross-linking with glutaraldehyde [Torres et al., 2000] or incorporation of short collagen fibers [Gentleman et al., 2004]. Increasing the cross-linking density of collagen in the scaffold has also been shown to decrease the degradation rate of the gels [Meinel, Hofmann et al., 2004; Meinel et al., 2004a,b; Hu et al., 2005].

Gelatin, a natural glycine-rich polymer derived from hydrolyzed collagen, is widely used in food industry as well as in pharmaceutical devices for the controlled release of growth factors. Besides its advantageous biodegradability and biocompatibility, the most attractive characteristic of gelatin is its ability to form polyion complexes. Depending on the manufacturing process, the isoelectric point (PI) of gelatin can be adjusted to yield a positively or negatively charged polymer. This flexibility makes gelatin a suitable matrix for controlled delivery of charged growth factors such as anionic basic fibroblast growth factor (bFGF) [Tabata et al., 1999] and BMP-2 [Yamamoto et al., 2003].

Different gelatin formulations have been studied to evaluate the drug loading capacity and release rate. Like the other hydrogels, drug release profiles obtained from gelatin hydrogels can be readily adjusted by changing the network cross-linking density. Several methods have been developed to cross-link gelatin hydrogels including glutaraldehyde, dehydrothermal treatment, UV or electron beam irradiation [Liang et al., 2004].

The preparation of gelatin hydrogels is rather easy compared to other natural polymers. For example, gelatin hydrogels can be formed by simply mixing gelatin solution with small amount of the cross-linker (e.g., glutaraldehyde) and left for several hours at 4°C. In addition to the conventional gelation methods such as freeze-drying, gelatin hydrogels can also be prepared via photopolymerization. For example, Matsuda and colleagues have synthesized photocurable, tissue-adhesive gelatin hydrogels for drug release [Okino et al., 2002, 2003; Manabe et al., 2004; Masuda et al., 2004], arterial repair [Li et al., 2003] as well as for nerve-guidance prosthetic scaffolds [Gamez et al., 2003]. The major advantage of these gelatin hydrogels is that they are photocurable and can be formed via *in situ* polymerization.

Burmania et al. [2003a,b] prepared interpenetrating networks (IPNs) containing gelatin and PEG-diacrylate and further characterized the protein release, fibroblast adhesion, and *in vivo* host response to these gels. The chemical and mechanical properties of these gelatin hydrogels can be controlled by changing the weight percentage of gelatin in the IPN or through chemical modifications of the gel precursors.

5.5.5 Dextran-Based Hydrogels

Dextran, with subunits consisting of α-1,6-linked D-glucopyranose, is another common, naturally occurring polymer used in the fabrication of degradable hydrogels. Dextran can be readily produced by bacteria or yeast via fermentation and is therefore an ideal polysaccharide for industrial as well as clinical usage. Dextran is also water-soluble and has been widely used in surgery owing to its antithrombotic effect.

Similar to other hydrogels, dextran hydrogels can be formed via physical or chemical cross-linking. Physically cross-linked dextran hydrogels can be fabricated via electrostatic interaction. Hennink and coworkers have fabricated microspheres with positive and negative charges by modifying dextran with dimethylaminoethyl methacrylate and methacrylic acid, respectively (Figure 5.13) [Van Tomme et al., 2005]. Dextran hydrogels are formed when microspheres with opposite charges were mixed together. One interesting characteristic of this physical hydrogel is that, when sufficient shear is applied, the viscosity of the gel decreases, rendering the gel injectable. After injection the shear is removed and the hydrogel spontaneously reforms.

Another method for fabricating physically cross-linked dextran hydrogels was developed by Hennink and coworkers. It is based on stereocomplexation between D-lactate and L-lactate oligomers [de Jong et al., 2000; Hennink et al., 2004]. Enantiomeric lactic acid oligomers were grafted to dextran and the dex-lactate

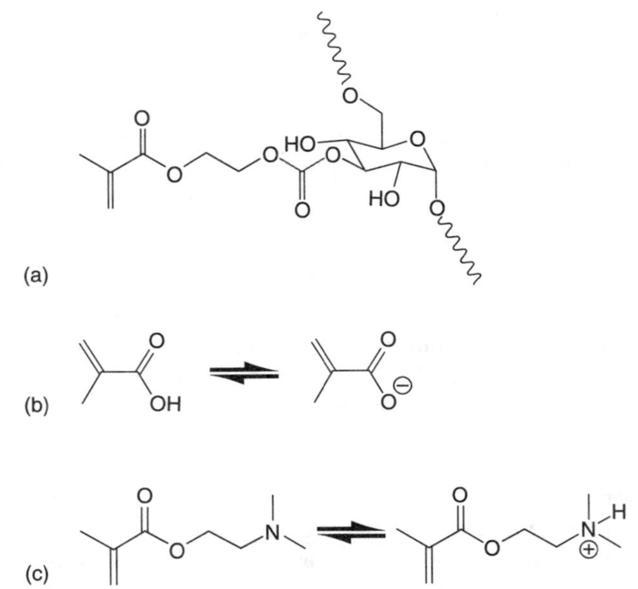

FIGURE 5.13 Chemical structures of dex-HEMA (a), methacrylic acid (b), and dimethylaminoethyl methacrylate (c). (From Van Tomme, S.R., van Steenbergen, M.J., et al. 2005. *Biomaterials* 26: 2129–2135. With permission.)

hydrogels were formed by stereocomplex formation between D-lactate and L-lactate. The sustained release of pharmaceutical proteins [Cadee et al., 2002; Hennink et al., 2004] as well as the biocompatibility [Cadee et al., 2000, 2001] have been demonstrated for these dextran-based hydrogels.

Several modification schemes have been proposed to fabricate degradable dextran hydrogels via chemical cross-linking. For instance, Hennink and coworkers have modified dextran with hydroxyethyl methacrylate (HEMA). The resulting dex-HEMA hydrogels can be formed via free-radical polymerization and cross-linking of the methacrylate side groups [van Dijk-Wolthuis et al., 1997]. Degradable hydrogels were formed by incorporating lactate into the modified dextran chains to form dex-lactate-HEMA hydrogels (Figure 5.14a). The degradation rate of these hydrogels can be tailored by varying the length of a spacer unit within the cross-link [Cadee et al., 1999, 2000, 2001, 2002]. Increasing the size of the cross-link lowers the overall cross-linking density, increases the degree of swelling and water content within the gel, and results in a faster rate of degradation.

PEGylation of dextran is another means of fabricating dextran hydrogels. Moriyama et al. have prepared multi-layered PEG-g-dextran hydrogels for pulsatile drug delivery. In this two-phase system, grafted-PEG chains act as an insulin depot, while dextran domains form the main matrix. Upon surface-limited degradation by dextranase, a pulsatile release profile appears due to the multi-layered structure of the polymer formulation [Moriyama and Yui, 1996; Moriyama et al., 1999].

Dextran hydrogels can also be prepared via photopolymerization. For example, Kim et al. [1999] synthesized dex-maleic acid macromers by reacting dextran with maleic anhydride (Figure 5.14b). GMA has also been reacted with dextran to form dex-GMA photo-cross-linkable hydrogels (Figure 5.14c) [Pitarresi et al., 2003; Li et al., 2004].

IPNs [Kurisawa et al., 1995, 1997; Yamamoto et al., 1996] or semi-IPNs [Kumashiro et al., 2002] consisting of dextran and other components provide another route for dextran hydrogel preparation. Kurisawa et al. have prepared a series of dextran-based IPNs including Dex-PEG [Kurisawa et al., 1995, 1997] and Dex-Gelatin [Yamamoto et al., 1996; Kurisawa and Yui, 1998]. These dextran-based hydrogels have been shown to exhibit a double-stimuli-response function and are degradable only when two enzymes, which independently degrade distinctly different substrates, are both present.

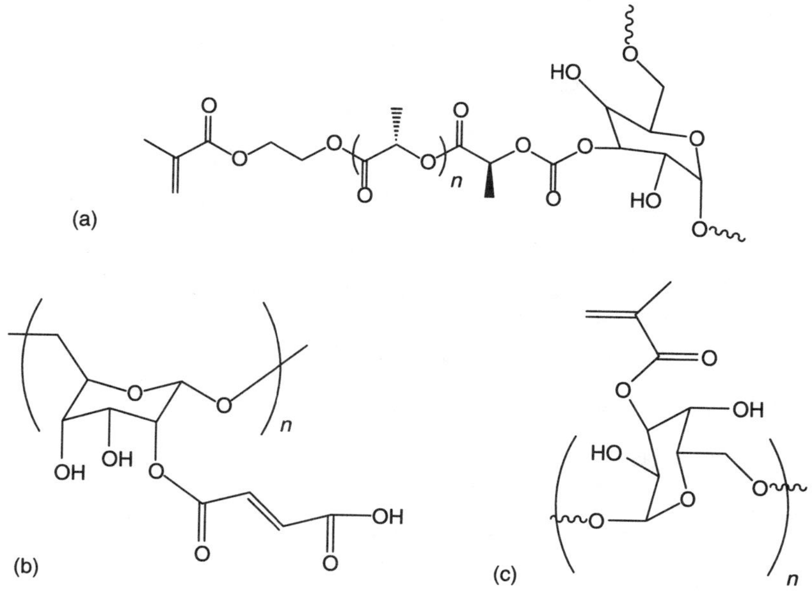

FIGURE 5.14   Structures of (a) dex-lactate-HEMA, (b) dex-maleic acid, and (c) dex-methacrylate.

5.5.6 Hyaluronic Acid

HA is another natural polymer derived from glycosaminoglycan composed of D-glucuronic acid and N-acetylglucosamine. Physiologically, HA is the backbone of connective tissues such as cartilage. In addition to its structural importance, HA also possesses many characteristics that make it suitable for biomedical application. For example, HA mediates angiogenesis, wound healing, metastasis, inflammation, and granulation [Chen and Abatangelo, 1999]. Because of these favorable properties, HA has been used in many clinical applications. However, disadvantages of using HA as tissue engineering scaffolds include its nonadhesive property, which prohibits cell adhesion. Furthermore, when HA is used as a scaffolding material, it does not provide enough mechanical strength, which largely limits its application. Fortunately, the first disadvantage can be overcome by grafting cell-adhesive peptides, such as arginine-glycine-aspartic acid (RGD), to HA molecules. By incorporating reactive side functionalities such as acrylate or methacrylate groups (Figure 5.15), linear HA chains can be cross-linked to form water-swellable hydrogels with adequate mechanical strengths.

Collagen, GMA, methacrylic anhydride, and gelatin have been used to cross-link HA chains and form cell-adhesive, mechanically stable hydrogels. For example, Park et al. [2003] have prepared methacrylated, RGD-containing HA-based hydrogels via photopolymerization. The cellular response as well as the gel swelling ratio and mechanical properties were extensively studied. Leach et al. [2003] have synthesized a photo-cross-linkable HA by reacting it with the epoxy group of GMA. The pendant methacrylate bonds present on the resulting GMA-modified HA (GMHA) molecules facilitate photopolymerization and cross-linking of the polymer chains. Furthermore, an acrylated PEG-peptide can be copolymerized into the GMHA hydrogels to render them cell adhesive [Leach et al., 2004].

When presented *in vivo*, HA hydrogels can be degraded by hyaluronidase [Menzel and Farr, 1998] or hydroxyl radicals [Yui et al., 1993; Hawkins and Davies, 1998] produced at the inflammation sites. Hyaluronidase-mediated degradation is relatively slow compared to degradation by hydroxyl radicals. This is mainly because of the low physiological concentration of hyaluronidase. On the other hand, hydroxyl radicals generated in the inflammation sites can effectively degrade HA via glycosidic cleavage [Yui et al., 1993; Hawkins and Davies, 1998]. As shown in Figure 5.16, the degradation rate of GMHA

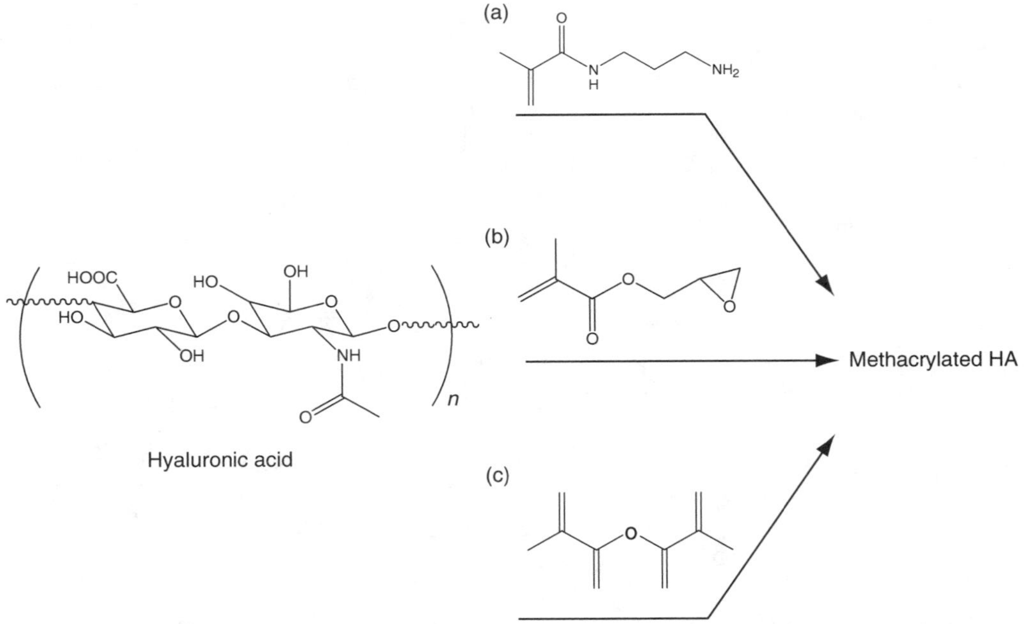

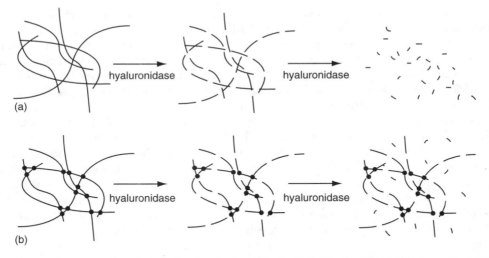

FIGURE 5.15 Reaction schemes for methacrylation of HA through (a) *N*-(3-aminopropyl)-methacrylamide, (b) glycidyl methacrylate, and (c) methacrylic anhydride.

FIGURE 5.16 Enzymatic degradation of HA hydrogels. (a) Native HA (shown schematically) is quickly degraded *in vivo* by the enzyme hyaluronidase. (b) Cross-linking the HA chains forms an insoluble hydrogel matrix that is more resistant to enzymatic degradation. (From Leach, J.B., Bivens, K.A., et al. 2003. *Biotechnol. Bioeng.* 82: 578–589. With permission.)

hydrogels is slower than native HA hydrogels [Leach et al., 2003]. Leach et al. [2003, 2004] have also tested the *in vitro* degradability of the GMHA hydrogels using hyaluronidase. Finally, Langer and coworkers were able to better control the resulting GMHA-hydrogel swelling and degradation behavior by partially oxidizing or grafting an oligomer onto the multifunctional HA chains prior to conjugation with GMA [Jia et al., 2004].

Methacrylic anhydride has also been reacted with HA to form methacrylated HA for subsequent gel formation [Masters et al., 2004, 2005; Burdick et al., 2005]. Applications of HA hydrogels based on this chemistry include the incorporation of specialized peptides to enhance the adhesion and proliferation of valvular interstitial cells [Masters et al., 2004, 2005] as well as the photoencapsulation of chondrocytes for cartilage regeneration [Burdick et al., 2005]. Smeds and Grinstaff have prepared photo-cross-linkable HA hydrogels by reacting HA with excess methacrylic anhydride [Smeds and Grinstaff, 2001]. The gel formation was rapid and its *in situ* photopolymerization enables minimally invasive implantation when used on an articular cartilage repair application [Nettles et al., 2004].

5.6   Degradable Hydrogels Derived from Synthetic Biomaterials

Degradable hydrogels made from synthetic polymers have recently gained considerable interest. The most common synthetic polymers used to construct hydrogel networks include PEG or poly(ethylene oxide), PVA, PNIPAAM), and others. The most apparent advantage of using synthetic polymers is the high degree of control over the polymer chemistry, architecture, and physical hydrogel properties including degradation behavior. Hydrogels made from synthetic polymers can be engineered with tailorable hydrophilicity, cross-linking density, and degradability. In addition, when degraded *in vivo*, there is no immune recognition associated with the low molecular weight degradation products if the polymer and hydrogel fabrication technique are carefully chosen.

5.6.1   Poly(ethylene glycol)-Based Hydrogels

PEG is a nonionic, hydrophilic polymer. Because of this, PEG is widely used to create nonfouling surfaces that repel nonspecific protein adsorption and cell adhesion. Many researchers have utilized the terminal hydroxyl groups on PEG molecules to modify PEG into a cross-linkable material. The most commonly used chemistry is the acrylation or methacrylation to endcap linear, branched, or star-shaped PEG molecules with reactive vinyl species (Figure 5.17). Free-radical polymerization of these methacrylate or acrylate groups results in cross-linked hydrogels that are stable over clinically relevant timescales.

Hydrolytically or enzymatically labile groups can be introduced into cross-linked PEG networks by modifying the soluble PEG precursors with ester-containing α-hydroxy acid oligomers or protease-sensitive peptide blocks prior to vinyl-group end-capping. Upon polymerization and cross-linking, these bonds will render the resulting hydrogels degradable. In particular, the incorporation of hydrolytically degradable PLA groups greatly increase the potential applications of PEG hydrogels (Figure 5.2). Sawhney et al. [1993] pioneered the acrylation of PLA-PEG-PLA diblock macromers for *in situ* formation of degradable hydrogels. Since that time, Bowman and Anseth and their colleagues have synthesized a series of degradable PLA-PEG-PLA hydrogels and extensively characterized their degradability via experimental observation as well as theoretical verification [Metters et al., 2000a,b; Metters, Bowman et al., 2000; Metters, Anseth et al., 2001; Metters, Bowman et al., 2001; Mason et al., 2001].

Mikos and colleagues have synthesized a series of PEG-containing, water-soluble block copolymers, including poly(proplyene fumarate)-co-ethylene glycol) [Suggs et al., 1998; Behravesh et al., 2002; Tanahashi et al., 2002] and oligo(poly(ethylene glycol) fumarate) (Figure 5.17) [Jo et al., 2001; Shin et al., 2002; Temenoff et al., 2002]. These hydrogels are biodegradable, injectable, *in situ* photo-cross-linkable, and biocompatible. Because of these properties, they hold great potential both in drug delivery [Holland et al., 2003; Kasper et al., 2005; Park et al., 2005] and tissue engineering applications [Shin et al., 2002, 2003, 2004; Behravesh and Mikos, 2003; Temenoff et al., 2003; Fisher et al., 2004]. The degradation of these hydrogels occurs at the ester linkage within the poly(propylene fumarate) block [Suggs et al., 1998] and can be accelerated by decreasing pH and cross-linking density [Timmer et al., 2003].

Recently, Hubbell and coworkers have used Michael-type addition reactions to form PEG-based hydrogels by reacting vinyl-sulfone functionalized PEG chains with thiol groups present on other macromers. This conjugation reaction can be carried out under physiological conditions without the use of initiators

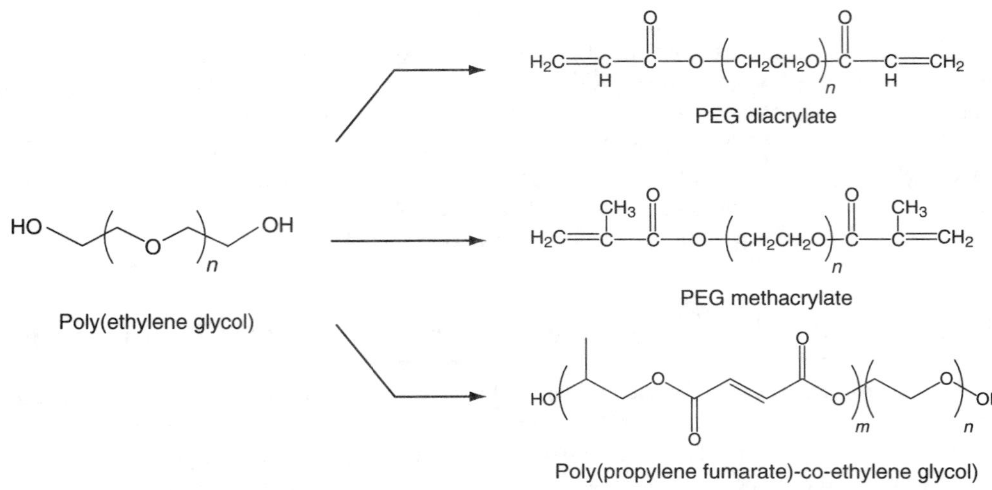

FIGURE 5.17   Poly(ethylene glycol) and its derivatives for PEG-based hydrogel synthesis.

[Elbert et al., 2001; Heggli et al., 2003; Lutolf and Hubbell, 2003; van de Wetering et al., 2005], which produce free radicals and often induce damage to the encapsulated protein [Lin and Metters, 2005], DNA [Quick and Anseth, 2003, 2004; Quick et al., 2004], or cells [Quick and Anseth, 2003].

West and colleagues have incorporated enzyme-sensitive peptides into PEG-based hydrogels. The resulting peptide-incorporated PEG-based hydrogels can be degraded by collagenase or plasmin [West and Hubbell, 1999; Mann et al., 2001; Gobin and West, 2002]. The incorporation of enzyme-sensitive peptides not only facilitates gel degradation, but also enhances cell migration [Gobin and West, 2002]. Hubbell and coworkers have also modified PEG hydrogels with recombinantly produced peptides that are proteolytically labile [Park et al., 2004]. The main advantage of these approaches is that the cellularly controlled gel-degradation rate self-adjusts to the rate of cell infiltration, making these gels extremely attractive for tissue regeneration applications.

5.6.2 Poly(vinyl alcohol)-Based Hydrogels

PVA hydrogels were one of the earliest hydrogels to play an important role in biomedical applications [Brinkman et al., 1991]. Unlike PEG, which has at most two derivatizable hydroxyl groups at the chain termini, PVA possesses numerous pendant hydroxyl groups that can be modified for cross-linking or ligand attachment. Both physical and chemical cross-linking methods can be used to fabricate PVA hydrogels. Peppas pioneered the fabrication of PVA hydrogels by the freeze–thaw process [Peppas and Scott, 1992; Stauffer and Peppas, 1992; Mongia et al., 1996]. The resulting PVA hydrogels were used for controlled release [Peppas and Scott, 1992] and wound healing applications [Mongia et al., 1996].

In addition to the freeze–thaw process, PVA hydrogels can be physically cross-linked using blend copolymers. Some examples of polymers that can be blended with PVA to create stable hydrogels are polysaccharides including chitosan and dextran. Cascone et al. [2001] reviewed the fabrication and applications of several PVA/polysaccharide blend hydrogels. These hydrogels could be used in drug-delivery systems and are capable of delivering human growth hormone in physiological amounts.

To create chemically cross-linked PVA hydrogels that can be formed *in situ*, Anseth and coworkers synthesized a series of photocurable PVA-based hydrogels and characterized their mechanical properties as well as degradability [Martens and Anseth, 2000; Nuttelman et al., 2002; Bryant et al., 2004]. Specifically, fibronectin-modified PVA hydrogels were found to enhance NIH3T3 cell adhesion, proliferation, and migration [Nuttelman et al., 2001]. PLA-g-PVA hydrogels were found to improve valve interstitial cell

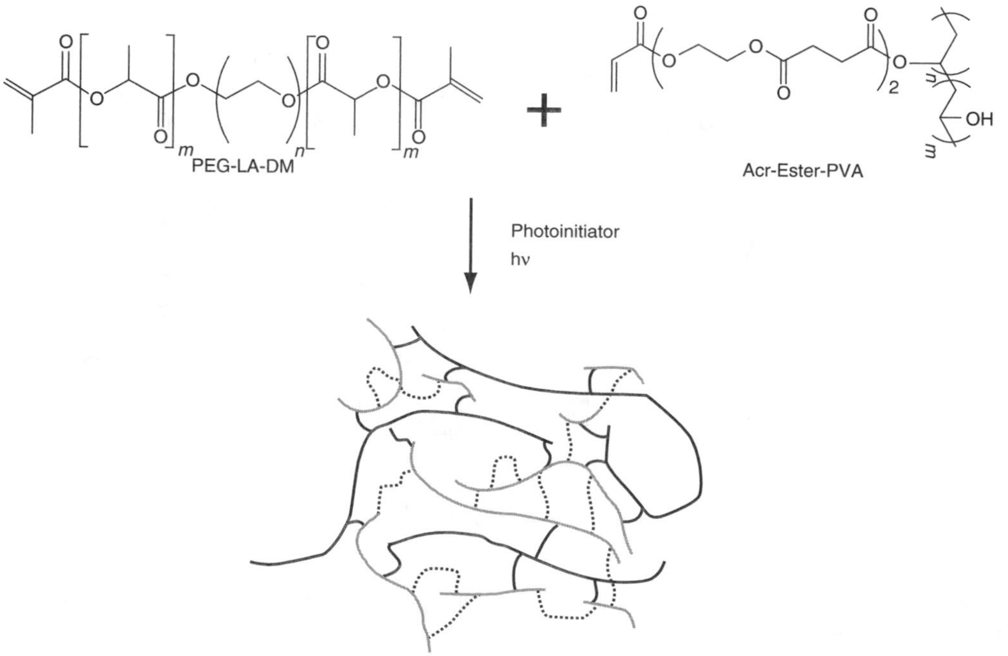

FIGURE 5.18 An idealized schematic of the structure of hydrogels formulated by copolymerizing poly(ethylene glycol)-lactic acid-dimethacrylate (PEG-LA-DM) with acrylate-ester-poly(vinyl alcohol) (Acr-Ester-PVA). The network consists of kinetic chains (light lines) connected via PEG cross-links (dotted lines) and the multifunctional PVA chains (bold lines). (From Martens, P.J., Bryant, S.J., et al. 2003. *Biomacromolecules* 4: 283–292. With permission.)

adhesion by increasing gel hydrophobicity [Nuttelman et al., 2002]. PEG-PVA hydrogels with tailorable characteristics were used in cartilage tissue engineering (Figure 5.18) [Martens et al., 2003]. West and coworkers also prepared photo-cross-linkable PVA hydrogels modified with cell-adhesive peptides for encouraging cell attachment [Schmedlen et al., 2002].

5.6.3 Polyacrylamide-Based Hydrogels

Polyacrylamide gels are widely used for protein separation (e.g., SDS–PAGE) and are prepared from free-radical polymerization of acrylamide monomer and N,N'-methylene bisacrylamide (BIS) cross-linker. The free radicals are typically generated via a redox reaction involving two components, ammonium persulfate (APS) and tetramethylenediamine (TEMED). To fulfill the desired protein separation requirements, the cross-linking density of these gels can be increased by increasing the monomer concentration or cross-linker to monomer ratio. While acrylamide bonds are considered to be stable in aqueous environments, the unique architecture of the bisacrylamide cross-linker enables it to be cleaved via hydrolytic attack. Hydrolytic cleavage of the cross-linker leads to eventual degradation of the polyacrylamide gel.

Hydrogels made from N-(2-hydroxypropyl)-methacrylamide (HPMA) have been prepared via radical polymerization (Figure 5.19) [Ulbrich et al., 1993, 1995]. The resulting poly(HPMA) hydrogels have been shown to be stable in acidic environments, but hydrolytically degradable at pH 7.4 [Ulbrich et al., 1995]. Applications of these degradable methacrylamide-based hydrogels include controlled drug delivery [Ulbrich et al., 1995; St'astny et al., 2002] and gene therapy [Howard et al., 2000; Oupicky et al., 2002]. Poly(HPMA) hydrogels have also been used to conjugate doxorubicin, an anticancer drug,

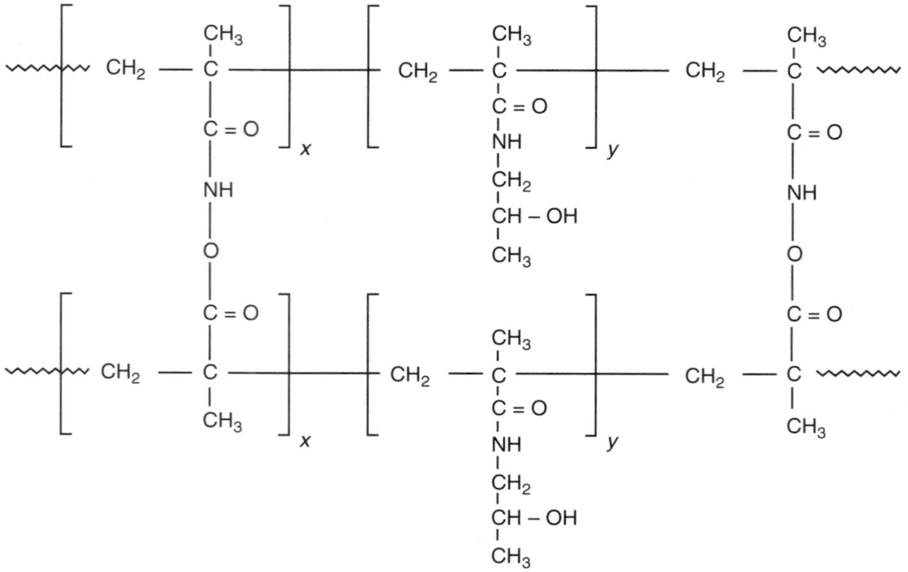

FIGURE 5.19 Structure of poly(HPMA) hydrogel. (From Ulbrich, K., Subr, V., et al. 1995. *J. Control. Release* 34: 155–165. With permission.)

via enzymatically or hydrolytically labile linkages [St'astny et al., 2002a,b]. Prolonged, degradation-controlled release of this drug from poly(HPMA) gels has been shown to effectively reduce tumor size as well as increase survival time in an animal model [St'astny et al., 2002].

Another well-known application of polyacrylamide-based polymers is the fabrication of thermally responsive PNIPAAM hydrogels. PNIPAAM, with a lower critical solution temperature (LCST) around 32°C, exhibits a reversible thermo-sensitive swelling behavior. Above the LCST, the PNIPAAM polymer collapses and precipitates from the surrounding solution due to strong intramolecular hydrophobic interactions. Below the LCST, the polymer chain swells and is soluble. This attractive feature allows PNIPAAM at a temperature below the LCST to be readily injected into the body. Immediately following injection, the system temperature increases to 37°C and the PNIPAAM gel forms *in situ*.

Although PNIPAAM itself is a nondegradable polymer, many efforts have been made to incorporate degradable functionalities to render degradable PNIPAAM hydrogels. These copolymers include dextran [Huh et al., 2000; Kumashiro et al., 2001, 2002; Huang and Lowe, 2005], acrylic acid [Kim and Healy, 2003], dimethylacrylamide [Kurisawa et al., 1998; Kurisawa and Yui, 1998], poly(amino acids) [Yoshida et al., 2003], and PLA [Xiao et al., 2004]. The temperature-responsiveness and biodegradability of these gels make them unique and in many cases allows better control over physical gel properties and function. For example, the degree of swelling, degradation rate, and release rate of drugs from hydrogels containing PNIPAAM can be modulated by small temperature changes [Kurisawa et al., 1998].

5.6.4 Polyphosphazene-Based Hydrogels

As a degradable synthetic hydrogel for biomedical applications, polyphosphazene has gained much attention for controlled drug release. Readers are referred to recent review articles for a more in-depth discussion of polyphosphazene polymers [Andrianov and Payne, 1998]. Unlike most of the other synthetic polymers with carbon–carbon backbones, polyphosphazene, on the other hand, possesses a backbone with alternative phosphorus and nitrogen atoms and two substitutive groups on phosphorus atoms. Polyphosphazenes can be synthesized from precursor macromers poly(dichlorophosphazene) into water soluble polyphosphazene polyacid. The resulting polyacid can then be cross-linked into ionic hydrogels (Figure 5.20)

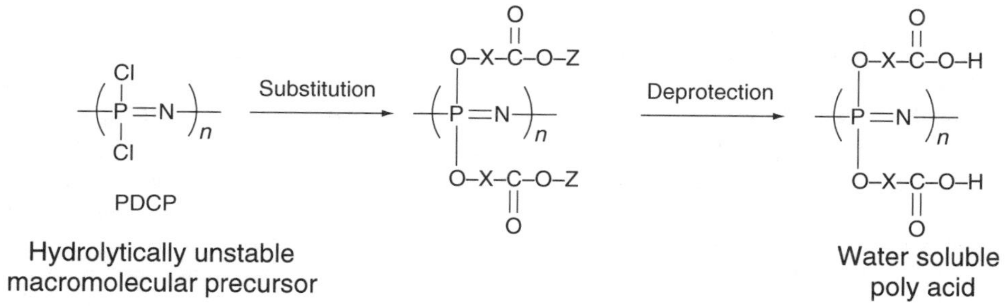

FIGURE 5.20 Synthetic pathway to polyphosphazene polyacids. (From Andrianov, A.K., Svirkin, Y.Y., et al. 2004. *Biomacromolecules* 5: 1999–2006. With permission.)

[Andrianov et al., 1998, 2004]. The degradation rate of polyphosphazene has been shown to depend on the degree of substitution of the pendant groups. Specifically, the degradation rate decreases with a decreasing degree of substitution [Andrianov et al., 2004].

Many modifications have been proposed in the fabrication of polyphosphazene hydrogels. For example, Allcock et al. have synthesized a series of phosphazene hydrogels for biomedical applications including, but not limited to, poly[(amino acid-ester) phosphazenes] [Allcock et al., 1994], poly(alkyl oxybenzoate) phosphazene [Greish et al., 2005], tyrosine-bearing polyphosphazenes [Allcock et al., 2003], and polyphosphazene blend copolymers [Ibim et al., 1997; Ambrosio et al., 2002]. Polyphosphazene can also be fabricated as thermo- or pH-responsive hydrogels.

5.6.5 Protein-Cross-Linked Hydrogels

Hydrogels made from synthetic polymers exhibit many favorable characteristics as discussed in the previous sections. However, one of the drawbacks of using pure synthetic hydrogels is that they do not possess biological recognition sites for cell–material interactions that may be advantageous for tissue engineering applications. For this reason, efforts have been made to develop hydrogels with synthetic polymer backbones and biologically derived cross-links. The use of proteins as cross-linkers for synthetic hydrogels can be traced back to the early 90s where albumin was used to cross-link poly(1-vinyl-2-pyrrolidinone) (PVP) hydrogels [Shalaby et al., 1990, 1991] The cross-linking density of the PVP hydrogels can be controlled by the degree of albumin functionality and the concentration of albumin. Furthermore, albumin-cross-linked hydrogels can be enzymatically degraded by pepsin via surface erosion or bulk degradation depending on the functionality of the albumin and the overall cross-linking density of the gel.

More recently, PEGylated fibrinogen has also been used to cross-link PEG hydrogels. Several advantages exist for using fibrinogen as a hydrogel cross-linking agent. For example, fibrinogen possesses inherent cell-recognition sites for cellular ingrowth as well as enzymatically degradable peptide sequences for gel degradation [Almany and Seliktar, 2005].

While protein-cross-linked hydrogels provide sites for cell–material interactions, genetically engineered or artificial proteins have emerged as alternative tools for cross-linking synthetic, polymer-based hydrogels. Recombinant DNA technology was used to create artificial proteins or peptides as building blocks within otherwise synthetic hydrogels. Hydrogels composed or cross-linked by genetically engineered proteins preserve all the favored characteristics of natural-protein cross-linkers while eliminating excessive recognition sites that may prove detrimental to successful application. Differing from conventional synthetic polymers and natural proteins, genetically engineered proteins have monodisperse molecular weights and precisely engineered functionalities.

Owing to this fine tuning over polymer size and chemistry, researchers are able to better predict and determine the physiological or biological fate of the artificial biopolymers [Haider et al., 2004]. Examples of genetically engineered proteins used for degradable hydrogels are elastin-like proteins (ELPs) [Megeed et al., 2004], silk-elastin-like proteins (SELPs) [Megeed et al., 2002], and coiled-coil proteins [Petka et al., 1998; Wang et al., 1999, 2001; Kopecek, 2003]. Hydrogels made from these materials have been used in many biomedical applications focusing on drug release [Kopecek, 2003; Haider et al., 2004], gene delivery [Megeed et al., 2002, 2004], and tissue engineering [Panitch et al., 1999].

Tirrell and coworkers pioneered the study of artificial-protein hydrogels in which the synthetically engineered proteins retain at least two domains, one for water retention and another for hydrogel network formation [Petka et al., 1998]. The resulting hydrogels can be delicately designed so that the gel is stimuli-sensitive. The mechanical properties of artificial-protein matrices were also determined and used to control cell and tissue behavior [Di Zio and Tirrell, 2003]. In addition to the stimuli-controlled swelling, another advantage of incorporating artificial proteins into synthetic hydrogels is that many key features of the extracellular matrix, including fibrinogen and elastin domains, can be engineered *in vitro* and used to promote tissue regeneration *in vivo* [Di Zio and Tirrell, 2003].

Kopecek and coworkers have developed genetically engineered stimuli-sensitive hydrogels based on coiled-coil proteins for controlled release and tissue engineering [Wang et al., 1999]. As shown in Figure 5.21, recombinant coiled-coil protein was used to cross-link HPMA hydrogels containing metal-chelating monomer N-(N',N'-dicarboxymethylaminopropyl) methacrylamide (DAMA) [Wang et al., 1999]. The resulting poly(HPMA-co-DAMA) hydrogels can swell or shrink in response to environmental changes and are very useful in stimuli-sensitive drug release [Xu et al., 2005]. On the other hand, the designed protein or peptide sequence can be used as an epitope when incorporated into hydrogels for cellular recognition [Tang et al., 2000, 2001].

Hubbell and coworkers have also devoted significant efforts to the design and fabrication of PEG-based hydrogels containing artificial proteins. They created several model synthetic polymeric systems containing molecularly engineered peptides or proteins for mimicking the natural ECM [Halstenberg et al., 2002; Lutof, Lauer-Fields et al., 2003a; Lutof, Weber et al., 2003b; Seliktar et al., 2004; Raeber et al., 2005]. In these systems, the proteins are incorporated as highly mobile pendant chains rather than as cross-links. The pendant protein systems are not stimuli-sensitive, but do facilitate the development of synthetic scaffolds for use in tissue engineering. They also provide a platform for investigating some of the fundamental yet critical cellular behavior such as migration, secretion, and proliferation. For example, a PEG-based hydrogel bearing a MMP inhibitor and tumor necrosis factor-alpha (TNF-α) was used to investigate cellular protease activity as well as cell migration [Raeber et al., 2005].

5.7 Conclusions

Degradable hydrogels have already been successfully employed in numerous biomedical applications. Because of their unique combination of properties, they have the potential to dramatically impact the future of biomaterials and biomedicine, especially in the fields of controlled drug delivery and tissue engineering. Irregardless of their chosen application, degradable hydrogels must meet a number of design criteria to function appropriately in complex biological environments. They must be biocompatible, mechanically resilient, selectively permeable, and degradable over appropriate timescales. Furthermore, the degradation behavior of all biodegradable hydrogels should be well defined, reproducible, and tunable via precursor chemistry and structure.

The chemistries of all hydrogel components are critical to maintaining sufficient biocompatibility, degradability, and hydrophilicity. Many polymers derived from biological sources are readily cross-linked into biocompatible, resorbable hydrogels. However, these materials suffer from concerns of reproducibility and engineerability. Synthetic polymers offer better engineerability in the chemistry, structure, and physicochemical properties of biodegradable hydrogels, but are not necessarily biodegradable nor biocompatible. Degradability is typically engineered into these systems via unique macromonomer designs (e.g., dimethacrylated PLA-PEG-PLA triblock copolymers) or specialized polymerization reactions that create labile bonds (e.g., Michael-type additions between an acrylate and thiol). Future designs of biodegradable

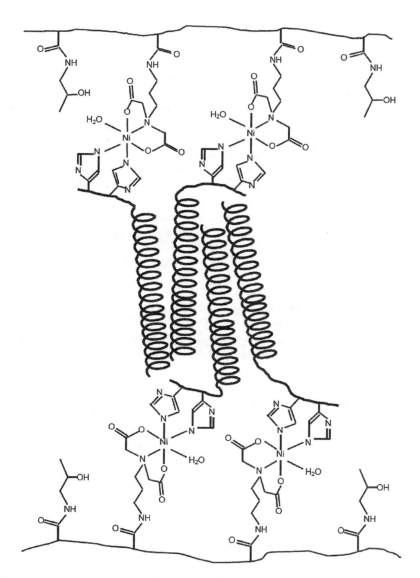

FIGURE 5.21 Structural representation of the hybrid hydrogel primary chains and the attachment of His-tagged coiled-coil proteins. (From Wang, C., Stewart, R.J., et al. 1999. Hybrid hydrogels assembled from synthetic polymers and coiled-coil protein domains. *Nature* 397: 417–420. With permission.)

hydrogels will rely on synthesizing gels exhibiting the advantages of both natural and synthetic polymers. Novel methods of genetic engineering and controlled polymerizations such as atom-transfer radical polymerizations are currently producing polymers with precisely defined molecular weights and functionalities that can be used to fabricate biodegradable hydrogels with exact physicochemical properties and degradation behaviors.

Finally, network structure also plays a key role in determining the elasticity, water content, permeability, and degradation behavior of biodegradable hydrogels. The significant majority of biodegradable gels currently used in biomedical applications degrade via cleavage of a small number of known hydrolytically or enzymatically labile bonds. However, the progression in observable hydrogel properties such as degree of swelling, modulus, and mass loss during degradation varies greatly due to differences in the location, arrangement, distribution, and total concentration of these labile bonds within the network structure. Network structure is a function of precursor chemistry, precursor architecture, and method of gel fabrication (e.g., chain vs. step-growth polymerization). In most cases, it cannot be experimentally characterized and is made more complex by interpolymer interactions as well as the occurrence of structural defects during fabrication. While physical cross-linking is readily accomplished in a number of systems, covalent cross-linking methods remove the need for interpolymer interactions and lead to the formation of homogeneous networks with relatively uniform chain distributions and cross-linking densities. In addition, fundamental thermodynamic principles can be used to relate experimentally observable gel properties during degradation of these gels to their time-dependent cross-linking density or average molecular weight between cross-links (M_c). These relationships have already been verified for covalently cross-linked PLA-PEG-PLA gels. Development of the next generation of biodegradable hydrogels with precise combinations of properties (e.g., in situ formation, cell-mediated degradation, high mechanical strength and solute permeability, etc.) will depend on furthering our understanding of hydrogel structure while also increasing our library of available biodegradable and biocompatible polymer chemistries.

References

Aklonis, J.J. and MacKnight, W.J. 1983. *Introduction to Polymer Viscoelasticity*. Wiley-Interscience, NY.

Allcock, H.R., Pucher, S.R., et al. 1994. Poly[(Amino-Acid-Ester)Phosphazenes] — synthesis, crystallinity, and hydrolytic sensitivity in solution and the solid-state. *Macromolecules* 27: 1071–1075.

Allcock, H.R., Singh, A., et al. 2003. Tyrosine-bearing polyphosphazenes. *Biomacromolecules* 4: 1646–1653.

Almany, L. and Seliktar, D. 2005. Biosynthetic hydrogel scaffolds made from fibrinogen and polyethylene glycol for 3D cell cultures. *Biomaterials* 26: 2467–2477.

Ambrosio, A.M.A., Allcock, H.R., et al. 2002. Degradable polyphosphazene/poly(alpha-hydroxyester) blends: Degradation studies. *Biomaterials* 23: 1667–1672.

Ameer, G.A., Mahmood, T.A., et al. 2002. A biodegradable composite scaffold for cell transplantation. *J. Orthopaed. Res.* 20: 16–19.

Amsden, B. 1998. Solute diffusion within hydrogels. Mechanisms and models. *Macromolecules* 31: 8382–8395.

Amsden, B.G., Stubbe, B.G., et al. 2004. Modeling the swelling pressure of degrading hydroxyethylmethacrylate-grafted dextran hydrogels. *J. Polym. Sci. Pol. Phys.* 42: 3397–3404.

Andrade, J.D. 1976. *Hydrogels for Medicine and Related Applications*. American Chemical Society, Washington, D.C.

Andrianov, A.K., Chen, J.P., et al. 1998. Preparation of hydrogel microspheres by coacervation of aqueous polyphosphazene solutions. *Biomaterials* 19: 109–115.

Andrianov, A.K. and Payne, L.G. 1998. Protein release from polyphosphazene matrices. *Adv. Drug Deliver. Rev.* 31: 185–196.

Andrianov, A.K., Svirkin, Y.Y., et al. 2004. Synthesis and biologically relevant properties of polyphosphazene polyacids. *Biomacromolecules* 5: 1999–2006.

Anseth, K.S., Bowman, C.N., et al. 1996. Mechanical properties of hydrogels and their experimental determination. *Biomaterials* 17: 1647–1657.

Anseth, K.S., Metters, A.T., et al. 2002. *In situ* forming degradable networks and their application in tissue engineering and drug delivery. *J. Control. Release* 78: 199–209.

Auerbach, R., Lewis, R., et al. 2003. Angiogenesis assays: A critical overview. *Clin. Chem.* 49: 32–40.

Behravesh, E., Jo, S., et al. 2002. Synthesis of *in situ* cross-linkable macroporous biodegradable poly(propylene fumarate-co-ethylene glycol) hydrogels. *Biomacromolecules* 3: 374–381.

Behravesh, E. and Mikos, A.G. 2003. Three-dimensional culture of differentiating marrow stromal osteoblasts in biomimetic poly(propylene fumarate-co-ethylene glycol)-based macroporous hydrogels. *J. Biomed. Mater. Res. A* 66A: 698–706.

Bell, E., Ivarsson, B., et al. 1979. Production of a tissue-like structure by contraction of collagen lattices by human fibroblasts of different proliferative potential *in vitro*. *Proc. Natl Acad. Sci. USA* 76: 1274–1278.

Benelli, R. and Albini, A. 1999. *In vitro* models of angiogenesis: The use of Matrigel. *Intl J. Biol. Marker* 14: 243–246.

Berger, J., Reist, M., et al. 2004. Structure and interactions in covalently and ionically crosslinked chitosan hydrogels for biomedical applications. *Eur. J. Pharm. Biopharm.* 57: 19–34.

Bird, R.B., Stewart, W.E., et al. 1960. *Transport Phenomena*. John Wiley & Sons, Toronto.

Bohl, K.S., Shon, J., et al. 1998. Role of synthetic extracellular matrix in development of engineered dental pulp. *J. Biomat. Sci.-Polym E* 9: 749–764.

Bouhadir, K.H., Hausman, D.S., et al. 1999. Synthesis of cross-linked poly(aldehyde guluronate) hydrogels. *Polymer* 40: 3575–3584.

Brinkman, E., Vanderdoes, L., et al. 1991. Poly(vinyl alcohol)-heparin hydrogels as sensor catheter membranes. *Biomaterials* 12: 63–70.

Bryant, S.J. and Anseth, K.S. 2002. Hydrogel properties influence ECM production by chondrocytes photoencapsulated in poly(ethylene glycol) hydrogels. *J. Biomed. Mater. Res.* 59: 63–72.

Bryant, S.J., Davis-Arehart, K.A., et al. 2004. Synthesis and characterization of photopolymerized multifunctional hydrogels: Water-soluble poly(vinyl alcohol) and chondroitin sulfate macromers for chondrocyte encapsulation. *Macromolecules* 37: 6726–6733.

Bueche, F. 1962. *Physical Properties of Polymers*. Interscience, New York.

Burdick, J.A., Chung, C., et al. 2005. Controlled degradation and mechanical behavior of photopolymerized hyaluronic acid networks. *Biomacromolecules* 6: 386–391.

Burmania, J.A., Martinez-Diaz, G.J., et al. 2003a. Synthesis and physicochemical analysis of interpenetrating networks containing modified gelatin and poly(ethylene glycol) diacrylate. *J. Biomed. Mater. Res. A* 67A: 224–234.

Burmania, J.A., Stevens, K.R., et al. 2003b. Cell interaction with protein-loaded interpenetrating networks containing modified gelatin and poly(ethylene glycol) diacrylate. *Biomaterials* 24: 3921–3930.

Cadee, J.A., Brouwer, L.A., et al. 2001. A comparative biocompatibility study of microspheres based on crosslinked dextran or poly(lactic-co-glycolic)acid after subcutaneous injection in rats. *J. Biomed. Mater. Res.* 56: 600–609.

Cadee, J.A., de Groot, C.J., et al. 2002. Release of recombinant human interleukin-2 from dextran-based hydrogels. *J. Control. Release* 78: 1–13.

Cadee, J.A., De Kerf, M., et al. 1999. Synthesis, characterization of 2-(methacryloyloxy)ethyl-(di-) L-lactate and their application in dextran-based hydrogels. *Polymer* 40: 6877–6881.

Cadee, J.A., van Luyn, M.J.A., et al. 2000. *In vivo* biocompatibility of dextran-based hydrogels. *J. Biomed. Mater. Res.* 50: 397–404.

Canal, T. and Peppas, N.A. 1989. Correlation between mesh size and equilibrium degree of swelling of polymeric networks. *J. Biomed. Mater. Res.* 23: 1183–1193.

Cascone, M.G., Barbani, N., et al. 2001. Bioartificial polymeric materials based on polysaccharides. *J. Biomat. Sci.-Polym. E* 12: 267–281.

Chen, W.Y.J. and Abatangelo, G. 1999. Functions of hyaluronan in wound repair. *Wound Repair Regen.* 7: 79–89.

Cho, J.H., Kim, S.H., et al. 2004. Chondrogenic differentiation of human mesenchymal stem cells using a thermosensitive poly(*N*-isopropylacrylamide) and water-soluble chitosan copolymer. *Biomaterials* 25: 5743–5751.

Cohen, M.H. and Turnbull, D. 1959. *J. Chem. Phys.* 31: 1164.

Cramer, N.B., Reddy, S.K., et al. 2004. Initiation and kinetics of thiol-ene photopolymerizations without photoinitiators. *J. Polym. Sci. Pol. Chem.* 42: 5817–5826.

Davis, K.A. and Anseth, K.S. 2002. Controlled release from crosslinked degradable networks. *Crit. Rev. Ther. Drug* 19: 385–423.

Davis, K.A., Burdick, J.A., et al. 2003. Photoinitiated crosslinked degradable copolymer networks for tissue engineering applications. *Biomaterials* 24: 2485–2495.

de Jong, S.J., De Smedt, S.C., et al. 2000. Novel self-assembled hydrogels by stereocomplex formation in aqueous solution of enantiomeric lactic acid oligomers grafted to dextran. *Macromolecules* 33: 3680–3686.

Di Zio, K. and Tirrell, D.A. 2003. Mechanical properties of artificial-protein matrices engineered for control of cell and tissue behavior. *Macromolecules* 36: 1553–1558.

DuBose, J.W., Cutshall, C., et al. 2005. Controlled release of tethered molecules via engineered hydrogel degradation: Model development and validation. *J. Biomed. Mater. Res. A* 74A: 104–116.

Dusek, K. and Duskov-Smrckova, M. 2000. Network structure formation during crosslinking of organic coating systems. *Prog. Polym. Sci.* 25: 1215–1260.

Ehrbar, M., Djonov, V.G., et al. 2004. Cell-demanded liberation of VEGF(121) from fibrin implants induces local and controlled blood vessel growth. *Circ. Res.* 94: 1124–1132.

Ehrbar, M., Metters, A., et al. 2005. Endothelial cell proliferation and progenitor maturation by fibrin-bound VEGF variants with differential susceptibilities to local cellular activity. *J. Control. Release* 101: 93–109.

Elbert, D.L., Pratt, A.B., et al. 2001. Protein delivery from materials formed by self-selective conjugate addition reactions. *J. Control. Release* 76: 11–25.

Elliott, J.E., Macdonald, M., et al. 2004. Structure and swelling of poly(acrylic acid) hydrogels: Effect of pH, ionic strength, and dilution on the crosslinked polymer structure. *Polymer* 45: 1503–1510.

Elliott, J.E., Nie, J., et al. 2003. The effect of primary cyclization on free radical polymerization kinetics: Experimental characterization. *Polymer* 44: 327–332.

Fisher, J.P., Jo, S., et al. 2004. Thermoreversible hydrogel scaffolds for articular cartilage engineering. *J. Biomed. Mater. Res. A* 71A: 268–274.

Flores-Ramirez, N., Elizalde-Pena, E.A., et al. 2005. Characterization and degradation of functionalized chitosan with glycidyl methacrylate. *J. Biomater. Sci.-Polym. E.* 16: 473–488.

Flory, P.J. 1953. *Principles of Polymer Chemistry*. Cornell University Press, Ithaca, NY.

Flory, P.J. and Rehner, B.D. 1943. Statistical mechanics of cross-linked polymer networks. *J. Chem. Phys.* 11: 521–526.

Gamez, E., Ikezaki, K., et al. 2003. Photoconstructs of nerve guidance prosthesis using photoreactive gelatin as a scaffold. *Cell Transplant.* 12: 481–490.

Gentleman, E., Nauman, E.A., et al. 2004. Short collagen fibers provide control of contraction and permeability in fibroblast-seeded collagen gels. *Tissue Eng.* 10: 421–427.

Gobin, A.S. and West, J.L. 2002. Cell migration through defined, synthetic extracellular matrix analogues. *FASEB J.* 16.

Gong, P., Zhang, L., et al. 1998. Synthesis and characterization of polyurethane-chitosan interpenetrating polymer networks. *J. Appl. Polym. Sci.* 68: 1321–1329.

Greish, Y.E., Bender, J.D., et al. 2005. Low temperature formation of hydroxyapatite-poly(alkyl oxybenzoate)phosphazene composites for biomedical applications. *Biomaterials* 26: 1–9.

Gupta, K.C. and Kumar, M.N.V. 2001. pH dependent hydrolysis and drug release behavior of chitosan/poly(ethylene glycol) polymer network microspheres. *J. Mater. Sci.-Mater. Med.* 12: 753–759.

Haider, M., Megeed, Z., et al. 2004. Genetically engineered polymers: Status and prospects for controlled release. *J. Control. Release* 95: 1–26.

Halstenberg, S., Panitch, A., et al. 2002. Biologically engineered protein-graft-poly(ethylene glycol) hydrogels: A cell adhesive and plasm in-degradable biosynthetic material for tissue repair. *Biomacromolecules* 3: 710–723.

Hawkins, C.L. and Davies, M.J. 1998. Degradation of hyaluronic acid, poly- and monosaccharides and model compounds by hypochlorite: Evidence for radical intermediates and fragmentation. *Free Radical Biol. Med.* 24: 1396–1410.

Heggli, M., Tirelli, N., et al. 2003. Michael-type addition as a tool for surface functionalization. *Bioconjugate Chem.* 14: 967–973.

Heller, J. 1980. Controlled release of biologically active compounds from bioerodible polymers. *Biomaterials* 1: 51–57.

Hennink, W.E., De Jong, S.J., et al. 2004. Biodegradable dextran hydrogels crosslinked by stereocomplex formation for the controlled release of pharmaceutical proteins. *Intl. J. Pharm.* 277: 99–104.

Hennink, W.E. and van Nostrum, C.F. 2002. Novel crosslinking methods to design hydrogels. *Adv. Drug Deliver. Rev.* 54: 13–36.

Holland, T.A., Tabata, Y., et al. 2003. *In vitro* release of transforming growth factor-beta 1 from gelatin microparticles encapsulated in biodegradable, injectable oligo(poly(ethylene glycol) fumarate) hydrogels. *J. Control. Release* 91: 299–313.

Homminga, G.N., Buma, P., et al. 1993. Chondrocyte behavior in fibrin glue *in vitro*. *Acta Orthop. Scand.* 64: 441–445.

Howard, K.A., Dash, P.R., et al. 2000. Influence of hydrophilicity of cationic polymers on the biophysical properties of polyelectrolyte complexes formed by self-assembly with DNA. *BBA-Gen. Subjects* 1475: 245–255.

Hu, Y.Q., Jiang, H.L., et al. 2005. Preparation and characterization of poly(ethylene glycol)-g-chitosan with water- and organosolubility. *Carbohyd. Polym.* 61: 472–479.

Huang, S. and Ingber, D.E. 1999. The structural and mechanical complexity of cell-growth control. *Nat. Cell Biol.* 1: E131–E138.

Huang, X. and Lowe, T.L. 2005. Biodegradable thermoresponsive hydrogels for aqueous encapsulation and controlled release of hydrophilic model drugs. *Biomacromolecules* 6: 2131–2139.

Hubbell, J.A. 1999. Bioactive biomaterials. *Curr. Opin. Biotech.* 10: 123–129.

Hubbell, J.A., Zisch, A.P., et al. 2003. Incorporation of engineered VEGF variants in fibrin cell ingrowth matrices. *FASEB J.* 17: A553–A553.

Huh, K.M., Hashi, J., et al. 2000. Synthesis and characterization of dextran grafted with poly (*N*-isopropylacrylamide-co-*N*,*N*-dimethyl-acrylamide). *Macromol. Chem. Physic.* 201: 613–619.

Ibim, S.E.M., Ambrosio, A.M.A., et al. 1997. Novel polyphosphazene/poly(lactide-co-glycolide) blends: Miscibility and degradation studies. *Biomaterials* 18: 1565–1569.

Jeong, B., Kim, S.W., et al. 2002. Thermosensitive sol-gel reversible hydrogels. *Adv. Drug Deliver. Rev.* 54: 37–51.

Jia, X.Q., Burdick, J.A., et al. 2004. Synthesis and characterization of *in situ* cross-linkable hyaluronic acid-based hydrogels with potential application for vocal fold regeneration. *Macromolecules* 37: 3239–3248.

Jo, S., Shin, H., et al. 2001. Synthesis and characterization of oligo(poly(ethylene glycol) fumarate) macromer. *Macromolecules* 34: 2839–2844.

Jockenhoevel, S., Zund, G., et al. 2001. Fibrin gel-advantages of a new scaffold in cardiovascular tissue engineering. *Eur. J. Cardio-Thorac.* 19: 424–430.

Johansson, L., Skantze, U., et al. 1991. Diffusion and interaction in gels and solutions. 2. experimental results on the obstruction effect. *Macromolecules* 24: 6019–6023.

Kasper, F.K., Kushibiki, T., et al. 2005. *In vivo* release of plasmid DNA from composites of oligo(poly (ethylene glycol)fumarate) and cationized gelatin microspheres. *J. Control. Release* 107: 547–561.

Kim, S. and Healy, K.E. 2003. Synthesis and characterization of injectable poly(*N*-isopropylacrylamide-co-acrylic acid) hydrogels with proteolytically degradable cross-links. *Biomacromolecules* 4: 1214–1223.

Kim, S.H., Won, C.Y., et al. 1999. Synthesis and characterization of dextran-maleic acid-based hydrogel. *J. Biomed. Mater. Res.* 46: 160–170.

Kohn, J. and Langer, R. 1996. Bioresorbable and bioerodible materials. In: *Biomaterials Science, An Introduction to Materials in Medicine*, Vol. 2. B.D. Ratner and A.S. Hoffman (Eds.), pp. 64–72. Academic, San Diego, CA.

Kong, H.J., Alsberg, E., et al. 2004. Controlling degradation of hydrogels via the size of cross-linked junctions. *Adv. Mater.* 16: 1917–1921.

Kong, H.J., Kaigler, D., et al. 2004. Controlling rigidity and degradation of alginate hydrogels via molecular weight distribution. *Biomacromolecules* 5: 1720–1727.

Kong, H.J., Lee, K.Y., et al. 2002. Decoupling the dependence of rheological/mechanical properties of hydrogels from solids concentration. *Polymer* 43: 6239–6246.

Kopecek, J. 2003. Smart and genetically engineered biomaterials and drug delivery systems. *Eur. J. Pharm. Sci.* 20: 1–16.

Koyano, T., Minoura, N., et al. 1998. Attachment and growth of cultures fibroblast cells on PVA/chitosan-blended hydrogels. *J. Biomed. Mater. Res.* 39: 486–490.

Kumar, M., Muzzarelli, R.A.A., et al. 2004. Chitosan chemistry and pharmaceutical perspectives. *Chem. Rev.* 104: 6017–6084.

Kumashiro, Y., Huh, K.M., et al. 2001. Modulatory factors on temperature-synchronized degradation of dextran grafted with thermoresponsive polymers and their hydrogels. *Biomacromolecules* 2: 874–879.

Kumashiro, Y., Lee, T.K., et al. 2002. Enzymatic degradation of semi-IPN hydrogels based on *N*-isopropylacrylamide and dextran at a specific temperature range. *Macromol. Rapid Comm.* 23: 407–410.

Kurisawa, M., Matsuo, Y., et al. 1998. Modulated degradation of hydrogels with thermo-responsive network in relation to their swelling behavior. *Macromol. Chem. Physic.* 199: 705–709.

Kurisawa, M., Terano, M., et al. 1995. Doublestimuli-responsive degradable hydrogels for drug-delivery — interpenetrating polymer networks composed of oligopeptide-terminated poly(ethylene glycol) and dextran. *Macromol. Rapid Comm.* 16: 663–666.

Kurisawa, M., Terano, M., et al. 1997. Double-stimuli-responsive degradation of hydrogels consisting of oligopeptide-terminated poly(ethylene glycol) and dextran with an interpenetrating polymer network. *J. Biomat. Sci.-Polym. E* 8: 691–708.

Kurisawa, M. and Yui, N. 1998a. Dual-stimuli-responsive drug release from interpenetrating polymer network-structured hydrogels of gelatin and dextran. *J. Control. Release* 54: 191–200.

Kurisawa, M. and Yui, N. 1998b. Gelatin/dextran intelligent hydrogels for drug delivery: Dual-stimuli-responsive degradation in relation to miscibility in interpenetrating polymer networks. *Macromol. Chem. Physic.* 199: 1547–1554.

Kurisawa, M. and Yui, N. 1998c. Modulated degradation of dextran hydrogels grafted with poly(*N*-isopropylacrylamide-co-*N*,*N*-dimethylacrylamide) in response to temperature. *Macromol. Chem. Physic.* 199: 2613–2618.

Langer, R. 1990. New methods of drug delivery. *Science* 249: 1527–1533.

Langer, R. 1991. Drug delivery systems. *Mrs Bull.* 16: 47–49.

Langer, R. and Peppas, N.A. 1981. Present and future application of biomaterials in controlled drug delivery systems. *Biomaterials* 2: 201–214.

Leach, J.B., Bivens, K.A., et al. 2003. Photocrosslinked hyaluronic acid hydrogels: Natural, biodegradable tissue engineering scaffolds. *Biotechnol. Bioeng.* 82: 578–589.

Leach, J.B., Bivens, K.A., et al. 2004. Development of photocrosslinkable hyaluronic acid-polyethylene glycol-peptide composite hydrogels for soft tissue engineering. *J. Biomed. Mater. Res. A* 70A: 74–82.

Lee, C.R., Grad, S., et al. 2005. Fibrin-polyurethane composites for articular cartilage tissue engineering: A preliminary analysis. *Tissue Eng.* 11: 1562–1573.

Lee, K.W., Yoon, J.J., et al. 2004. Sustained release of vascular endothelial growth factor from calcium-induced alginate hydrogels reinforced by heparin and chitosan. *Transplant. P* 36: 2464–2465.

Lee, K.Y., Alsberg, E., et al. 2001. Degradable and injectable poly(aldehyde guluronate) hydrogels for bone tissue engineering. *J. Biomed. Mater. Res.* 56: 228–233.

Lee, K.Y., Bouhadir, K.H., et al. 2000. Degradation behavior of covalently cross-linked poly(aldehyde guluronate) hydrogels. *Macromolecules* 33: 97–101.

Lee, K.Y., Kong, H.J., et al. 2003. Hydrogel formation via cell crosslinking. *Adv. Mater.* 15: 1828–1832.

Lee, K.Y. and Mooney, D.J. 2001. Hydrogels for tissue engineering. *Chem. Rev.* 101: 1869–1879.

Lee, S.B., Ha, D.I., et al. 2004. Temperature/pH-sensitive comb-type graft hydrogels composed of chitosan and poly(*N*-isopropylacrylamide). *J. Appl. Polym. Sci.* 92: 2612–2620.

Li, C.L., Sajiki, T., et al. 2003. Novel visible-light-induced photocurable tissue adhesive composed of multiply styrene-derivatized gelatin and poly(ethylene glycol) diacrylate. *J. Biomed. Mater. Res. B-Appl. Biomater.* 66B: 439–446.

Li, Q., Williams, C.G., et al. 2004. Photocrosslinkable polysaccharides based on chondroitin sulfate. *J. Biomed. Mater. Res. A* 68A: 28–33.

Liang, H.C., Chang, W.H., et al. 2004. Crosslinking structures of gelatin hydrogels crosslinked with genipin or a water-soluble carbodiimide. *J. Appl. Polym. Sci.* 91: 4017–4026.

Lin, C.C. and Metters, A.T. 2006. Enhanced protein delivery from photopolymerized hydrogels using a pseudospecific metal chelating ligand. *Phar. Res.* 23: 614–622.

Lin, Y.H., Liang, H.F., et al. 2005. Physically crosslinked alginate/*N*,*O*-carboxymethyl chitosan hydrogels with calcium for oral delivery of protein drugs. *Biomaterials* 26: 2105–2113.

Lovestead, T.M., Burdick, J.A., et al. 2002. Coupling GPC and modeling to investigate kinetic chain lengths in multivinyl photopolymerized degradable networks. *Abstr. Pap. Am. Chem. Soc.* 224: U467–U467.

Lowman, A.M. and Peppas, N.A. 1999. Hydrogels. In: *Encyclopedia of Controlled Drug Delivery.* E. Mathiowitz (Ed.), pp. 397–418. Wiley, New York.

Lu, H., Carioscia, J.A., et al. 2005. Investigations of step-growth thiol-ene polymerizations for novel dental restoratives. *Dent. Mater.* 21: 1129–1136.

Lu, S.X. and Anseth, K.S. 2000. Release behavior of high molecular weight solutes from poly(ethylene glycol)-based degradable networks. *Macromolecules* 33: 2509–2515.

Lustig, S.R. and Peppas, N.A. 1988. Solute diffusion in swollen membranes.9. scaling laws for solute diffusion in gels. *J. Appl. Polym. Sci.* 36: 735–747.

Lutolf, M.P. and Hubbell, J.A. 2003. Synthesis and physicochemical characterization of end-linked poly(ethylene glycol)-co-peptide hydrogels formed by Michael-type addition. *Biomacromolecules* 4: 713–722.

Lutolf, M.P. and Hubbell, J.A. 2005. Synthetic biomaterials as instructive extracellular microenvironments for morphogenesis in tissue engineering. *Natl. Biotechnol.* 23: 47–55.

Lutolf, M.P., Lauer-Fields, J.L., et al. 2003. Synthetic matrix metalloproteinase-sensitive hydrogels for the conduction of tissue regeneration: Engineering cell-invasion characteristics. *Proc. Natl Acad. Sci.USA* 100: 5413–5418.

Lutolf, M.R., Weber, F.E., et al. 2003. Repair of bone defects using synthetic mimetics of collagenous extracellular matrices. *Natl. Biotechnol.* 21: 513–518.

Mackie, J.S. and Meares, P. 1955. *Proc. R. Soc. London* A232: 498.

Mallapragada, S.K. and Narasimhan, B. 1999. Drug delivery systems. In: *Handbook of Biomaterial Evaluation, Scientific, Technical and Clinical Testing of Implant Materials*, Vol. 27. A. F. von Recum (Ed.), pp. 425–437. Edwards, Ann Arbor, MI.

Manabe, T., Okino, H., et al. 2004. *In situ*-formed, tissue-adhesive co-gel composed of styrenated gelatin and styrenated antibody: Potential use for local anti-cytokine antibody therapy on surgically resected tissues. *Biomaterials* 25: 5867–5873.

Mann, B.K., Gobin, A.S., et al. 2001. Smooth muscle cell growth in photopolymerized hydrogels with cell adhesive and proteolytically degradable domains: Synthetic ECM analogs for tissue engineering. *Biomaterials* 22: 3045–3051.

Mao, C. and Kisaalita, W.S. 2004. Characterization of 3-D collagen hydrogels for functional cell-based biosensing. *Biosens. Bioelectron.* 19: 1075–1088.

Martens, P. and Anseth, K.S. 2000. Characterization of hydrogels formed from acrylate modified poly(vinyl alcohol) macromers. *Polymer* 41: 7715–7722.

Martens, P., Metters, A.T., et al. 2001. A generalized bulk-degradation model for hydrogel networks formed from multivinyl cross-linking molecules. *J. Phys. Chem. B* 105: 5131–5138.

Martens, P.J., Bowman, C.N., et al. 2004. Degradable networks formed from multi-functional poly(vinyl alcohol) macromers: Comparison of results from a generalized bulk-degradation model for polymer networks and experimental data. *Polymer* 45: 3377–3387.

Martens P.J., Bryant, S.J., et al. 2003. Tailoring the degradation of hydrogels formed from multivinyl poly(ethylene glycol) and poly(vinyl alcohol) macromers for cartilage tissue engineering. *Biomacromolecules* 4: 283–292.

Masaro, L. and Zhu, X.X. 1999. Physical models of diffusion for polymer solutions, gels and solids. *Prog. Polym. Sci.* 24: 731–775.

Mason, M.N., Metters, A.T., et al. 2001. Predicting controlled-release behavior of degradable PLA-b-PEG-b-PLA hydrogels. *Macromolecules* 34: 4630–4635.

Masters, K.S., Shah, D.N., et al. 2004. Designing scaffolds for valvular interstitial cells: Cell adhesion and function on naturally derived materials. *J. Biomed. Mater. Res. A* 71A: 172–180.

Masters, K.S., Shah, D.N., et al. 2005. Crosslinked hyaluronan scaffolds as a biologically active carrier for valvular interstitial cells. *Biomaterials* 26: 2517–2525.

Masuda, T., Furue, M., et al. 2004. Photocured, styrenated gelatin-based microspheres for de novo adipogenesis through corelease of basic fibroblast growth factor, insulin, and insulin-like growth factor I. *Tissue Eng.* 10: 523–535.

Megeed, Z., Cappello, J., et al. 2002. Controlled release of plasmid DNA from a genetically engineered silk-elastinlike hydrogel. *Pharm. Res.* 19: 954–959.

Megeed, Z., Haider, M., et al. 2004. *In vitro* and *in vivo* evaluation of recombinant silk-elastinlike hydrogels for cancer gene therapy. *J. Control. Release* 94: 433–445.

Meinel, L., Hofmann, S., et al. 2004. Engineering cartilage-like tissue using human mesenchymal stem cells and silk protein scaffolds. *Biotechnol. Bioeng.* 88: 379–391.

Meinel, L., Karageorgiou, V., et al. 2004a. Bone tissue engineering using human mesenchymal stem cells: Effects of scaffold material and medium flow. *Ann. Biomed. Eng.* 32: 112–122.

Meinel, L., Karageorgiou, V., et al. 2004b. Engineering bone-like tissue *in vitro* using human bone marrow stem cells and silk scaffolds. *J. Biomed. Mater. Res. A* 71A: 25–34.

Menzel, E.J. and Farr, C. 1998. Hyaluronidase and its substrate hyaluronan: Biochemistry, biological activities and therapeutic uses. *Cancer Lett.* 131: 3–11.

Metters, A. and Hubbell, J. 2005. Network formation and degradation behavior of hydrogels formed by Michael-type addition reactions. *Biomacromolecules* 6: 290–301.

Metters, A.T., Anseth, K.S., et al. 2000a. Fundamental studies of a novel, biodegradable PEG-b-PLA hydrogel. *Polymer* 41: 3993–4004.

Metters, A.T., Anseth, K.S., et al. 2000b. Predicting degradation behavior of PLA-b-PEG-b-PLA hydrogels. *Abstr. Pap. Am. Chem. Soc.* 220: U289–U289.

Metters, A.T., Anseth, K.S., et al. 2001. A statistical kinetic model for the bulk degradation of PLA-b-PEG-b-PLA hydrogel networks: Incorporating network non-idealities. *J. Phys. Chem. B* 105: 8069–8076.

Metters, A.T., Bowman, C.N., et al. 2000. A statistical kinetic model for the bulk degradation of PLA-b-PEG-b-PLA hydrogel networks. *J. Phys. Chem. B* 104: 7043–7049.

Metters, A.T., Bowman, C.N., et al. 2001. Verification of scaling laws for degrading PLA-b-PEG-b-PLA hydrogels. *Aiche J.* 47: 1432–1437.

Mongia, N.K., Anseth, K.S., et al. 1996. Mucoadhesive poly(vinyl alcohol) hydrogels produced by freezing/thawing processes: Applications in the development of wound healing systems. *J. Biomater. Sci.-Polym. E* 7: 1055–1064.

Moriyama, K., Ooya, T., et al. 1999. Pulsatile peptide release from multi-layered hydrogel formulations consisting of poly(ethylene glycol)-grafted and ungrafted dextrans. *J. Biomater. Sci.-Polym. E* 10: 1251–1264.

Moriyama, K. and Yui, N. 1996. Regulated insulin release from biodegradable dextran hydrogels containing poly(ethylene glycol). *J. Control. Release* 42: 237–248.

Muhr, A.H. and Blanshard, J.M.V. 1982. Diffusion in gels. *Polymer* 23: 1012–1026.

Murata, Y., Kontani, Y., et al. 2002. Behavior of alginate gel beads containing chitosan salt prepared with water-soluble vitamins. *Eur. J. Pharm. Biopharm.* 53: 249–251.

Nakagawa, S., Pawelek, P., et al. 1989. Long-term culture of fibroblasts in contracted collagen gels — effects on cell-growth and biosynthetic activity. *J. Invest. Dermatol.* 93: 792–798.

Navarro, R.R. and Tatsumi, K. 2001. Improved performance of a chitosan-based adsorbent for the sequestration of some transition metals. *Water Sci. Technol.* 43: 9–16.

Nettles, D.L., Vail, T.P., et al. 2004. Photocrosslinkable hyaluronan as a scaffold for articular cartilage repair. *Ann. Biomed. Eng.* 32: 391–397.

Nuttelman, C.R., Henry, S.M., et al. 2002. Synthesis and characterization of photocrosslinkable, degradable poly(vinyl alcohol)-based tissue engineering scaffolds. *Biomaterials* 23: 3617–3626.

Nuttelman, C.R., Mortisen, D.J., et al. 2001. Attachment of fibronectin to poly(vinyl alcohol) hydrogels promotes NIH3T3 cell adhesion, proliferation, and migration. *J. Biomed. Mater. Res.* 57: 217–223.

Odian, G. 1991. *Principles of Polymerization.* John Wiley & Sons, Inc., New York.

Okay, O. and Bowman, C.N. 2005. Kinetic modeling of thiol-ene reactions with both step and chain growth aspects. *Macromol. Theor. Simul.* 14: 267–277.

Okay, O., Reddy, S.K., et al. 2005. Molecular weight development during thiol-ene photopolymerizations. *Macromolecules* 38: 4501–4511.

Okino, H., Manabe, T., et al. 2003. Novel therapeutic strategy for prevention of malignant tumor recurrence after surgery: Local delivery and prolonged release of adenovirus immobilized in photocured, tissue-adhesive gelatinous matrix. *J. Biomed. Mater. Res. A* 66A: 643–651.

Okino, H., Nakayama, Y., et al. 2002. *In situ* hydrogelation of photocurable gelatin and drug release. *J. Biomed. Mater. Res.* 59: 233–245.

Oupicky, D., Ogris, M., et al. 2002. Importance of lateral and steric stabilization of polyelectrolyte gene delivery vectors for extended systemic circulation. *Mol. Ther.* 5: 463–472.

Panitch, A., Yamaoka, T., et al. 1999. Design and biosynthesis of elastin-like artificial extracellular matrix proteins containing periodically spaced fibronectin CS5 domains. *Macromolecules* 32: 1701–1703.

Park, H., Temenoff, J.S., et al. 2005. Delivery of TGF-beta 1 and chondrocytes via injectable, biodegradable hydrogels for cartilage tissue engineering applications. *Biomaterials* 26: 7095–7103.

Park, I.K., Kim, T.H., et al. 2001. Galactosylated chitosan-graft-poly(ethylene glycol) as hepatocyte-targeting DNA carrier. *J. Control. Release* 76: 349–362.

Park, Y., Lutolf, M.P., et al. 2004. Bovine primary chondrocyte culture in synthetic matrix metalloproteinase-sensitive poly(ethylene glycol)-based hydrogels as a scaffold for cartilage repair. *Tissue Eng.* 10: 515–522.

Park, Y.D., Tirelli, N., et al. 2003. Photopolymerized hyaluronic acid-based hydrogels and interpenetrating networks. *Biomaterials* 24: 893–900.

Peppas, N.A. 1986. *Hydrogels in Medicine and Pharmacy.* CRC Press, Boca Raton, FL.

Peppas, N.A., Huang, Y., et al. 2000. Physicochemical, foundations and structural design of hydrogels in medicine and biology. *Annu. Rev. Biomed. Eng.* 2: 9–29.

Peppas, N.A. and Scott, J.E. 1992. Controlled release from poly(vinyl alcohol) gels prepared by freezing–thawing processes. *J. Control. Release* 18: 95–100.

Petka, W.A., Harden, J.L., et al. 1998. Reversible hydrogels from self-assembling artificial proteins. *Science* 281: 389–392.

Pitarresi, G., Palumbo, F.S., et al. 2003. Biodegradable hydrogels obtained by photocrosslinking of dextran and polyaspartamide derivatives. *Biomaterials* 24: 4301–4313.

Pratt, A.B., Weber, F.E., et al. 2004. Synthetic extracellular matrices for *in situ* tissue engineering. *Biotechnol. Bioeng.* 86: 27–36.

Quick, D.J. and Anseth, K.S. 2003. Gene delivery in tissue engineering: A photopolymer platform to coencapsulate cells and plasmid DNA. *Pharm. Res.* 20: 1730–1737.

Quick, D.J. and Anseth, K.S. 2004. DNA delivery from photocrosslinked PEG hydrogels: Encapsulation efficiency, release profiles, and DNA quality. *J. Control. Release* 96: 341–351.

Quick, D.J., Macdonald, K.K., et al. 2004. Delivering DNA from photocrosslinked, surface eroding polyanhydrides. *J. Control. Release* 97: 333–343.

Raeber, G.P., Lutolf, M.P., et al. 2005. Molecularly engineered PEG hydrogels: A novel model system for proteolytically mediated cell migration. *Biophys. J.* 89: 1374–1388.

Reddy, S.K., Anseth, K.S., et al. 2005. Modeling of network degradation in mixed step-chain growth polymerizations. *Polymer* 46: 4212–4222.

Reddy, S.K., Cramer, N.B., et al. 2004. Rate mechanisms of a novel thiol-ene photopolymerization reaction. *Macromol. Symp.* 206: 361–374.

Rydholm, A.E., Bowman, C.N., et al. 2005. Degradable thiol-acrylate photopolymers: Polymerization and degradation behavior of an *in situ* forming biomaterial. *Biomaterials* 26: 4495–4506.

Sakiyama-Elbert, S.E. and Hubbell, J.A. 2000a. Controlled release of nerve growth factor from a heparin-containing fibrin-based cell ingrowth matrix. *J. Control. Release* 69: 149–158.

Sakiyama-Elbert, S.E. and Hubbell, J.A. 2000b. Development of fibrin derivatives for controlled release of heparin-binding growth factors. *J. Control. Release* 65: 389–402.

Sawhney, A.S., Pathak, C.P., et al. 1993. Bioerodible hydrogels based on photopolymerized poly(ethylene glycol)-co-poly(alpha-hydroxy acid) diacrylate macromers. *Macromolecules* 26: 581–587.

Schmedlen, K.H., Masters, K.S., et al. 2002. Photocrosslinkable polyvinyl alcohol hydrogels that can be modified with cell adhesion peptides for use in tissue engineering. *Biomaterials* 23: 4325–4332.

Segura, T., Anderson, B.C., et al. 2005. Crosslinked hyaluronic acid hydrogels: A strategy to functionalize and pattern. *Biomaterials* 26: 359–371.

Segura, T., Chung, P.H., et al. 2005. DNA delivery from hyaluronic acid-collagen hydrogels via a substrate-mediated approach. *Biomaterials* 26: 1575–1584.

Seliktar, D., Zisch, A.H., et al. 2004. MMP-2 sensitive, VEGF-bearing bioactive hydrogels for promotion of vascular healing. *J. Biomed. Mater. Res. A* 68A: 704–716.

Shalaby, W.S.W., Blevins, W.E., et al. 1990. Enzyme-induced degradation behavior of albumin-cross-linked hydrogels. *Abstr. Pap. Am. Chem. Soc.* 200: 73.

Shalaby, W.S.W., Blevins, W.E., et al. 1991. Enzyme-degradable hydrogels — properties associated with albumin-cross-linked polyvinylpyrrolidone hydrogels. *ACS Sym. Ser.* 467: 484–492.

Shin, H., Jo, S., et al. 2002. Modulation of marrow stromal osteoblast adhesion on biomimetic oligo[poly(ethylene glycol) fumarate] hydrogels modified with Arg-Gly-Asp peptides and a poly(ethylene glycol) spacer. *J. Biomed. Mater. Res.* 61: 169–179.

Shin, H., Ruhe, P.Q., et al. 2003. *In vivo* bone and soft tissue response to injectable, biodegradable oligo(poly(ethylene glycol) fumarate) hydrogels. *Biomaterials* 24: 3201–3211.

Shin, H., Zygourakis, K., et al. 2004. Modulation of differentiation and mineralization of marrow stromal cells cultured on biomimetic hydrogels modified with Arg-Gly-Asp containing peptides. *J. Biomed. Mater. Res. A* 69A: 535–543.

Shin, M.S., Kim, S.I., et al. 2002. Characterization of hydrogels based on chitosan and copolymer of poly(dimethylsiloxane) and poly(vinyl alcohol). *J. Appl. Polym. Sci.* 84: 2591–2596.

Silva, S.S., Menezes, S.M.C., et al. 2003. Synthesis and characterization of polyurethane-g-chitosan. *Eur. Polym. J.* 39: 1515–1519.

Simmons, C.A., Alsberg, E., et al. 2004. Dual growth factor delivery and controlled scaffold degradation enhance *in vivo* bone formation by transplanted bone marrow stromal cells. *Bone* 35: 562–569.

Smeds, K.A. and Grinstaff, M.W. 2001. Photocrosslinkable polysaccharides for *in situ* hydrogel formation. *J. Biomed. Mater. Res.* 54: 115–121.

St'astny, M., Plocova, D., et al. 2002a. HPMA-hydrogels result in prolonged delivery of anticancer drugs and are a promising tool for the treatment of sensitive and multidrug resistant leukaemia. *Eur. J. Cancer* 38: 602–608.

St'astny, M., Plocova, D., et al. 2002b. HPMA-hydrogels containing cytostatic drugs — Kinetics of the drug release and *in vivo* efficacy. *J. Control. Release* 81: 101–111.

Stauffer, S.R. and Peppas, N.A. 1992. Poly(vinyl alcohol) hydrogels prepared by freezing-thawing cyclic processing. *Polymer* 33: 3932–3936.

Stupp, S.I. 2005. Biomaterials for regenerative medicine. *Mrs Bull.* 30: 546–553.

Suggs, L.J., Kao, E.Y., et al. 1998. Preparation and characterization of poly(propylene fumarate-co-ethylene glycol) hydrogels. *J. Biomater. Sci.-Polym. E.* 9: 653–666.

Suggs, L.J., Krishnan, R.S., et al. 1998. *In vitro* and *in vivo* degradation of poly(propylene fumarate-co-ethylene glycol) hydrogels. *J. Biomed. Mater. Res.* 42: 312–320.

Sun, A.M. 1988. Microencapsulation of pancreatic-islet cells — a bioartificial endocrine pancreas. *Methods Enzymol.* 137: 575–580.

Tabata, Y., Hijikata, S., et al. 1999. Neovascularization effect of biodegradable gelatin microspheres incorporating basic fibroblast growth factor. *J. Biomater. Sci.-Polym. E.* 10: 79–94.

Tanahashi, K., Jo, S.B., et al. 2002. Synthesis and characterization of biodegradable cationic poly(propylene fumarate-co-ethylene glycol) copolymer hydrogels modified with agmatine for enhanced cell adhesion. *Biomacromolecules* 3: 1030–1037.

Tang, A., Wang, C., et al. 2001. The coiled coils in the design of protein-based constructs: Hybrid hydrogels and epitope displays. *J. Control. Release* 72: 57–70.

Tang, A.J., Wang, C., et al. 2000. Self-assembled peptides exposing epitopes recognizable by human lymphoma cells. *Bioconjugate Chem.* 11: 363–371.

Taub, M., Wang, Y., et al. 1990. Epidermal growth-factor or transforming growth factor-alpha is required for kidney tubulogenesis in matrigel cultures in serum-free medium. *Proc. Natl Acad. Sci. USA* 87: 4002–4006.

Taylor, S.J., McDonald, J.W., et al. 2004. Controlled release of neurotrophin-3 from fibrin gels for spinal cord injury. *J. Control. Release* 98: 281–294.

Temenoff, J.S., Athanasiou, K.A., et al. 2002. Effect of poly(ethylene glycol) molecular weight on tensile and swelling properties of oligo(poly(ethylene glycol) fumarate) hydrogels for cartilage tissue engineering. *J. Biomed. Mater. Res.* 59: 429–437.

Temenoff, J.S., Shin, H., et al. 2003. *In vitro* cytotoxicity of redox radical initiators for cross-linking of oligo(poly(ethylene glycol) fumarate) macromers. *Biomacromolecules* 4: 1605–1613.

Thompson, D.F., Letassy, N.A., et al. 1988. Fibrin glue — a review of its preparation, efficacy, and adverse-effects as a topical hemostat. *Drug Intel. Clin. Pharm.* 22: 946–952.

Timmer, M.D., Ambrose, C.G., et al. 2003. *In vitro* degradation of polymeric networks of poly(propylene fumarate) and the crosslinking macromer poly(propylene fumarate)-diacrylate. *Biomaterials* 24: 571–577.

Torres, D.S., Freyman, T.M., et al. 2000. Tendon cell contraction of collagen-GAG matrices *in vitro*: Effect of cross-linking. *Biomaterials* 21: 1607–1619.

Treloar, L.R.G. 1975. *Physics of Rubber Elasticity.* Clarendon Press, Oxford.

Ulbrich, K., Subr, V., et al. 1993. Novel biodegradable hydrogels prepared using the divinylic cross-linking agent *N,O*-dimethacryloylhydroxylamine.1. synthesis and characterization of rates of gel degradation, and rate of release of model-drugs, *in vitro* and *in vivo. J. Control. Release* 24: 181–190.

Ulbrich, K., Subr, V., et al. 1995. Synthesis of novel hydrolytically degradable hydrogels for controlled drug-release. *J. Control. Release* 34: 155–165.

Urech, L., Bittermann, A.G., et al. 2005. Mechanical properties, proteolytic degradability and biological modifications affect angiogenic process extension into native and modified fibrin matrices *in vitro. Biomaterials* 26: 1369–1379.

van de Wetering, P., Metters, A.T., et al. 2005. Poly(ethylene glycol) hydrogels formed by conjugate addition with controllable swelling, degradation, and release of pharmaceutically active proteins. *J. Control. Release* 102: 619–627.

Van Tomme, S.R., van Steenbergen, M.J., et al. 2005. Self-gelling hydrogels based on oppositely charged dextran microspheres. *Biomaterials* 26: 2129–2135.

van Dijk-Wolthuis, W.N.E., Tsang, S.K.Y., et al. 1997. A new class of polymerizable dextrans with hydrolyzable groups: Hydroxyethyl methacrylated dextran with and without oligolactate spacer. *Polymer* 38: 6235–6242.

Voet, D. and Voet, J.G. 1995. *Biochemistry.* John Wiley & Sons, New York.

Wang, C., Kopecek, J., et al. 2001. Hybrid hydrogels cross-linked by genetically engineered coiled-coil block proteins. *Biomacromolecules* 2: 912–920.

Wang, C., Stewart, R.J., et al. 1999. Hybrid hydrogels assembled from synthetic polymers and coiled-coil protein domains. *Nature* 397: 417–420.

Wang, T., Turhan, M., et al. 2004. Selected properties of pH-sensitive, biodegradable chitosan-poly(vinyl alcohol) hydrogel. *Polym. Int.* 53: 911–918.

West, J.L. and Hubbell, J.A. 1999. Polymeric biomaterials with degradation sites for proteases involved in cell migration. *Macromolecules* 32: 241–244.

Wichterle, O. and Lim, D. 1960. Hydrophilic gels for biological use. *Nature* 185: 117.

Xiao, H., Nayak, B.R., et al. 2004. Synthesis and characterization of novel thermoresponsive-co-biodegradable hydrogels composed of *N*-isopropylacrylamide, poly(L-lactic acid), and dextran. *J. Polym. Sci. A-Polym. Chem.* 42: 5054–5066.

Xu, C.Y., Breedveld, V., et al. 2005. Reversible hydrogels from self-assembling genetically engineered protein block copolymers. *Biomacromolecules* 6: 1739–1749.

Yamamoto, M., Takahashi, Y., et al. 2003. Controlled release by biodegradable hydrogyels enhances the ectopic bone formation of bone morphogenetic protein. *Biomaterials* 24: 4375–4383.

Yamamoto, N., Kurisawa, M., et al. 1996. Double-stimuli-responsive degradable hydrogels: Interpenetrating polymer networks consisting of gelatin and dextran with different phase separation. *Macromol. Rapid Commun.* 17: 313–318.

Yasuda, H., Lamaze, C.E., et al. 1968. Permeability of solutes through hydrated polymer membranes. I. Diffusion of sodium chloride. *Makromol. Chem.* 118: 19–35.

Ye, Q., Zund, G., et al. 2000. Fibrin gel as a three dimensional matrix in cardiovascular tissue engineering. *Eur. J. Cardio-Thorac.* 17: 587–591.

Yoshida, T., Aoyagi, T., et al. 2003. Newly designed hydrogel with both sensitive thermoresponse and biodegradability. *J. Polym. Sci. A — Polym. Chem.* 41: 779–787.

Yui, N., Nihira, J., et al. 1993. Regulated release of drug microspheres from inflammation responsive degradable matrices of cross-linked hyaluronic-acid. *J. Control. Release* 25: 133–143.

Zhao, X. and Harris, J.M. 1997. Novel degradable poly(ethylene glycol) esters for drug delivery. *Poly(Ethylene Glycol)* 680: 458–472.

Zhao, X. and Harris, J.M. 1998. Novel degradable poly(ethylene glycol) hydrogels for controlled release of protein. *J. Pharm. Sci.* 87: 1450–1458.

Zisch, A.H., Lutolf, M.P., et al. 2003. Biopolymeric delivery matrices for angiogenic growth factors. *Cardiovasc. Pathol.* 12: 295–310.

Zisch, A.H., Schenk, U., et al. 2001. Covalently conjugated VEGF-fibrin matrices for endothelialization. *J. Control. Release* 72: 101–113.

6

Biodegradable Polymeric Biomaterials: An Updated Overview

Chih-Chang Chu
Cornell University

6.1 Introduction

The term **biodegradation** is loosely associated with materials that could be broken down by nature either through hydrolytic mechanisms without the help of enzymes and/or enzymatic mechanism. Other terms like *absorbable, erodible, and resorbable* have also been used in the literature to indicate biodegradation.

TABLE 6.1   Properties of Commercially Important Synthetic Absorbable Polymers

Polymer	Crystallinity	T_m (°C)	T_g (°C)	T_{dec} (°C)	Fiber Strength MPa	Fiber Modulus GPa	Fiber Elongation (%)
PGA	High	230	36	260	890	8.4	30
PLLA	High	170	56	240	900	8.5	25
PLA	None	—	57	—	—	—	—
Polyglactin910[a]	High[c]	200	40	250	850	8.6	24
Polydioxanone	High	106	<20	190	490	2.1	35
Polyglyconate[b]	High[c]	213	<20	260	550	2.4	45
Poliglecaprone25[d]	—	<220	−36~15		91,100[e]	113,000[e]	39

[a] Glycolide per lactide = 9/1.
[b] Glycolide per trimethylene carbonate = 9/1.
[c] Depending on the copolymer composition.
[d] 2/0 size Monocryl (glycolide-ε-caprolactone copolymer).
[e] PSI unit.

Source: Kimura, Y., 1993. *Biomedical Applications of Polymeric Materials*, T. Tsuruta, T. Hayashi, K. Kataoka, K. Ishihara, and Y. Kimura, Eds. CRC Press, Boca Raton, FL and Chu, C.C., von Fraunhofer, J.A., and Greisler, H.P., 1996. *Wound Closure Biomaterials and Devices*. CRC Press, Boca Raton, FL.

The interests in biodegradable polymeric biomaterials for biomedical engineering use have increased dramatically during the past decade. This is because this class of biomaterials has two major advantages that non-biodegradable biomaterials do not have. First, they do not elicit permanent chronic foreign-body reactions due to the fact that they are gradually absorbed by the human body and do not permanently leave traces of residual in the implantation sites. Second, some of them have recently been found to be able to regenerate tissues, so called **tissue engineering**, through the interaction of their biodegradation with immunologic cells like macrophages. Hence, surgical implants made from biodegradable biomaterials could be used as a temporary scaffold for tissue regeneration. This approach toward the reconstruction of injured, diseased, or aged tissues is one of the most promising fields in the next century.

Although the earliest and most commercially significant biodegradable polymeric biomaterials were originated from linear aliphatic polyesters like polyglycolide and polylactide from poly(α-hydroxyacetic acids), recent introduction of several new synthetic and natural biodegradable polymeric biomaterials extends the domain beyond this family of simple polyesters. These new commercially significant biodegradable polymeric biomaterials include poly(orthoesters), polyanhydrides, polysaccharides, poly(ester-amides), tyrosine-based polyarylates or polyiminocarbonates or polycarbonates, poly(D,L-lactide-urethane), poly(β-hydroxybutyrate), poly(ε-caprolactone), poly[*bis*(carboxylatophenoxy) phosphazene], poly(amino acids), pseudo-poly(amino acids), and copolymers derived from amino acids and non-amino acids.

All the above biodegradable polymeric biomaterials could be generally divided into eight groups based on their chemical origin: (1) Biodegradable linear aliphatic polyesters (e.g., polyglycolide, polylactide, polycaprolactone, polyhydroxybutyrate) and their copolymers within the aliphatic polyester family like poly(glycolide-L-lactide) copolymer and poly(glycolide-ε-caprolactone) copolymer; (2) Biodegradable copolymers between linear aliphatic polyesters in (1) and monomers other than linear aliphatic polyesters like, poly(glycolide-trimethylene carbonate) copolymer, poly(L-lactic acid-L-lysine) copolymer, Tyrosine-based polyarylates or polyiminocarbonates or polycarbonates, poly(D,L-lactide-urethane), and poly(ester-amide); (3) Polyanhydrides; (4) Poly(orthoesters); (5) Poly(ester-ethers) like poly-p-dioxanone; (6) Biodegradable polysaccharides like hyaluronic acid, chitin, and chitson; (7) polyamino acids like poly-L-glutamic acid and poly-L-lysine; (8) Inorganic biodegradable polymers like polyphosphazene and poly[*bis*(carboxylatophenoxy)phosphazene] which have a nitrogen-phosphorus backbone instead of ester linkage. Recently, there is a new approach of making new biodegradable polymers through melt-blending of highly accepted biodegradable polymers like those of glycolide and lactide base [Shalaby, 1994].

The earliest, most successful, and frequent biomedical applications of biodegradable polymeric biomaterials have been in wound closure [Chu et al., 1996]. All biodegradable wound closure biomaterials are based upon the glycolide and lactide families. For example, polyglycolide (Dexon from American Cyanamid), poly(glycolide-L-lactide) random copolymer with 90 to 10 — ratio (Vicryl from Ethicon), poly(ester-ether) (PDS from Ethicon), poly(glycolide-trimethylene carbonate) random block copolymer (Maxon from American Cyanamid), and poly(glycolide-ε-caprolactone) copolymer (Monocryl from Ethicon). This class of biodegradable polymeric biomaterials is also the one most studied for their chemical, physical, mechanical, morphological, and biological properties and their changes with degradation time and environment. Some of the above materials like Vicryl have been commercially used as surgical meshes for repair of a hernia or the body wall.

The next largest biomedical application of biodegradable polymeric biomaterials that are commercially satisfactory is drug control/release devices. Some well-known examples in this application are polyanhydrides and poly(ortho-ester). Biodegradable polymeric biomaterials, particularly totally resorbable composites, have also been experimentally used in the field of orthopedics, mainly as components for internal bone fracture fixation like PDS pins. However, their wide acceptance in other parts of orthopaedic implants may be limited due to their inherent mechanical properties and their biodegradation rate. Besides the commercial uses described above, biodegradable polymeric biomaterials have been experimented with as (1) vascular grafts, (2) vascular stents, (3) vascular couplers for vessel anastomosis, (4) nerve growth conduits, (5) augmentation of defected bone, (6) ligament/tendon prostheses, (7) intramedullary plug during total hip replacement, (8) anastomosis ring for intestinal surgery, and (9) stents in ureteroureterostomies for accurate suture placement.

Due to space limitation, the emphasis of this chapter will be on the commercially most significant and successful biomedical biodegradable polymers based on (1) linear aliphatic polyesters, (2) some very recent research and development of important classes of synthetic biodegradable polymers, (3) a new theoretical approach to modeling the hydrolytic degradation of glycolide/lactide based biodegradable polymers, (4) the effects of some new extrinsic factors on the degradation of the most commercially significant biodegradable polymers, and (5) the new biomedical applications of this class of synthetic biodegradable polymers in tissue engineering and regeneration. The details of the applications of this family and other biodegradable polymeric biomaterials and their chemical, physical, mechanical, biological, and biodegradation properties can be found in other recent reviews [Barrows, 1986; Vert et al., 1992; Kimura, 1993; Park et al., 1993; Shalaby, 1994; Hollinger, 1995; Chu et al., 1996].

6.2 Glycolide/Lactide Based Biodegradable Linear Aliphatic Polyesters

This class of biodegradable polymers is the most successful, important, and commercially widely used biodegradable biomaterials in surgery. It is also the class of biodegradable biomaterials that were most extensively studied in terms of degradation mechanisms and structure–property relationships. Among them, polyglycolide or polyglycolic acid (PGA) is the most important one because most other biodegradable polymers are derived from PGA either through copolymerization, for example, poly(glycolide-L-lactide) copolymer or modified glycolide monomer, for example, poly-p-dioxanone.

6.2.1 Glycolide Based Biodegradable Homopolymers Polyesters

PGA can be polymerized either directly or indirectly from glycolic acid. The direct polycondensation produces a polymer of M_n less than 10,000 because of the requirement of a very high degree of dehydration (99.28% up) and the absence of monofunctional impurities. For PGA of molecular weight higher than 10,000 it is necessary to proceed through the ring-opening polymerization of the cyclic dimers

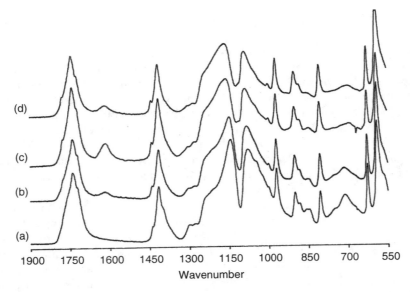

FIGURE 6.1 FTIR spectra of polyglycolic acid disks as a function of *in vitro* hydrolysis time in phosphate buffer of pH 7.44 at 37°C. (a) 0 day; (b) 55 h; (c) 7 days; (d) 21 days.

of glycolic acid. Numerous catalysts are available for this ring-opening polymerization. They include organometallic compounds and Lewis acids [Chujo et al., 1967; Wise et al., 1979]. For biomedical applications, stannous chloride dihydrate or trialkyl aluminum are preferred. PGA was found to exhibit an orthorhombic unit cell with dimensions $a = 5.22$ Å, $b = 6.19$ Å, and c(fiber axis) $= 7.02$ Å. The planar zigzag-chain molecules form a sheet structure parallel to the ac plane and do not have the polyethylene type arrangement [Chatani et al., 1968]. The molecules between two adjacent sheets orient in opposite directions. The tight molecular packing and the close approach of the ester groups might stabilize the crystal lattice and contribute to the high melting point, T_m, of PGA (224 to 230°C). The glass transition temperature, T_g, ranges from 36 to 40°C. The specific gravities of PGA are 1.707 for a perfect crystal and 1.50 in a completely amorphous state [Chujo et al., 1967a]. The heat of fusion of 100% crystallized PGA is reported to be 12 kJ/mol (45.7 cal/g) [Brandrup et al., 1975]. A recent study of injection molded PGA disks reveals their IR spectroscopic characteristics [Chu et al., 1995]. As shown in Figure 6.1, the four bands at 850, 753, 713, and 560 cm^{-1} are associated with the amorphous regions of the PGA disks and could be used to assess the extends of hydrolysis. Peaks associated with the crystalline phase included those at 972, 901, 806, 627, and 590 cm^{-1}. Two broad, intense peaks at 1142 and 1077 cm^{-1} can be assigned to C—O stretching modes in the ester and oxymethylene groups, respectively. These two peaks are associated mainly with ester and oxymethylene groups originating in the amorphous domains. Hydrolysis could cause both of these C—O stretching modes to substantially decrease in intensity.

6.2.2 Glycolide-Based Biodegradable Copolyesters Having Aliphatic Polyester Based Co-Monomers

Other commercially successful glycolide-based biodegradable polymeric biomaterials are the copolymers of glycolide with other monomers within linear aliphatic polyesters like lactides, carbonates, and ε-caprolactone. The glycolide-lactide random copolymers are the most studied and have a wide range of properties and applications, depending on the composition ratio of glycolide to lactide. Figure 6.2 illustrates the dependence of biodegradation rate on the composition of glycolide to lactide in the copolymer. For wound closure purposes, a high concentration of glycolide monomer is required for achieving proper mechanical and degradation properties. Vicryl sutures, sometime called polyglactin 910,

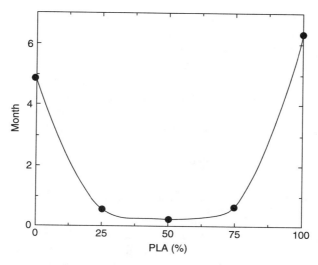

FIGURE 6.2 The effect of poly(L-lactide) composition in polyglycolide on the time required for 50% mass loss implanted under the dorsal skin of rat. (From Miller, R.A., Brady, J.M., and Cutright, D.E., 1977. *J. Biomed. Mater. Res.*, 11:711. With permission.)

contain a 90/10 molar ratio of glycolic to L-lactide and this molar ratio is important for the Vicryl suture to retain crystalline characteristics. For biomedical use, Lewis acid catalysts are preferred for the copolymers [Wise et al., 1979]. If D,L-instead of L-lactide is used as the co-monomer, the U-shape relationship between the level of crystallinity and glycolide composition disappears. This is because polylactide from 100% D,L-lactide composition is totally amorphous. IR bands associated with Vicryl molecules in the amorphous domains are 560, 710, 850, and 888 cm^{-1}, while 590, 626, 808, 900, and 972 cm^{-1} are associated with the crystalline domains [Frederick et al., 1984]. Like PGA, these IR bands could be used to assess the extent of hydrolysis.

A relatively new block copolymer of glycolide and carbonates, such as trimethylene carbonate, has been commercialized. Maxon is made from a block copolymer of glycolide and 1,3-dioxan-2-one (trimethylene carbonate or GTMC) and consists of 32.5% by weight (or 36 mol%) of trimethylene carbonate [Casey et al., 1984; Katz et al., 1985]. Maxon is a poly(ester-carbonate). The polymerization process of Maxon is divided into two stages. The first stage is the formation of a middle block which is a random copolymer of glycolide and 1,3-dioxan-2-one. Diethylene glycol is used as an initiator and stannous chloride dihydrate (SnCl$_2 \cdot$ 2H$_2$O) serves as the catalyst. The polymerization is conducted at about 180°C. The weight ratio of glycolide to trimethylene carbonate in the middle block is 15 : 85. After the synthesis of the middle block, the temperature of the reactive bath is raised to about 220°C to prevent the crystallization of the copolymer and additional glycolide monomers as the end blocks are added into the reaction bath to form the final triblock copolymer.

The latest glycolide-based copolymer that has become commercially successful is Monocryl suture. It is a segmented block copolymer consisting of both soft and hard segments. The purpose of having soft segments in the copolymer is to provide good handling properties like pliability, while the hard segments are used to provide adequate strength. The generic copolymerization process between glycolic acid and ε-caprolactone was recently reported by Fukuzaki et al. in Japan [1989, 1991]. The resulting copolymers were low molecular weight biodegradable copolymers of glycolic acid and various lactones for potential drug delivery purposes. The composition of lactone ranged from as low as 15 to as high as 50 mol% and the weight average for molecular weight ranged from 4,510 to 16,500. The glass transition temperature ranged from 18 to −43°C, depending on the copolymer composition and molecular weight.

Monocryl is made from two stages of the polymerization process [Bezwada et al., 1995]. In the first stage, soft segments of prepolymer of glycolide and ε-caprolactone are made. This soft segmented prepolymer is further polymerized with glycolides to provide hard segments of polyglycolide. Monocryl has a composition of 75% glycolide and 25% ε-caprolactone and should have a higher molecular weight than those glycolide/ε-caprolactone copolymers reported by Fukuzaki et al., for adequate mechanical properties required by sutures. The most unique aspect of Monocryl monofilament suture is its pliability as claimed by Ethicon [Bezwada et al., 1995]. The force required to bend a 2/0 suture is only about 2.8×10^4 lb-in^2 for Monocryl, while the same size PDSII and Maxon monofilament sutures require about 3.9 and 11.6×10^4 lb-in^2 force, respectively. This inherent pliability of Monocryl is due to the presence of soft segments and T_g resulting from the ε-caprolactone co-monomer unit. Its T_g is expected to be between 15 and $-36°C$.

6.2.3 Glycolide-Based Biodegradable Copolyesters with Non-Aliphatic Polyester-Based Co-Monomers

In this category, the most important one is the glycolide copolymer consisting of poly(ethylene 1,4-phenylene-bis-oxyacetate) (PEPBO) (Jamiokowski and Shalaby, 1991). The development of this type of glycolide-based copolymer was initiated because of the adverse effect of γ-irradiation on the mechanical properties of glycolide-based synthetic absorbable sutures. There is a great desire to develop γ-irradiation sterilizable, synthetic, absorbable polymers to take advantage of the highly convenient and reliable method of sterilization. Shalaby et al. recently reported that the incorporation of about 10 mol% of a polymeric radiostabilizer like PEPBO into PGA backbone chains would make the copolymer sterilizable by γ-irradiation without a significant accelerated loss of mechanical properties upon hydrolysis when compared with the unirradiated copolymer control (MPG) [Jamiokowski et al., 1991]. The changes in tensile breaking force of both MPG and PGA sutures implanted intramuscularly and subcutaneously in rats for various periods show the great advantage of such copolymers. MPG fibers γ-irradiated at 2.89 Mrads did not show any loss in tensile breaking force during the first 14 days postimplantation when compared with unimplanted samples. On the contrary, PGA sutures γ-irradiated at 2.75 Mrads lost 62% of the tensile breaking force of their unimplanted samples. There was no tensile breaking force remaining for the irradiated PGA at the end of 21 days, while both 2.89 and 5 Mrads irradiated MPG retained 72 and 55% of their corresponding 0 day controls, respectively. The inherent more hydrolytic resistance of MPG must be attributed to the presence of an aromatic group in the backbone chains. This aromatic polyester component is also responsible for the observed γ-irradiation stability. It is not known at this time whether the new γ-irradiation-resistant MPG is biocompatible with biologic tissues due to the lack of published histologic data.

6.2.4 Glycolide-Derived Biodegradable Polymers Having Ether Linkage

Poly-p-dioxanone (PDS) is derived from the glycolide family with better flexibility. It is polymerized from ether-containing lactones, 1,4-dioxane-2,5-dione (i.e., p-dioxanone) monomers with a hydroxylic initiator and tin catalyst [Shalaby, 1994]. The resulting polymer is semi-crystalline with T_m about 106–115°C and T_g −10–0°C. The improved flexibility of PDS relative to PGA as evidenced in its lower T_g is due to the incorporation of an ether segment in the repeating unit which reduces the density of ester linkages for intermolecular hydrogen bonds. Because of the less dense ester linkages in PDS when compared with PGA or glycolide-L-lactide copolymers, PDS is expected and has been shown to degrade at a slower rate *in vitro* and *in vivo*. PDS having an inherent viscosity of 2.0 dl/g in hexafluoroisopropanol is adequate for making monofilament sutures. Recently, an advanced version of PDS, PDSII, was introduced. PDSII was achieved by subjecting the melt-spun fibers to a high temperature (128°C) for a short period of time. This additional treatment partially melts the outermost surface layer of PDS fibers and leads to a distinctive skin-core morphology. The heat employed also results in larger crystallites in the core of the fiber than the untreated PDS fiber. The tensile strength-loss profile of PDSII sutures is better than that of PDS sutures.

A variety of copolymers having high molar ratios of PDS compared to other monomers within the same linear aliphatic polyester family have been reported for the purpose of improving the mechanical and biodegradation properties [Shalaby, 1994]. For example, copolymer of PDS (80%) and PGA (up to 20%) has an absorption profile similar to Dexon and Vicryl sutures but it has compliance similar to PDS. Copolymer of PDS (85%) and PLLA (up to 15%) results in a more compliant (low modulus) suture than homopolymer PDS but with absorption profiles similar to PDS [Bezwada et al., 1990].

Copolymer fibers made from PDS and monomers other than linear aliphatic polyester like morpholine-2,5-dione (MD) exhibit rather interesting biodegradation properties. This copolymer fiber was absorbed 10 to 25% earlier than PDS. The copolymer, however, retained a tensile breaking strength profile similar to PDS with a slightly faster strength loss during the earlier stage, that is, the first 14 days [Shalaby, 1994]. This ability to break the inherent fiber structure–property relationship through copolymerization is a major improvement in biodegradation properties of absorbable sutures. It is interesting to recognize that a small % (3%) of MD in the copolymer suture is sufficient to result in a faster mass loss profile without the expense of its tensile strength-loss profile. The ability to achieve this ideal biodegradation property might be attributed to both an increasing hydrophilicity of the copolymer and the disruption of crystalline domains due to MD moiety. As described later, the loss of suture mass is mainly due to the destruction of crystalline domains, while the loss of tensile breaking strength is chiefly due to the scission of tie-chain segments located in the amorphous domains. The question is why MD–PDS copolymeric suture retains its strength-loss similar to PDS. The possible explanation is that the amide functional groups in MD could form stronger intermolecular hydrogen bonds than ester functional groups. This stronger hydrogen bond contributes to the strength retention of the copolymer of PDS and MD during *in vivo* biodegradation. The incorporation of MD moiety into PDS also lowers the unknot and knot strength of unhydrolyzed specimens, but increases elongation at break. This suggests that the copolymer of PDS and MD should have a lower level of crystallinity than PDS which is consistent with its observed faster mass loss *in vivo*.

To improve γ-irradiation stability of PDS, radiostabilizers like PEPBO have been copolymerized with PDS to form segmented copolymers the same way as PEPBO with glycolide described above [Koelmel et al., 1991; Shabaly, 1994]. The incorporation of 5 to 10% of such stabilizer in PDS has been shown not only to improve γ-irradiation resistance considerably but to also increase the compliance of the material. For example, PEPBO-PDS copolymer retained 79, 72, and 57% of its original tensile breaking strength at 2, 3, and 4 weeks in *in vivo* implantation, while PDS homopolymer retained only 43, 30, and 25% at the corresponding periods. It appears that an increasing (CH_2) group between the two ester functional groups of the radiation stabilizers improves the copolymer resistance toward γ-irradiation.

6.2.5 Lactide Biodegradable Homopolymers and Copolymers

Polylactides, particularly poly-L-lactide (PLLA), and copolymers having >50% L- or DL-lactide have been explored for medical use without much success mainly due to their much slower absorption and difficulty in melt processing. PLLAs are prepared in solid state through ring-opening polymerization due to their thermal instability and should be melt-processed at the lowest possible temperature [Shalaby, 1994]. Other methods like solution spinning, particularly for high molecular weight, and suspension polymerization have been reported as better alternatives. PLLA is a semi-crystalline polymer with $T_m = 170°C$ and $T_g = 56°C$. This high T_g is mainly responsible for the extremely slow biodegradation rate reported in the literature. The molecular weight of lactide-based biodegradable polymers suitable for medical use ranges from 1.5 to 5.0 dl/g inherent viscosity in chloroform. Ultra high molecular weight of polylactides have been reported [Tunc, 1983; Leenslag et al., 1984]. For example, an intrinsic viscosity as high as 13 dl/g was reported by Leenslag et al. High strength PLLA fibers from this ultra high molecular weight polylactide was made by hot-drawing fibers from solutions of good solvents. The resulting fibers had tensile breaking strength close to 1.2 GPa [Gogolewski et al., 1983]. Due to a dissymmetric nature of lactic acid, the polymer made from the optically inactive racemic mixture of D and L enantiomers, poly-DL-lactide, however, is an amorphous polymer.

Lactide-based copolymers having a high percentage of lactide have recently been reported, particularly those copolymerized with aliphatic polycarbonates like trimethylene carbonate (TMC) or 3,3-dimethyltrimethylene carbonate (DMTMC) [Shieh et al., 1990]. The major advantage of incorporating TMC or DMTMC units into lactide is that the degradation products from TMC or DMTMC are largely neutral pH and hence are considered to be advantageous. Both *in vitro* toxicity and *in vivo* non-specific foreign body reactions like sterile sinuses have been reported in orthopaedic implants made from PGA and/or PLLA [Eitenmuller et al., 1989; Bostman et al., 1990; Daniels et al., 1992; Hofmann, 1992; Winet et al., 1993]. Several investigators indicated that the glycolic or lactic-acid rich-degradation products have the potential to significantly lower the local pH in a closed and less body-fluid buffered regions surrounded by bone [Sugnuma et al., 1992]. This is particularly true if the degradation process proceeds with a burst mode (i.e., a sudden and rapid release of degradation products). This acidity tends to cause abnormal bone resorption and/or demineralization. The resulting environment may be cytotoxic [Daniels et al., 1992]. Indeed, inflammatory foreign body reactions with a discharging sinus and osteolytic foci visible on x-ray have been encountered in clinical studies [Eitenmuller et al., 1989]. Hollinger et al. recently confirmed the problem associated with PGA and/or PLLA orthopaedic implants [Winet et al., 1993]. A rapid degradation of a 50:50 ratio of glycolide-lactide copolymer in bone chambers of rabbit tibias has been found to inhibit bone regeneration. However, emphasis has been placed on the fact that extrapolation of *in vitro* toxicity to *in vivo* biocompatibility must consider microcirculatory capacity. The increase in the local acidity due to a faster accumulation of the highly acidic degradation products is also known to lead to an accelerated acid-catalyzed hydrolysis in the immediate vicinity of the biodegradable device. This acceleration in hydrolysis could lead to a faster loss of mechanical property of the device than we expect. This finding suggests the need to use components in totally biodegradable composites so that degradation products with less acidity would be released into the surrounding area. A controlled slow release rather than a burst release of degradation products at a level that the surrounding tissue could timely metabolize them would also be helpful in dealing with the acidity problem. Copolymers of composition ratio of 10DMTMC/90LLA or 10TMC/90LLA appear to be a promising absorbable orthopaedic device. Other applications of this type of copolymers include nerve growth conduits, tendon prostheses, and coating materials for biodegradable devices.

Another unique example of L-lactide copolymer is the copolymer of L-lactide and 3-(S)[(alkyloxycarbonyl) methyl]-1,4-dioxane-2,5-dione, a cyclic diester [Kimura, 1993]. The most unique aspect of this new biodegradable copolymer is the carboxyl acid pendant group which obviously would make the new polymer not only more hydrophilic and hence faster biodegradation but also more reactive toward future chemical modification through the pendant carboxyl group. The availability of these carboxyl reactive pendant sites could be used to chemically bond antimicrobial agents or other biochemicals like growth factors for making future wound closure biomaterials having new and important biological functions. Unfortunately, there are no reported data to evaluate the performance of this new absorbable polymer for biomedical engineering use up to the present time.

Block copolymers of PLLA with poly(amino acids) have also been reported as a potential controlled drug delivery system [Nathan et al., 1994]. This new class of copolymers consists of both ester and amide linkages in the backbone molecules and is sometimes referred as poly(depsipeptides) or poly(esters-amides). Poly(depsipeptides) could also be synthesized from ring-opening polymerization of morpholine-2,5-dione and its derivatives [Helder et al., 1986]. Barrows has also made a series of poly(ester-amides) from polyesterification of diols that contain preformed amide linkages, such as amidediols [Barrows, 1994]. Katsarava and Chu et al. just reported the synthesis of high-molecular-weight poly(ester-amides) of M_w from 24,000 to 167,000 with narrow polydispersity ($M_w/M_n = 1.20 - 1.81$) via solution polycondensation of di-p-toluenesulfonic acid salts of *bis*-(α-amino acid) α, ω-alkylene diesters and di-p-nitrophenyl esters of diacids [Katsarava et al., In press]. These poly(ester-amide)s consisted of naturally occurring and non-toxic building blocks and had excellent film forming properties. These polymers were mostly amorphous materials with T_g from 11 to 59°C. The rationale for making poly(ester-amides) is to combine the well-known absorbability and biocompatibility of linear aliphatic polyesters with the high performance and the flexibility of potential chemical reactive sites of amide of polyamides. Poly(ester-amides) could be degraded either by enzyme and/or nonenzymatic mechanisms. There is no commercial use of this class of copolymers at the present time.

The introduction of poly(ethylene oxide) (PEO) into PLLA in order to modulate the hydrophilicity and degradability of PLLA for drug control/release biomaterials has been reported and an example is the triblock copolymer of PLA/PEO/PLA [Li et al., 1998a]. Biomaterials having an appropriate PLLA and PEO block length were found to have a hydrogel property that could deliver hydrophilic drugs as well as hydrophobic ones like steroids and hormones. Another unique biodegradable biomaterial consisting of a star-block copolymer of PLLA, PGA, and PEO was also reported for protein drug delivery devices [Li et al., 1998b]. This star-shaped copolymer has 4 or 8 arms made of PEO, PLLA, and PGA. The glass transition temperature and the crystallinity of this star-shaped block copolymer were significantly lower than the corresponding linear PLLA and PGA.

Because of the characteristic of very slow biodegradation rate of PLLA and the copolymers having a high composition ratio of PLLA, their biomedical applications have been mainly limited to (1) orthopaedic surgery, (2) drug control/release devices, (3) coating materials for suture, (4) vascular grafts, and (5) surgical meshes to facilitate wound healing after dental extraction.

6.3 Non-Glycolide/Lactide Based Linear Aliphatic Polyesters

All glycolide/lactide based linear aliphatic polyesters are based on poly(α-hydroxy acids). Recently, there are two unique groups of linear aliphatic polyesters based on poly(ω-hydroxy acids) and the most famous ones are poly(ε-caprolactone) [Kimura, 1993], poly(β-hydroxybutyrate) (PHB), poly(β-hydroxyvalerate) (PHV) and the copolymers of PHB/PHV (Gross, 1994). Poly(ε-caprolactone) has been used as a comonomer with a variety of glycolide/lactide based linear aliphatic polyesters described earlier. PHB and PHV belong to the family of poly(hydroxyalkanoates) and are mainly produced by prokaryotic types of microorganisms like *Pseudomonas olevorans* or *Alcaligenes eutrophus* through biotechnology. PHB and PHV are the principal energy and carbon storage compounds for these microorganisms and are produced when there are excessive nutrients in the environment. These naturally produced PHB and PHV are stereochemically pure and are isotactic. They could also be synthesized in labs, but the characteristics of steroregularity is lost.

This family of biodegradable polyesters is considered to be environmentally friendly because they are produced from propionic acid and glucose and could be completely degraded to water, biogas, biomass, and humic materials [Gross, 1994]. Their biodegradation requires enzymes. Hence, PHB, PHV, and their copolymers are probably the most important biodegradable polymers for environmental use. However, the biodegradability of this class of linear aliphatic polyesters in human or animal tissues has been questionable. For example, high molecular weight PHB or PHB/PHV fibers do not degrade in tissues or simulated environments over periods of up to six months [Williams, 1990]. The degradability of PHB could be accelerated by γ-irradiation or copolymerization with PHV.

An interesting derivative of PHB, poly(β-malic acid) (PMA), has been synthesized from β-benzyl malolactonate followed by catalytic hydrogenolysis. PMA differs from PHB in that the β-(CH_3) substituent is replaced by $-COOH$ [Kimura, 1993]. The introduction of pendant carboxylic acid group would make PMA more hydrophilic and easier to be absorbed.

6.4 Non-Aliphatic Polyesters Type Biodegradable Polymers

6.4.1 Aliphatic and Aromatic Polycarbonates

The most significant aliphatic polycarbonates are based upon DMTMC and TMC. They are made by the same ring-opening polymerization as glycolide-based biodegradable polyesters. The homopolymers are biocompatible with a controllable rate of biodegradation. Pellets of poly(ethylene carbonate) were absorbed completely in two weeks in the peritoneal cavity of rats. A slight variation of this polycarbonate, that is, poly(propylene carbonate), however, did not show any sign of absorption after two months [Barrows, 1986]. Copolymers of DMTMC/ε-caprolactone and DMTMC/TMC have been reported to have adequate properties for wound closure, tendon prostheses, and vascular grafts. The most important advantage of aliphatic polycarbonates is the neutral pH of the degradation products.

Poly(BPA-carbonates) made from bisphenol A (BPA) and phosgene is non-biodegradable, but an analog of poly(BPA-carbonate) like poly(iminocarbonates) have been shown to degrade in about 200 days [Barrows, 1986]. In general, this class of aromatic polycarbonates takes an undesirably long period to degrade, presumably due to the presence of an aromatic ring which could protect adjacent ester bonds to be hydrolyzed by water or enzymes. Different types of degradation products of this polymer under different pH environments are produced. At pH >7.0, the degradation products of this polymer are BPA, and ammonia and CO_2, while insoluble poly(BPA-carbonate) oligomers were produced with pH <7.0 [Barrows, 1986]. The polymer had good mechanical properties and acceptable tissue biocompatibility. Unfortunately, there is currently no commercial use of this class of polymer in surgery.

6.4.2 Poly(alkylene oxalates) and Copolymers

This class of high crystalline biodegradable polymers was initially developed [Shalaby, 1994] for absorbable sutures and their coating. They consist of $[-ROOC-COO-]_n$ repeating unit where R is $(CH_2)_x$ with x ranging from 4 to 12. R could also be cyclic (1,4-*trans*-cyclohexanedimethanol) or aromatic (1,4-benzene, 1,3-benzene dimethanol) for achieving higher melting temperature. The biodegradation properties depend on the number of (CH_2) group, x, and the type of R group (i.e., acyclic vs. cyclic or aromatic). In general, a higher number of methylene group and/or the incorporation of cyclic or aromatic R group would retard the biodegradation rate and hence make the polymer absorbed slower. For example, there was no mass of the polymer with $x = 4$ remaining *in vivo* (rats) after 28 days, while the polymer with $x = 6$ retained 80% of its mass after 42 days *in vivo*. An isomorphic copolyoxalate consisting of 80% cyclic R group like 1,4-*trans*-cyclohexanedimethanol and 20% with acyclic R group like 1,6-hexanediol retained 56% of its original mass after 180 days *in vivo*. By varying the ratio of cyclic to acyclic monomers, copolymers with a wide range of melting temperatures could be made, for example, copolymer of 95/5 ratio of cyclic (i.e., 1,4-*trans*-cyclohexanedimethanol)/acyclic (i.e., 1,6-hexanediol) monomers had a $T_m = 210°C$, while the copolymer with 5/95 ratio had a $T_m = 69°C$. Poly(alkylene oxalates) with $x = 3$ or 6 had been experimented with as drug control/release devices. The tissue reaction to this class of biodegradable polymers has been minimal.

6.5 Biodegradation Properties of Synthetic Biodegradable Polymers

The reported biodegradation studies of a variety of biodegradable polymeric biomaterials have mainly focused on their tissue biocompatibility, the rate of drug release, or loss of strength and mass. Recently, the degradation mechanisms and the effects of intrinsic and extrinsic factors, such as pH [Chu, 1981, 1982], enzymes [Williams et al., 1977, 1984; Williams, 1979; Chu et al., 1983], γ-irradiation [Campbell et al., 1981; Chu et al., 1982, 1983; Williams et al., 1984; Zhang et al., 1993], electrolytes [Pratt et al., 1993a], cell medium [Chu et al., 1992], annealing treatment [Chu et al., 1988], plasma surface treatment [Loh et al., 1992], external stress [Miller et al., 1984; Chu, 1985a], and polymer morphology [Chu et al., 1989] and on a chemical means to examine the degradation of PGA fibers [Chu et al., 1985] have been systemically examined and the subject has been recently reviewed [Chu, 1985b, 1991, 1995a, b; Hollinger, 1995; Chu et al., 1996]. Table 6.2 is an illustration of structural factors of polymers that could control their degradation [Kimura, 1993]. Besides these series of experimental studies of a variety of factors that could

TABLE 6.2 Structural Factors to Control the Polymer Degradability

Factors	Methods of control
Chemical structure of main chain and side groups	Selection of chemical bonds and functional groups
Aggregation state	Processing, copolymerization
Crystalline state	Polymer blend
Hydrophilic/hydrophobic balance	Copolymerization, introduction of functional groups
Surface area	Micropores
Shape and morphology	Fiber, film, composite

Source: Kimura, Y., 1993, in *Biomedical Applications of Polymeric Materials*, T. Tsuruta, T. Hayashi, K. Kataoka, K. Ishihara, and Y. Kimura, Eds. pp. 164–190. CRC Press, Boca Raton, FL.

affect the degradation of biodegradable polymeric biomaterials, there are two new areas that broaden the above traditional study of biodegradation properties of biodegradable polymers into the frontier of science. They are: theoretical modeling and the role of free radicals.

6.5.1 Theoretical Modeling of Degradation Properties

The most systematic theoretical modeling study of degradation properties of biodegradable biomaterials was reported by Pratt and Chu who used computational chemistry to theoretically model the effects of a variety of substituents which could exert either steric effect and/or inductive effect on the degradation properties of glycolide/lactide based biodegradable polymers [Pratt et al., 1993b, 1994a, b]. This new approach could provide scientists with a better understanding of the relationship between the chemical structure of biodegradable polymers and their degradation behavior at a molecular level. It also could help the future research and development of this class of polymers through the intelligent prediction of structure–property relationships. In those studies, Pratt and Chu examined the affect of various derivatives of linear aliphatic polyester (PGA) and a naturally occurring linear polysaccharide (hyaluronic acid) on their hydrolytic degradation phenomena and mechanisms.

The data showed a decrease in the rate of hydrolysis by about a factor of 106 with isopropyl α-substituents, but nearly a six-fold increase with *t*-butyl α-substituents [Pratt et al., 1993b]. The role of electron donating and electron withdrawing groups on the rate of hydrolytic degradation of linear aliphatic polyesters was also theoretically modeled by Pratt and Chu [Pratt et al., 1994a]. Electron withdrawing substituents a to the carbonyl group would be expected to stabilize the tetrahedral intermediate resulting from hydroxide attack, that is, favoring hydroxide attack but disfavoring alkoxide elimination. Electron releasing groups would be expected to show the opposite effect. Similarly, electronegative substituents on the alkyl portion of the ester would stabilize the forming alkoxide ion and favor the elimination step. Pratt and Chu found that the rate of ester hydrolysis is greatly affected by halogen substituents due primarily to charge delocalization. The data suggest that the magnitude of the inductive effect on the hydrolysis of glycolic esters decreases significantly as the location of the substituent is moved further away from the α-carbon because the inductive effect is very distance-sensitive. In all three locations of substitutions (α, β, and γ), Cl and Br substituents exhibited the largest inductive effect compared to other halogen elements.

Therefore, Pratt and Chu concluded that the rate of ester hydrolysis is greatly affected by both alkyl and halogen substituents due primarily to either steric hindrance or charge delocalization. In the steric effect, alkyl substituents on the glycolic esters cause an increase in activation enthalpies, and a corresponding decrease in reaction rate, up to about three carbon sizes, while bulkier alkyl substituents other than isopropyl make the rate-determining elimination step more facile. It appears that aliphatic polyesters containing a isopropyl groups, or slightly larger linear alkyl groups, such as n-butyl, n-pentyl, etc., would be expected to show a longer strength retention, given the same fiber morphology. In the inductive effect, α-substituents on the acyl portion of the ester favor the formation of the tetrahedral intermediate through charge delocalization, with the largest effect seen with Cl substitution, but retard the rate-determining alkoxide elimination step by stabilizing the tetrahedral intermediate. The largest degree of stabilization is caused by the very electronegative F substituent.

6.5.2 The Role of Free Radicals in Degradation Properties

Salthouse et al. had demonstrated that the biodegradation of synthetic absorbable sutures is closely related to macrophage activity through the close adhesion of macrophage onto the surface of the absorbable sutures [Matlaga et al., 1980]. It is also known that inflammatory cells, particularly leukocytes and macrophages are able to produce highly reactive oxygen species like superoxide ($\cdot O_2^-$) and hydrogen peroxide during inflammatory reactions toward foreign materials [Badwey et al., 1980; Devereux et al., 1991]. These highly reactive oxygen species participate in the biochemical reaction, frequently referred to as a respiratory burst, which is characterized by the one electron reduction of O_2 into superoxide via either NADPH or NADH oxidase as shown below. The reduction of O_2 results in an increase in O_2 uptake and the consumption of glucose.

$$2O_2 + NADPH \xrightarrow{\text{(NADPH Oxidase)}} 2\,\cdot O_2^- + NADP^+ + H^+ \qquad (6.1)$$

The resulting superoxide radicals are then neutralized to H_2O_2 via cytoplasmic enzyme superoxide dismutase (SOD).

$$2\,\cdot O_2^- + 2H^+ \xrightarrow{\text{(SOD)}} H_2O_2 + O_2 \qquad (6.2)$$

Very recently, Williams et al. suggested that these reactive oxygen species may be harmful to polymeric implant surfaces through their production of highly reactive, potent, and harmful hydroxyl radicals $\cdot OH$ in the presence of metals like iron as shown in the following series of redox reactions [Williams et al., 1991; Ali et al., 1993; Zhong et al., 1994].

$$\cdot O_2 + M^{+n} \rightarrow O_2 + M^{+(n-1)} \qquad (6.3)$$

$$H_2O_2 + M^{+(n-1)} \rightarrow \cdot OH + HO^- + M^{+n} \qquad (6.4)$$

The net reaction will be:

$$\cdot O_2^- + H_2O_2 \rightarrow \cdot OH + HO^- + O_2 \qquad (6.5)$$

and is often referred to as the metal-catalyzed Haber–Weiss reaction [Haber et al., 1934].

Although the role of free radicals in the hydrolytic degradation of synthetic biodegradable polymers is largely unknown, a very recent study using absorbable sutures like Vicryl in the presence of an aqueous free radical solution prepared from H_2O_2 and ferrous sulfate, $FeSO_4$, raised the possibility of the role of free radicals in the biodegradation of synthetic absorbable sutures [Williams et al., 1991; Zhong et al., 1994]. As shown below, both $\cdot OH$ radicals and OH^- are formed in the process of oxidation of Fe^{+2} by H_2O_2 and could exert some influence on the subsequent hydrolytic degradation of Vicryl sutures.

$$Fe^{+2} + H_2O_2 \rightarrow Fe^{+3} + \cdot OH + OH^-$$

SEM results indicated that Vicryl sutures in the presence of free radical solutions exhibited many irregular surface cracks at both 7 and 14 days *in vitro*, while the same sutures in the two controls (H_2O_2 or $FeSO_4$ solutions) did not have these surface cracks. Surprisingly, the presence of surface cracks of Vicryl sutures treated in the free radical solutions did not accelerate the tensile breaking strength-loss as would be expected. Thermal properties of Vicryl sutures under the free radical and 3% H_2O_2 media showed the classical well-known maximum pattern of the change of the level of crystallinity with hydrolysis time. The level of crystallinity of Vicryl sutures peaked at 7 days in both media (free radical and 3% H_2O_2). The time for peak appearance in these two media was considerably earlier than Vicryl sutures in conventional physiological buffer media. Based on the Chu's suggestion of using the time of the appearance of the crystallinity peak as an indicator of degradation rate, it appears that these two media accelerated the degradation of Vicryl sutures when compared with regular physiological buffer solution. Based on their findings, Williams et al. proposed the possible routes of the role of ·OH radicals in the hydrolytic degradation of Vicryl sutures [Zhong et al., 1994]. Unfortunately, the possible role of OH^-, one of the byproducts of Fenton reagents ($H_2O_2/FeSO_4$), was not considered in the interpretation of their findings. OH^- species could be more potent than OH toward hydrolytic degradation of synthetic absorbable sutures. This is because hydroxyl anions are the sole species which attack carbonyl carbon of the ester linkages during alkaline hydrolysis. Since an equal amount of ·OH and OH^- are generated in Fenton reagents, the observed changes in morphological, mechanical, and thermal properties could be partially attributed to OH^- ions as well as ·OH radicals.

Besides hydroxyl radicals, the production of superoxide ions and singlet oxygen during phagocytosis has been well documented [Babior et al., 1973]. Although the role of superoxide in simple organic ester hydrolysis has been known since the 1970s [Forrester et al., 1984, 1987; Johnson, 1976; Mango et al., 1976; San Fillipo et al., 1976], its role in the hydrolytic degradation of synthetic biodegradable polyester-based biomaterials has remained largely unknown. Such an understanding of the superoxide ion role during the biodegradation of foreign materials has become increasing desirable because of the advanced understanding of how the human immune system reacts to foreign materials and the increasing use of synthetic biomaterials for human body repair.

Lee and Chu very recently examined the reactivity of the superoxide ion towards biodegradable biomaterials having an aliphatic polyester structure at different reaction conditions such as temperature, time, and superoxide ion concentration [Lee et al., 1996a]. Due to the extreme reactivity of the superoxide ion, it has been observed that the effect of superoxide ion-induced hydrolytic degradation of PDLLA and PLLA was significant in terms of changes in molecular weights and thermal properties. The superoxide ion-induced fragmentation of PDLLA would result in a mixture of various species with different chain lengths. A combined GPC method with a chemical tagging method revealed that the structure of oligomer species formed during the superoxide-induced degradation of PDLLA and PLLA was linear. The significant reduction in molecular weight of PDLLA by superoxide ion was also evident in the change of thermal properties like T_g. The linear low molecular species (oligomer, trimers, and dimers) in the reaction mixture could act as an internal plasticizer to provide the synergetic effects of lowering T_g by increasing free volume. The effect of the superoxide ion-induced hydrolytic degradation on molecular weight of PLLA was similar to PDLLA but with a much smaller magnitude. The mechanism of simple hydrolysis of ester by superoxide ion proposed by Forrester et al. was subsequently modified to interpret the data obtained from the synthetic biodegradable polymers.

In addition to the PDLLA and PLLA, superoxide ions also have a significant adverse effect on the hydrolytic degradation of synthetic absorbable sutures [Lee et al., 1996c]. A significant reduction in molecular weight has been found along with mechanical and thermal properties of these sutures over a wide range of superoxide ion concentrations, particularly during the first few hours of contact with superoxide ions. For example, the PGA suture lost almost all of its mass at the end of 24 h contact with superoxide ions at $25°C$, while the same suture would take at least 50 days in an *in vitro* buffer for a complete mass loss. The surface morphology of these sutures was also altered drastically. The exact mechanism, however, is not fully known yet; Lee et al. suggested the possibility of simultaneous occurrence of several main-chain scissions by three different nucleophilic species.

Lee and Chu also reported that the addition of Fenton agent or hydrogen peroxide to the degradation medium would retard the well-known adverse effect of the conventional γ-irradiation sterilization of synthetic absorbable sutures [Lee et al., 1996a]. They found that these γ-irradiated sutures retained better tensile breaking strength in the Fenton medium than in the regular buffer media. Chu et al. postulated that the γ-irradiation induced α-carbon radicals in these sutures react with the hydroxyl radicals from the Fenton agent medium and hence neutralize the adverse effect of α-carbon radicals on the backbone chain scission. This mechanism is supported by the observed gradual loss of ESR signal of the sutures in the presence of the Fenton agent in the medium.

Instead of the adverse effect of free radicals on the degradation properties of synthetic biodegradable polyesters, Lee and Chu described an innovative approach of covalent bonding nitroxyl radicals onto these biodegradable polymers so that the nitroxyl radical attached polymers would have biological functions similar to nitric oxide [Lee et al., 1996b, 1998]. A preliminary *in vitro* cell culture study of these new biologically active biodegradable polymers indicated that they could retard the proliferation of human smooth muscle cells as native nitric oxide. The full potential of this new class of biologically active biodegradable polymers is currently under investigation by Chu for a variety of therapeutic applications.

6.6 The Role of Linear Aliphatic Biodegradable Polyesters in Tissue Engineering and Regeneration

The use of biodegradable polymers as the temporary scaffolds either to grow cells/tissues *in vitro* for tissue engineering applications or to regenerate tissues *in vivo* has very recently become a highly important aspect of research and development that broadens this class of biodegradable polymers beyond their traditional use in wound closure and drug control/release biomaterials. The scaffolds used in either tissue engineering or regeneration are to provide support for cellular attachment and subsequent controlled proliferation into a predefined shape or form. Obviously, a biodegradable scaffold would be preferred because of the elimination of chronic foreign body reaction and the generation of additional volume for regenerated tissues.

Although many other biodegradable polymers of natural origin like alginate [Atala et al., 1994], hyaluronate [Benedetti et al., 1993; Larsen et al., 1993], collagen [Hirai et al., 1995] and laminin [Dixit, 1994] have been experimented with for such a purpose, synthetic biodegradable polymers of linear aliphatic polyesters like PGA, PLA, and their copolymers [Bowald et al., 1979, 1980; Greisler, 1982; Greisler et al., 1985, 1987a, b, 1988a, b, c, 1991a; Freed et al., 1993; Mikos et al., 1993; Yu et al., 1993, 1994; Mooney et al., 1994, 1995, 1996a, b, c; Kim et al., 1998a, b] have received more attention because of their consistent sources, reproducible properties, means to tailor their properties, and versatility in manufacturing processes.

Biodegradable polymers must be fabricated into stable textile structures before they can be used as the scaffold for tissue engineering or regeneration. The stability of the scaffold structure is important during tissue engineering and regeneration in order to maintain its proper size, shape, or form upon the shear force imposed by the circulating culture media in a bioreactor, the contractile force imposed by the growing cells on the scaffold surface, and other forces like the compression from surrounding tissues.

Kim et al. reported that, although ordinary non-woven PGA matrices have very good porosity (to facilitate diffusion of nutrients) with a high surface to volume ratio (to promote cell attachment and proliferation) and have been used to engineer dental pulp and smooth muscle tissues having comparable biological contents as the native tissues [Kim et al., 1998b; Mooney et al., 1996c], these non-woven PGA matrices could not maintain their original structure during tissue engineering due to the relatively weak non-woven textile structure and stronger contractile force exerted by the attached and proliferated cells/tissues [Kim et al., 1998a]. This led to deformed engineered tissues that may have undesirable properties; for example, the smooth muscle engineered on collagen gels exhibited significant contraction over time [Zeigler et al., 1994; Hirai et al., 1995].

Because of this shortcoming of the existing non-woven PGA matrices, Kim et al. very recently reported the use of PLLA to stabilize the PGA matrices [Kim et al., 1998a]. A 5% w/v PLLA solution in chloroform was sprayed onto PGA non-woven matrices (made of 12 μm diameter PGA fibers) of 97% porosity and either 3 mm or 0.5 mm thickness. The PLLA-impregnated PGA non-wovens could be subjected to additional heat treatment at 195°C to enhance their structural stability further. Figure 6.3 shows the

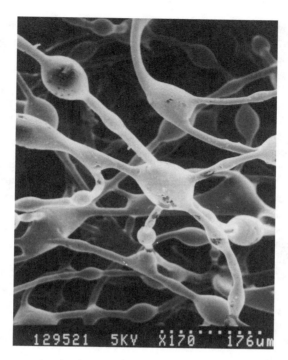

FIGURE 6.3   Scanning electron micrograph of the exterior of PLLA-impregnated and annealed PGA matrix. (From Kim, B.S. and Mooney, D.J., 1998a. *J. Biomed. Mater. Res.*, 41: 322–332. With permission.)

morphology of such a heat annealed PLLA-impregnated PGA non-woven matrix [Kim et al., 1998]. The PLLA was deposited mainly on the crosspoints of PGA fibers and hence interlocked the possible sliding of PGA fibers upon external force. Depending on the amount of PLLA used and subsequent heat treatment, the resulting PLLA-impregnated PGA non-woven matrices had an increase in compressive modulus of 10- to 35-fold when compared with the original PGA non-woven. The PLLA-impregnated PGA non-woven matrices also retained their initial volume (101 ± 4%) and about same shape as the original during the seven weeks in culture, while the untreated PGA non-woven exhibited severe distortion in shape and contracted about 5% of its original volume. Since PLLA is well-known to degrade at a much slower rate than PGA, its presence on the PGA fibers surface would be expected to make the treated PGA non-woven matrices degrade at a much slower rate than the untreated PGA non-woven. For example, the PLLA treated PGA non-woven retained about 80% of its initial mass, while the untreated PGA control had only 10% at the end of the seven week culture.

Linear aliphatic polyesters like PGA, its lactide copolymer, and poly-p-dioxanone have also been fabricated into both woven and knitted forms for the *in vivo* regeneration of blood vessels in animals [Bowald et al., 1979, 1980; Greisler, 1982; Greisler et al., 1985, 1987a, 1988c, 1991b; Yu et al., 1993, 1994]. The published results from a variety of animals like dogs and rabbits indicate that full-wall healing with pseudo-endothelial lining was observed. This class of synthetic biodegradable polymers are promising candidates for the regeneration of vascular tissue.

These encouraging findings were believed to be associated with the intense macrophage/biomaterial interactions. [Greisler, 1988a; Greisler et al., 1989]. This interaction leads to a differential activation of the macrophage which, in turn, yields different macrophage products being released into the microenvironment [Greisler et al., 1991b]. Greisler et al. [1988b] have documented active stimulatory or inhibitory effects of various bioresorbable and non-resorbable materials on myofibroblast, vascular smooth muscle cell, and endothelial cell regeneration, and has shown a transinterstitial migration to be their source when lactide/glycolide copolymeric prostheses are used. The rate of tissue ingrowth parallels the kinetics of macrophage mediated prosthetic resorption in all lactide/glycolides studied [Gresler, 1982; Greisler et al., 1985, 1987a, 1988a]. Macrophage phagocytosis of the prosthetic material is observed histologically as early as one week following implantation of a rapidly resorbed material, such as PGA or polyglactin 910 (PG910), and is followed by an extensive increase in the myofibroblast population and neovascularization of the inner capsules [Greisler, 1982; Greisler et al., 1985, 1986]. Autoradiographic analyses using tritiated thymidine demonstrated a significantly increased mitotic index within these inner capsular cells, that mitotic index paralleling the course of prosthetic resorption [Greisler et al., 1991a]. Polyglactin 910, for example, resulted in a mitotic index of $20.1 \pm 16.6\%$ three weeks following implantation, progressively decreasing to $1.2 \pm 1.3\%$ after 12 weeks. The more slowly resorbed polydioxanone prostheses demonstrated a persistently elevated mitotic index, $7.1 \pm 3.8\%$, 12 weeks after implantation, a time in which the prosthetic material was still being resorbed. By contrast Dacron never yielded greater than a $1.2 \pm 1.3\%$ mitotic index [Greisler et al., 1991a]. These mitotic indices correlated closely with the slopes of the inner capsule thickening curves suggesting that myofibroblast proliferation contributed heavily to this tissue deposition.

Therefore, the degradation property of synthetic biodegradable polymers somehow relates to macrophage activation which subsequently leads to the macrophage production of the required growth factors that initiate tissue regeneration. Different degradation properties of synthetic biodegradable polymers would thus be expected to result in different levels of macrophage activation, i.e., different degrees of tissue regeneration.

Defining Terms

Biodegradation: Materials that could be broken down by nature either through hydrolytic mechanisms without the help of enzymes and/or enzymatic mechanism. It is loosely associated with absorbable, erodable, resorbable.

Tissue Engineering: The ability to regenerate tissue through the help of artifical materials and devices.

References

Ali, S.A.M., Zhong, S.P., Doherty, P.J., and Williams, D.F., 1993. Mechanisms of polymer degradation in implantable devices. I. Poly(caprolactone). *Biomaterials*, 14: 648.

Atala, A., Kim, W., Paige, K.T., Vancanti, C.A., and Retil, A., 1994. Endoscopic treatment of vesicoureterall reflux with a chondrocye-alginate suspension. *J. Urol.*, 152: 641–643.

Babior, B.M., Kipnes R.S., and Cumutte, J.T., 1973. Biological defense mechanisms. The production by leukocytes of superoxide, A potential bactercidal agent. *J. Clin. Invest.*, 52: 741.

Badwey, J.A. and Kamovsky, M.L., 1980. Active oxygen species and the functions of phagocytic leucocytes. *Ann. Rev. Biochem.*, 49: 695.

Barrows, T.H., 1986, Degradable implant materials: a review of synthetic absorbable polymers and their applications. *Clin. Mater.*, 1: 233–257.

Barrows, T.H., 1994. Bioabsorbable poly(ester-amides). In: *Biomedical Polymers: Designed-to-Degrade Systems*, S.W. Shalaby, Ed., Hanser, New York, chap. 4.

Benedetti, L., Cortivo, R., Berti, T., Berti, A., and Pea, F., 1993. Biocompatibility and biodegradation of different hyaluronan derivatives (Hyaff) implanted in rats. *Biomaterials*, 14: 1154–1160.

Bezwada, R.S., Jamiolkowski, D.D., Lee, I.Y., Agarwal, V., Persivale, J., Trenka-Benthin, S., Erneta, M., Suryadevara, J., Yang, A., and Liu, S., 1995. Monocryl suture: a new ultra-pliable absorbable monofilament suture, *Biomaterials*, 16: 1141–1148.

Bezwada, R.S., Shalaby, S.W., Newman, H.D. Jr., and Kafrawy, A., 1990. Bioabsorbable copolymers of p-dioxanone and lactide for surgical devices. *Trans. Soc. for Biomater.* vol. XIII, p. 194.

Bostman, O., Hirvensalo, E., Vainionpaa, S. et al., 1990. Degradable polyglycolide rods for the internal fixation of displaced bimalleolar fractures. *Intern. Orthop. (Germany)*, 14: 1–8.

Bowald, S., Busch, C., and Eriksson, I., 1979. Arterial regeneration following polyglactin 910 suture mesh grafting, *Surgery*, 86: 722–729.

Bowald, S., Busch, C., and Eriksson, I., 1980. Absorbable material in vascular prosthesis. *Acta. Chir. Scand.*, 146: 391–395.

Brandrup, J., and Immergut, E.H., 1975. *Polymer Handbook*, 2nd ed., John Wiley & Sons, New York.

Campbell, N.D., and Chu, C.C., 1981. The effect of γ-irradiation on the biodegradation of polyglycolic acid synthetic sutures, the Tensile Strength Study. *27th International Symposium on Macromolecules*, Abstracts of Communications, Vol. II, pp. 1348–1352, Strasbourg, France, July 6–9, 1981.

Casey, D.J. and Roby, M.S., 1984. Synthetic copolymer surgical articles and method of manufacturing the same. US Patent 4,429,080, American Cyanamid.

Chatani, Y., Suehiro, K., Okita, Y., Tadokoro, H., and Chujo, K., 1968. Structural studies of polyesters, I. Crystal structure of polyglycolide. *Die Makromol. Chem.*, 113: 215–229.

Chu, C.C., 1981. The *In-vitro* degradation of poly(glycolic acid) sutures: effect of pH. *J. Biomed. Mater. Res.*, 15: 795–804.

Chu, C.C., 1982. The effect of pH on the *in vitro* degradation of poly(glycolide lactide) copolymer absorbable sutures. *J. Biomed. Mater. Res.*, 16: 117–124.

Chu, C.C., 1985a. Strain-accelerated hydrolytic degradation of synthetic absorbable sutures. In: *Surgical Research Recent Development*, C.W. Hall, Ed., Pergamon Press, San Antonio, Texas.

Chu, C.C., 1985b. The Degradation and biocompatibility of suture materials. In: *CRC Critical Reviews in Biocompat.*, D.F. Williams, Ed., Vol. 1 (3), CRC Press, Boca Raton, FL, pp. 261–322.

Chu, C.C., 1991. Recent advancements in suture fibers for wound closure. In: *High-Tech Fibrous Materials: Composites, Biomedical Materials, Protective Clothing, and Geotextiles*, T.L. Vigo and A.F. Turbak, Eds., ACS Symposium Series #457, American Chemical Society, Washington, D.C. pp. 167–213.

Chu, C.C., 1995a. Biodegradable suture materials: intrinsic and extrinsic factors affecting biodegradation phenomena. In: *Handbook of Biomaterials and Applications*, D.L. Wise, D.E. Altobelli, E.R. Schwartz, M. Yszemski, J.D. Gresser, and D.J. Trantolo, Eds., Marcel Dekker, New York.

Chu, C.C. 1995b. Biodegradable suture materials: intrinsic and extrinsic factors affecting biodegradation. In: *Encyclopedic Handbook of Biomaterials and Applications*, Part A: Materials, Vol. 1, D.L. Wise, Ed., Marcel Dekker, chap. 17, pp. 543–688.

Chu, C.C. and Browning, A., 1988. The study of thermal and gross morphologic properties of polyglycolic acid upon annealing and degradation treatments. *J. Biomed. Mater. Res.*, 22: 699–712.

Chu, C.C. and Campbell, N.D., 1982. Scanning electron microscope study of the hydrolytic degradation of poly(glycolic acid) suture. *J. Biomed. Mater. Res.*, 16: 417–430.

Chu, C.C., Hsu, A., Appel, M., and Beth, M. 1992. The effect of macrophage cell media on the *in vitro* hydrolytic degradation of synthetic absorbable sutures. *4th World Biomaterials Congress*, April 27–May 1, 1992, Berlin, Germany.

Chu, C.C. and Kizil, Z., 1989. The effect of polymer morphology on the hydrolytic degradation of synthetic absorbable sutures. *3rd International ITV Conference on Biomaterials — Medical Textiles*, Stuttgart, W. Germany, June 14–16, 1989.

Chu, C.C. and Louie, M., 1985. A chemical means to study the degradation phenomena of polyglycolic acid absorbable polymer. *J. Appl. Polym. Sci.*, 30: 3133–3141.

Chu, C.C., von Fraunhofer, J.A., and Greisler, H.P., 1996. *Wound Closure Biomaterials and Devices*. CRC Press, Boca Raton, FL.

Chu, C.C. and Williams, D.F., 1983. The effect of γ-irradiation on the enzymatic degradation of polyglycolic acid absorbable sutures. *J. Biomed. Mater. Res.*, 17: 1029.

Chu, C.C., Zhang, L., and Coyne, L., 1995. Effect of irradiation temperature on hydrolytic degradation properties of synthetic absorbable sutures and polymers. *J. Appl. Polym. Sci.* 56: 1275–1294.

Chujo, K., Kobayashi, H., Suzuki, J., Tokuhara, S., and Tanabe, M., 1967a. Ring-opening polymerization of glycolide. *Die Makromol. Chemi.*, 100: 262–266.

Chujo, K., Kobayashi, H., Suzuki, J., and Tokuhara, S., 1967b. Physical and chemical characteristics of polyglycolide. *Die Makromol. Chem.*, 100: 267–270.

Daniels, A.U., Taylor, M.S., Andriano, K.P., and Heller, J., 1992. Toxicity of absorbable polymers proposed for fracture fixation devices. *Trans. 38th Ann. Mtg. Orthop. Res. Soc.*, 17:88.

Devereux, D.F., O'Connell, S.M., Liesch, J.B., Weinstein, M., and Robertson, F.M., 1991. Induction of leukocyte activation by meshes surgically implanted in the peritoneal cavity. *Am. J. Surg.*, 162:243.

Dixit, V., 1994. Development of a bioartificial liver using isolated hepatocytes. *Artif. Organs*, 18:371–384.

Eitenmüller, K.L., Schmickal, G.T., and Muhr, G., 1989. Die versorgung von sprunggelenksfrakturen unter verwendung von platten und schrauben aus resorbierbarem polymer material. Presented at Jahrestagung der Deutschen Gesellschaft für Unfallheilkunde, Berlin, November 1989.

Forrester, A.R. and Purushotham, V., 1987. Reactions of carboxylic acid derivatives with superoxide. *J. Chem. Soc. Perkin Trans.* 1,945.

Forrester, A.R. and Purushotham, V., 1984. Mechnism of hydrolysis of esters by superoxide. *J. Chem. Soc., Chem. Commun.*, 1505.

Fredericks, R.J., Melveger, A.J., and Dolegiewitz, L.J., 1984. Morphological and structural changes in a copolymer of glycolide and lactide occurring as a result of hydrolysis. *J. Polym. Sci. Phy. Ed.* 22: 57–66.

Freed, L.E., Marquis, J.C., Nohia, A., Emmanual, J., Mikos, A.G., and Langer, R., 1993. Neocartilage formation *in vitro* and *in vivo* using cells cultured on synthetic biodegradable polymers. *J. Biomed. Mater. Res.*, 27: 11–23.

Fukuzaki, H., Yoshida, M., Asano, M., Aiba, Y., and Kumakura, M., 1989. Direct copolymerization of glycolic acid with lactones in the absence of catalysts. *Eur. Polym. J.*, 26: 457–461.

Fukuzaki, H., Yoshida, M., Asano, M., Kumakura, M., Mashimo, T., Yuasa, H., Imai, K., Yamandka, H., Kawaharada, U., and Suzuki, K., 1991. A new biodegradable copolymer of glycolic acid and lactones with relatively low molecular weight prepared by direct copolycondensation in the absence of catalysts. *J. Biomed. Mater. Res.*, 25: 315–328.

Gogolewski, S. and Pennings, A.J., 1983. Resorbable materials of poly(L-lactide). II. Fibres spun from solutions of poly(L-lactide) in good solvents. *J. Appl. Polym. Sci.*, 28: 1045–1061.

Greisler, H.P., 1982. Arterial regeneration over absorbable prostheses. *Arch. Surg.*, 117: 1425-1431.

Greisler, H.P., 1988a. Macrophage-biomaterial interactions with bioresorbable vascular prostheses. *Transactions of ASAIO*, 34: 1051–1059.

Greisler, H.P., 1991a. Macrophage activation in bioresorbable vascular grafts. In: *Vascular Endothelium: Physiological Basis of Clinical Problems*. J.D. Catravas, A.D. Callow, C.N. Gillis, and U. Ryan, Eds., Plenum Publishing, New York, NATO Advanced Study Institute. pp. 253–254.

Greisler, H.P., Dennis, J.W., Endean, E.D., Ellinger, J., Friesel, R., and Burgess, W., 1989. Macrophage/Biomaterial interactions: The stiulation of endotherlialization. *J. Vasc. Surg.*, 9: 588–593.

Greisler, H.P., Dennis, J.W., Endean, E.D., and Kim, D.U., 1988b. Derivation of neointima of vascular grafts. *Circ. Suppl. I*, 78: I6–I12.

Greisler, H.P., Ellinger, J., Schwarcz, T.H., Golan, J., Raymond, R.M., and Kim, D.U., 1987a. Arterial regeneration over polydioxanone prostheses in the rabbit. *Arch. Surg.*, 122: 715–721.

Greisler, H.P., Endean, E.D., Klosak, J.J., Ellinger, J., Dennis, J.W., Buttle, K., and Kim, D.U., 1988c. Polyglactin 910/polydioxanone bicomponent totally resorbable vascular prostheses. *J. Vasc. Surg.*, 7: 697–705.

Greisler, H.P., Kim, D.U., Dennis, J.W., Klosak, J.J., Widerborg, K.A., Endean, E.D., Raymond, R.M., and Ellinger, J., 1987b. Compound polyglactin 910/polypropylene small vessel prostheses. *J. Vasc. Surg.*, 5: 572–583.

Greisler, H.P., Kim, D.U., Price, J.B., and Voorhees, A.B., 1985. Arterial regenerative activity after prosthetic implantation. *Arch. Surg.*, 120: 315–323.

Greisler, H.P., Schwarcz, T.H., Ellinger, J., and Kim, D.U., 1986. Dacron inhibition of arterial regenerative activity. *J. Vasc. Surg.*, 747–756.

Greisler, H.P., Tattersall, C.W., Kloask, J.J. et al., 1991b. Partially bioresorbable vascular grafts in dogs. *Surgery*, 110: 645–655.

Gross, R.A., 1994. Bacterial polyesters: structural variability in microbial synthesis. In: *Biomedical Polymers: Designed-to-Degrade Systems*, S.W. Shalaby, Ed., chap. 7, Hanser, New York.

Haber, F. and Weiss, J., 1934. The catalytic decomposition of hydrogen peroxide by iron salts. *Proc. R. Soc. Lond.*, A, 147: 332.

Helder, J., Feijen, J., Lee, S.J., and Kim, W., 1986. Copolyemrs of DL-lactic acid and glycine. *Makromol. Chem. Rapid. Commun.*, 7: 193.

Hirai, J. and Matsuda, T., 1995. Self-organized, tubular hybrid vascular tissue composed of vascular cells and collagen for low pressure-loaded venous system, *Cell Transpl.*, 4: 597–608.

Hofmann, G.O., 1992. Biodegradable implants in orthopaedic surgery-A review of the state of the art. *Clin. Mater.*, 10: 75.

Hollinger, J.O., 1995. *Biomedical Applications of Synthetic Biodegradable Polymers*, CRC Press, Boca Raton, FL.

Jamiokowski, D.D. and Shalaby, S.W., 1991. A polymeric radiostabilizer for absorbable polyesters. In: Radiation Effect of Polymers, R.L. Clough and S.W. Shalaby, Eds., chap. 18, pp. 300–309. ACS Symposium Series # 475, ACS, Washington, D.C.

Johnson, R.A., 1976. *Tetrahedron Lett.*, 331.

Katsarava, R., Beridze, V., Arabuli, N. Kharadze, D., Chu, C.C., and Won, C.Y., Amino acid based bioanalogous polymers. Synthesis and study of regular poly(ester amide)s based on bis(α-amino acid) α, ω-alkylene diesters and aliphatic dicarboxylic acids. *J. Polym. Sci. Chem.* (in press).

Katz, A., Mukherjee, D.P., Kaganov, A.L., and Gordon, S., 1985. A new synthetic monofilament absorbable suture made from polytrimethylene carbonate. *Surg. Gynecol. Obstet.*, 161: 213–222.

Kim, B.S. and Mooney, D.J., 1998a. Engineering smooth muscle tissue with a predefined structure. *J. Biomed. Mater. Res.*, 41: 322.

Kim, B.S., Putman, A.J., Kulik, T.J., and Mooney, D.J., 1998b. Optimizing seeding and culture methods to engineer smooth muscle tissue on biodegradable polymer matrices. *Biotechnol. Bioeng.*, 57: 64–54.

Kimura, Y., 1993. Biodegradable polymers. In: *Biomedical Applications of Polymeric Materials*, T. Tsuruta, T. Hayashi, K. Kataoka, K. Ishihara, and Y. Kimura, Eds., pp. 164–190. CRC Press, Inc., Boca Raton, FL.

Koelmel, D.F., Jamiokowski, D.D., Shalaby, S.W., and Bezwada, R.S., 1991. Low modulus radiation sterilizable monofilament sutures. *Polym. Prepr.*, 32: 235–236.

Larsen, N.E., Pollak, C.T., Reiner, K., Leshchiner, E., and Balazs, E.A., 1993. Hylan gel biomaterial: Dermal and immunologic compatibility. *J. Biomed. Meter. Res.*, 27: 1129–1134.

Lee, K.H. and Chu, C.C., 1996a. The role of free radicals in hydrolytic degradation of absorable polymeric biomaterials. *5th World Biomaterials Congress*, Toronto, Canada, May 29–June 2.

Lee, K.H. and Chu, C.C., 1998. Molecular design of biologically active biodegradable polymers for biomedical applications. *Macromol. Symp.*, 130: 71.

Lee, K.H., Chu, C.C., and Fred, J., 1996b. Aminoxyl-containing radical spin in polymers and copolymers. U.S. Patent 5,516,881, May 16.

Lee, K.H., Won, C.Y., and Chu, C.C., 1996c. Hydrolysis of absorable polymeric biomaterials by superoxide. *5th World Biomaterials Congress*, Toronto, Canada, May 29–June 2.

Leenslag, J.W. and Pennings, A.J., 1984. Synthesis of high-molecular weight poly(L-lactide) initiated with tin 2-ethylhexanoate. *Makromol. Chem.*, 188: 1809–1814.

Li, S., Anjard, S., Tashkov, I., and Vert, M., 1998a. Hydrolytic degradation of PLA/PEO/PLA triblock copolymers prepared in the presence of Zn metal or CaH_2, *Polymer*, 39: 5421–5430.

Li, Y. and Kissel, T., 1998b. Synthesis, characteristics and *in vitro* degradation of star-block copolymers consisting of L-lactide, glycolide and branched multi-arm poly(ethylene oxide). *Polymer*, 39: 4421–4427.

Loh, I.H., Chu, C.C., and Lin, H.L., 1992. Plasma surface modification of synthetic absorbable fibers for wound closure. *J. Appl. Biomater.* 3: 131–146.

Magno, F. and Bontempelli, G., 1976. *J. Electroanal. Chem.*, 68: 337.

Matlaga, V.F. and Salthouse, T.N., 1980. Electron microscopic observations of polyglactin 910 suture sites, In *First World Biomaterials Congress*, Abstr., Baden, Austria, April 8–12, 2.

Mikos, A.G., Sarakinos, G., Leite, S.M., Vacanti, J.P., and Langer, R., 1993. Laminated three-dimensional biodegradable forms for use in tissue engineering. *Biomaterials*, 14: 323–330.

Miller, N.D. and Williams, D.F., 1984. The *in vivo* and *in vitro* degradation of poly(glycolic acid) suture material as a function of applied strain. *Biomaterials*, 5: 365–368.

Mooney, D.J., Baldwin, D.F., Vacanti, J.P., and Langer, R., 1996a. Novel approach to fabricate porous sponges of poly(D,L-lactic-co-glycolic acid) without the use of organic solvents. *Biomaterials*, 17: 1417–1422.

Mooney, D.J., Breuer, C., McNamara, K., Vacanti, J.P., and Langer, R., 1995. Fabricating tubular devices from polymers of lactic and glycolic acid for tissue engineering. *Tissue Eng.*, 1: 107–118.

Mooney, D.J., Mazzoni, C.L., Breuer, K., McNamara, J.P., Vacanti, J.P., and Langer, R., 1996b. Stabilized polyglycolic acid fibre-based tubes for tissue engineering. *Biomaterials*, 17: 115–124.

Mooney, D.J., Organ, G., Vacanti, J.P., and Langer, R., 1994. Design and fabrication of biodegradable polymer devices to engineer tubular tissue. *Cell Transplant*, 3: 203–210.

Mooney, D.J, Powell, C., Piana, J., and Rutherford, B., 1996c. Engineering dental pulp-like tissue *in vitro*. *Biotechnol. Prog.*, 12: 865–868.

Nathan, A. and Kohn, J., 1994. Amino acid derived polymers. In: *Biomedical Polymers: Designed-to-Degrade Systems*, S.W. Shalaby, chap. 5. Hanser Publishers, New York.

Park, K., Shalaby, W.S.W., and Park. H., 1993. *Biodegradable Hydrogels for Drug Delivery*, Technomic Publishing, Lancaster, PA.

Pratt, L., Chu, A., Kim, J., Hsu, A., and Chu, C.C., 1993a. The effect of electrolytes on the *in vitro* hydrolytic degradation of synthetic biodegradable polymers: mechanical properties, thermodynamics and molecular modeling. *J. Polym Sci. Chem. Ed.*, 31: 1759–1769.

Pratt, L. and Chu, C.C., 1993b. Hydrolytic degradation of α-substituted polyglycolic acid: a semi-empirical computational study. *J. Comput. Chem.*, 14: 809–817.

Pratt, L. and Chu, C.C., 1994a. The effect of electron donating and electron withdrawing substituents on the degradation rate of bioabsorbable polymers: a semi-empirical computational study. *J. Mol. Struct.*, 304: 213–226.

Pratt, L. and Chu, C.C., 1994b. A computational study of the hydrolysis of degradable polysaccharide biomaterials: substituent effects on the hydrolytic mechanism. *J. Comput. Chem.*, 15: 241–248.

Puelacher, W.C., Mooney, D., Langer, R., Upton, J., Vacanti, J.P., and Vananti, C.A., 1994. Design of nasoseptal cartilage replacements sunthesized from biodegradable polymers and chondrocytes. *Biomaterials*, 15: 774–778.

San Fillipo, Jr, J., Romano, L.J., Chem, C.I., and Valentine, J.S., 1976. Cleavage of esters by superoxide. *J. Org. Chem.*, 4: 586.

Shalaby, S.W., 1994. *Biomedical Polymers: Designed-to-Degrade Systems*, Hanser Publishers, New York.

Shieh, S.J., Zimmerman, M.C., and Parsons, J.R., 1990. Preliminary characterization of bioresorbable and nonresorbable synthetic fibers for the repair of soft tissue injuries. *J. Biomed. Mater. Res.*, 24: 789–808.

Sugnuma, J., Alexander, H., Traub, J., and Ricci, J.L., 1992. Biological response of intramedullary bone to poly-l-lactic acid. In: *Tissue-Inducing Biomater*, L.G. Cima and E.S. Ron, Eds., *Mater. Res. Soc. Sump. Proc.*, 252: 339–343.

Tunc, D.C., 1983. A high strength absorbable polymer for internal bone fixation. *Trans. Soc. Biomater.*, 6: 47.

Vert, M., Feijen, J., Albertsson, A., Scott, G., and Chiellini, E., 1992. *Biodegradable Polymers and Plastics*, Royal Society of Chemistry, Cambridge, England.

Williams, D.F., 1979. Some observations on the role of cellular enzymes in the In vivo degradation of polymers. *ASTM Spec. Tech. Publ.*, 684: 61–75.

Williams, D.F., 1990. Biodegradation of medical polymers. In: *Concise Encyclopedia of Medical and Dental Materials*, D.F. Williams, Ed., pp. 69–74. Pergamon Press, New York.

Williams, D.F. and Mort, E., 1977. Enzyme-accelerated hydrolysis of polyglycolic acid. *J. Bioeng.* 1: 231–238.

Williams, D.F. and Chu, C.C., 1984. The effects of enzymes and gamma irradiation on the tensile strength and morphology of poly(p-dioxanone) fibers. *J. Appl. Polym. Sci.*, 29: 1865–1877.

Williams, D.F. and Zhong, S.P., 1991. Are free radicals involved in the biodegradation of implanted polymers. *Advanced Materials*, 3: 623.

Winet, H. and Hollinger, J.O., 1993. Incorporation of polylactide–polyglycolide in a cortical defect: neoosteogenesis in a bone chamber. *J. Biomed. Mater. Res.*, 27: 667–676.

Wise, D.L., Fellmann, T.D., Sanderson, J.E., and Wentworth, R.L., 1979. Lactic/Glycolic acid polymers. In: *Drug Carriers in Biology and Medicine*, G. Gregoriadis, Ed., pp. 237–270. Academic Press, New York.

Yu, T.J. and Chu, C.C., 1993. Bicomponent vascular grafts consisting of synthetic biodegradable fibers. Part I. *In vitro* study. *J. Biomed. Mater. Res.*, 27: 1329–1339.

Yu, T.J., Ho, D.M., and Chu, C.C., 1994. Bicomponent vascular grafts consisting of synthetic biodegradable fibers. Part II. In vivo Healing Response. *J. Investigative Surg.* 7: 195–211.

Zhang, L., Loh, I.H., and Chu, C.C., 1993. A combined γ-irradiation and plasma deposition treatment to achieve the ideal degradation properties of synthetic absorbable polymers. *J. Biomed. Mater. Res.*, 27: 1425–1441.

Zhong, S.P., Doherty, P.J., and Williams, D.F., 1994. A preliminary study on the free radical degradation of glycolic acid/lactic acid copolymer. *Lastics, Rubber and Composites Processing and Application*, 21: 89.

Ziegler, T. and Nerem, R.M., 1994. Tissue engineering a blood vessel: Regulation of vascular biology by mechanical stress. *J. Cell. Biochem.*, 56: 204–209.

Further Information

Several recent books have very comprehensive descriptions of a variety of biodegradable polymeric biomaterials, their synthesis, physical, chemical, mechanical, biodegradable, and biological properties.

Barrows, T.H., 1986, Degradable implant materials: a review of synthetic absorbable polymers and their applications, *Clin. Mater.*, 1: 233–257.

Chu, C.C., Biodegradable suture materials: intrinsic and extrinsic factors affecting biodegradation phenomena, In: *Handbook of Biomaterials and Applications*, D.L. Wise, D.E. Altobelli, E.R. Schwartz, M. Yszemski, J.D. Gresser, and D.J. Trantolo, Eds., Marcel Dekker, New York (1995).

Chu, C.C., von Fraunhofer, J.A., and Greisler, H.P., 1996. *Wound Closure Biomaterials and Devices*. CRC Press, Boca Raton, FL.

Hollinger, J.O., Ed., 1995. *Biomedical Applications of Synthetic Biodegradable Polymers*, CRC Press, Boca Raton, FL.

Kimura, Y., 1993, Biodegradable polymers, In: *Biomedical Applications of Polymeric Materials*, T. Tsuruta, T. Hayashi, K. Kataoka, K. Ishihara, and Y. Kimura, Eds., pp. 164–190. CRC Press, Inc., Boca Raton, FL.

Park, K., Shalaby, W.S.W., and Park. H., 1993. *Biodegradable Hydrogels for Drug Delivery*, Technomic Publishing, Lancaster, PA.

Shalaby, S.W., 1994. *Biomedical Polymers: Designed-to-Degrade Systems*, Hanser Publishers, New York.

Vert, M., Feijen, J., Albertsson, A., Scott, G. and Chiellini, E., 1992. *Biodegradable Polymers and Plastics*, Royal Society of Chemistry, Cambridge, England.

The review by Barrows is brief with an emphasis on their applications with an extensive lists of patents. The book and chapter by Chu et al. focuses on the most successful use of biodegradable polymers in medicine, namely wound closure biomaterials like sutures. It is so far the most comprehensive review of all aspects of biodegradable wound closure biomaterials with very detailed chemical, physical, mechanical, biodegradable, and biological information. The chapter by Kimura is an overview of the subject with some interesting new polymers. The chapter includes both enzymatically degradable natural polymers and non-enzymatically degradable synthetic polymers. The biodegradable hydrogel book by Park et al. is the only book available that focuses on hydrogel. Probably the most broad coverage of biodegradable polymeric biomaterials is the very recent book edited by Shalaby. It has 8 chapters and covers almost all commercially and experimentally available biodegradable polymers. The book edited by Vert et al. is based on the *Proceedings of the 2nd International Scientific Workshop on Biodegradable Polymers and Plastics* held in Montpellier, France in November 1991. The book covers both medical and non-medical applications of biodegradable polymers. It has broader coverage of biodegradable polymers with far more chapters than Shalaby's book, but its chapters are shorter and less comprehensive than Shalaby's book which is far more focused on biomedical use.

Topics of biodegradable polymeric biomaterials can also be found in *Journal of Biomedical Materials Research, Journal of Applied Biomaterials, Biomaterials, Journal of Biomaterials Science: Polymer Ed., Journal of Applied Polymer Science, Journal of Materials Science,* and *Journal Polymer Science.*

7

Biologic Biomaterials: Tissue-Derived Biomaterials (Collagen)

Shu-Tung Li
Collagen Matrix, Inc.

7.1 Structure and Properties of Collagen and Collagen-Rich Tissues

7.1.1 Structure of Collagen

Collagen is a multifunctional family of proteins of unique structural characteristics. It is the most abundant and ubiquitous protein in the body, its functions ranging from serving crucial biomechanical functions in bone, skin, tendon, and ligament to controlling cellular gene expressions in development [Nimni and Harkness, 1988]. Collagen molecules like all proteins are formed *in vivo* by enzymatic regulated step-wise polymerization reaction between amino and carboxyl groups of amino acids, where R is a side group of an amino acid residue.

$$
(-\overset{\overset{\displaystyle O}{\|}}{C} - \overset{\overset{\displaystyle H}{|}}{N} - \underset{\underset{\displaystyle R}{|}}{\overset{\overset{\displaystyle H}{|}}{C}} -)_n
$$

(7.1)

The simplest amino acid is **glycine** (Gly) (R=H), where a hypothetical flat sheet organization of polyglycine molecules can form and be stabilized by intermolecular hydrogen bonds (Figure 7.1a). However, when R is a large group as in most other amino acids, the stereochemical constraints frequently force the **polypeptide** chain to adapt a less constraining conformation by rotating the bulky R groups away from the crowded interactions, forming a helix, where the large R groups are directed toward the surface of the helix (Figure 7.1b). The hydrogen bonds are allowed to form within a helix between the hydrogen attached to nitrogen in one amino acid residue and the oxygen attached to a second amino acid residue. Thus, the final conformation of a protein, which is directly related to its function, is governed primarily by the amino acid sequence of the particular protein.

Collagen is a protein comprised of three polypeptides (α chains), each having a general amino acid sequence of $(-Gly-X-Y-)_n$, where X is any other amino acid and is frequently **proline** (Pro) and Y is any other amino acid and is frequently *hydroxyproline* (Hyp). A typical amino acid composition of collagen is shown in Table 7.1. The application of helical diffraction theory to high-angle collagen x-ray

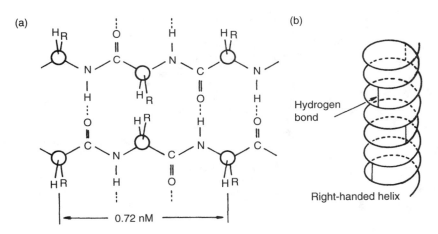

FIGURE 7.1 (a) Hypothetical flat sheet structure of a protein. (b) Helical arrangement of a protein chain.

TABLE 7.1 Amino Acid Content of Collagen

Amino Acids	Content, residues/1000 residues[a]
Gly	334
Pro	122
Hyp	96
Acid polar (Asp, Glu, Asn)	124
Basic polar (Lys, Arg, His)	91
Other	233

[a] Reported values are average values of 10 different determinations for tendon tissue.
Source: Eastoe, J.E. (1967). *Treatise on Collagen*, G.N. Ramachandran (Ed.), pp. 1–72, Academic Press, New York. With permission.

diffraction pattern [Rich and Crick, 1961] and the stereochemical constraints from the unusual amino acid composition [Eastoe, 1967] led to the initial triple-helical model and subsequent modified triple helix of the collagen molecule. Thus, collagen can be broadly defined as a protein which has a typical triple helix extending over the major part of the molecule. Within the triple helix, glycine must be present as every third amino acid, and proline and hydroxyproline are required to form and stabilize the triple helix.

To date, 19 proteins can be classified as collagen [Fukai et al., 1994]. Among the various collagens, type I collagen is the most abundant and is the major constituent of bone, skin, ligament, and tendon. Due to the abundance and ready accessibility of these tissues, they have been frequently used as a source for the preparation of collagen. This chapter will not review the details of the structure of the different collagens. The readers are referred to recent reviews for a more in-depth discussion of this subject [Nimni, 1988; van der Rest et al., 1990; Fukai et al., 1994; Brodsky and Ramshaw, 1997]. It is, however, of particular relevance to review some salient structural features of the type I collagen in order to facilitate the subsequent discussions of properties and its relation to biomedical applications.

A type I collagen molecule (also referred to as *tropocollagen*) isolated from various tissues has a molecular weight of about 283,000 daltons. It is comprised of three left-handed helical polypeptide chains (Figure 7.2a) which are intertwined forming a right-handed helix around a central molecular axis (Figure 7.2b). Two of the polypeptide chains are identical (α_1) having 1056 amino acid residues, and the third polypeptide chain (α_2) has 1029 amino acid residues [Miller, 1984]. The triple-helical structure has a rise per residue of 0.286 nm and a unit twist of 108°, with 10 residues in three turns and a **helical pitch** (repeating distance within a single chain) of 30-residues or 8.68 nm [Fraser et al., 1983]. Over 95% of the amino acids have the sequence of Gly–X–Y. The remaining 5% of the molecule does not have the sequence of Gly–X–Y and is therefore not triple-helical. These nonhelical portions of the molecule are located at the N- and C-terminal ends and are referred to as **telopeptides** (9 to 26 residues) [Miller, 1984]. The whole molecule has a length of about 280 nm and a diameter of about 1.5 nm and has a conformation similar to a rigid rod (Figure 7.2c).

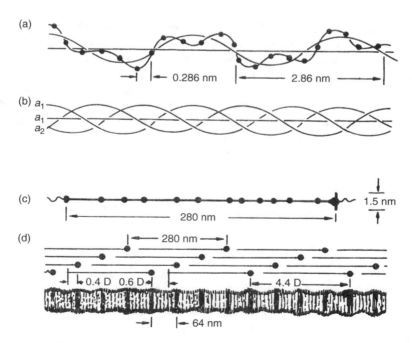

FIGURE 7.2   Diagram depicting the formation of collagen, which can be visualized as taking place in several steps: (a) single chain left-handed helix; (b) three single chains intertwined into a triple stranded helix; (c) a collagen (tropocollagen) molecule; (d) collagen molecules aligned in D staggered fashion in a fibril producing overlap and hole regions.

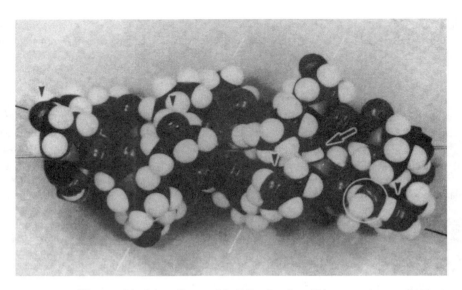

FIGURE 7.3 A space-filling model of the collagen triple helix, showing all the atoms in a ten-residue segment of repeating triplet sequence $(Gly–Pro–Hyp)_n$. The arrow shows an interchain hydrogen bond. The arrow heads identify the hydroxy groups of hydroxyproline in one chain. The circle shows a hydrogen-bonded water molecule. The short white lines identify the ridge of amino acid chains. The short black lines indicate the supercoil of one chain [Piez, 1984].

The triple-helical structure of a collagen molecule is stabilized by several factors (Figure 7.3): (1) a tight fit of the amino acids within the triple-helix — this geometrical stabilization factor can be appreciated from a space-filling model constructed from a triple helix with (Gly–Pro–Hyp) sequence (Figure 7.3); (2) the interchain hydrogen bond formation between the backbone carbonyl and amino hydrogen interactions; and (3) the contribution of water molecules to the interchain hydrogen bond formation.

The telopeptides are regions where **intermolecular crosslinks** are formed *in vivo*. A common inter-molecular crosslinks is formed between an **allysine** (the ε-amino group of **lysine** or hydroxy-lysine has been converted to an aldehyde) of one telopeptide of one molecule and an ε-amino group of a lysine or **hydroxylysine** in the triple helix or a second molecule (6.2). Thus the method commonly used to solubil-ize the collagen molecules from crosslinked **fibrils** with **proteolytic enzymes** such as **pepsin** removes the telopeptides (cleaves the intermolecular crosslinks) from the collagen molecule. The pepsin solubilized collagen is occasionally referred to as **atelocollagen** [Stenzl, 1974].

$$
\begin{array}{c}
 OH \\
 | \\
Pr - CH_2 - CH_2 - CH_2 - CHO \quad + \quad H_2N - CH_2 - CH - CH_2 - CH_2 - Pr \\
\text{Allysine} \text{Hydroxylysine} \\
OH \\
| \\
\rightarrow Pr - CH_2 - CH_2 - CH_2 - CH = N - CH_2 - CH - CH_2 - CH_2 - Pr \\
\text{Dehydrohydroxylysinononorleucine}
\end{array}
$$

$$(7.2)$$

Since the presence of hydroxyproline is unique in collagen **elastin** contains a small amount), the determ-ination of collagen content in a collagen-rich tissue is readily done by assaying the hydroxyproline content.

Collagen does not appear to exist as isolated molecules in the extracellular space in the body. Instead, collagen molecules aggregate into *fibrils*. Depending on the tissue and age, a collagen fibril varies from about 50 to 300 nm in diameter with indeterminate length and can be easily seen under electron microscopy (Figure 7.4). The fibrils are important structural building units for large **fibers** (Figure 7.5). Collagen molecules are arranged in specific orders both longitudinally and in cross-section, and the organization

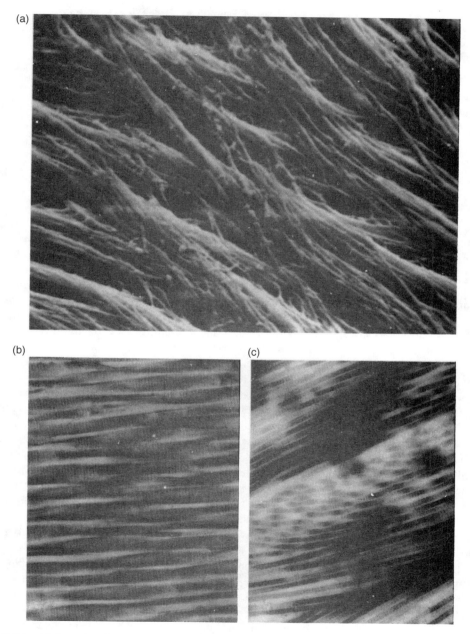

FIGURE 7.4 (a) Scanning electron micrograph of the surface of an adult rabbit bone matrix, showing how the collagen fibrils branch and interconnect in an intricate, woven pattern ($\times 4800$) (Tiffit, 1980). (b) Transmission electron micrographs of ($\times 24,000$) parallel collagen fibrils in tendon [Fung, 1992]. (c) Transmission electron micrographs of ($\times 24,000$) mesh work of fibrils in skin [Fung, 1993].

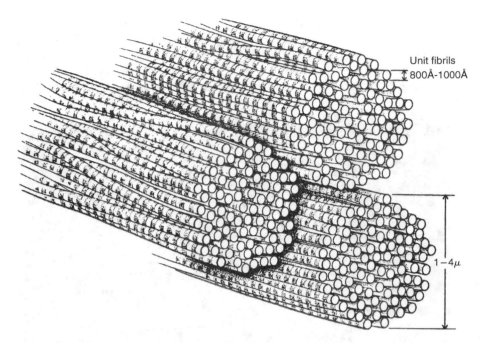

FIGURE 7.5 Diagram showing the collagen fibers of the connective tissue in general which are composed of unit collagen fibrils.

of collagen molecules in a fibril is tissue-specific [Katz and Li, 1972, 1973b]. The two-dimensional structure (the projection of a three-dimensional structure onto a two-dimensional plane) of a type I collagen fibril has been unequivocally defined both by an analysis of small-angle x-ray diffraction pattern along the meridian of a collagenous tissue [Bear, 1944] and by examination of the transmission electron micrographs of tissues stained with negative or positive stains [Hodge and Petruska, 1963]. In this structure (Figure 7.2d) the collagen molecules are staggered with respect to one another by a distance of D (64 to 67 nm) or multiple of D, where D is the fundamental repeat distance seen in the small-angle x-ray diffraction pattern, or the repeating distance seen in the electron micrographs. Since a collagen molecule has a length of about $4.4D$, this staggering of collagen molecules creates overlap regions of about $0.4D$ and hole or defect regions of about $0.6D$.

One interesting and important structural aspect of collagen is its approximate equal number of acidic (**aspartic** and **glutamic acids**) and basic (lysines and **arginines**) side groups. Since these groups are charged under physiological conditions, the collagen is essentially electrically neutral [Li and Katz, 1976]. The packing of collagen molecules with a D staggering results in clusters of regions where the charged groups are located [Hofmann and Kuhn, 1981]. These groups therefore are in close proximity to form intra- and intermolecular hydrogen-bonded **salt-linkages** of the form $(Pr - COO^{-} {}^{+}H_3N - Pr)$ [Li et al., 1975]. In addition, the side groups of many amino acids are nonpolar [**alanine (Ala)**, **valine** (Val), **leucine** (Leu), **isoleucine** (Ile), *proline* (Pro), and *phenolalanine* (Phe)] in character and hence *hydrophobic*; therefore, chains with these amino acids avoid contact with water molecules and seek interactions with the nonpolar chains of amino acids. In fact, the result of molecular packing of collagen in a fibril is such that the nonpolar groups are also clustered, forming hydrophobic regions within collagen fibrils [Hofmann and Kuhn, 1981]. Indeed, the packing of the collagen molecules in various tissues is believed to be a result of intermolecular interactions involving both the electrostatic and hydrophobic interactions [Hofmann and Kuhn, 1981; Katz and Li, 1981; Li et al., 1975].

The three-dimensional organization of type I collagen molecules within a fibril has been the subject of extensive research over the past 40 years [Fraser et al., 1983; Katz and Li, 1972, 1973a, b, 1981; Miller, 1976; Ramachandran, 1967; Yamuchi et al., 1986]. Many structural models have been proposed based on an analysis of equatorial and off-equatorial x-ray diffraction patterns of rat-tail-tendon collagen [Miller, 1976; North et al., 1954], *intrafibrillar volume* determination of various collagenous tissues [Katz and Li, 1972, 1973a, b], intermolecular side chain interactions [Hofmann and Kuhn, 1981; Katz and Li, 1981; Li et al., 1981], and intermolecular crosslinking patterns studies [Yamuchi et al., 1986]. The general understanding of the three-dimensional molecular packing in type I collagen fibrils is that the collagen molecules are arranged in hexagonal or near hexagonal arrays [Katz and Li, 1972, 1981; Miller, 1976]. Depending on the tissue, the intermolecular distance varies from about 0.15 nm in rat tail tendon to as large as 0.18 nm in bone and dentin [Katz and Li, 1973b]. The axial staggering of the molecules by $1 \sim 4D$ with respect to one another is tissue-specific and has not yet been fully elucidated.

There are very few interspecies differences in the structure of type I collagen molecule. The extensive homology of the structure of type I collagen may explain why this collagen obtained from animal species is acceptable as a material for human implantation.

7.1.2 Properties of Collagen-Rich Tissue

The function of collagenous tissue is related to its structure and properties. This section reviews some important properties of collagen-rich tissues.

7.1.2.1 Physical and Biomechanical Properties

The physical properties of tissues vary according to the amount and structural variations of the collagen fibers. In general, a collagen-rich tissue contains about 75 to 90% of collagen on a dry weight basis. Table 7.2 is a typical composition of a collagen-rich soft tissue such as skin. Collagen fibers (bundles of collagen fibrils) are arranged in different configurations in different tissues for their respective functions at specific anatomic sites. For example, collagen fibers are arranged in parallel in tendon (Figure 7.4b) and ligament for their high-tensile strength requirements, whereas collagen fibers in skin are arranged in random arrays (Figure 7.4c) to provide the resiliency of the tissue under stress. Other structure-supporting functions of collagen such as transparency for the lens of the eye and shaping of the ear or tip of the nose can also be provided by the collagen fiber. Thus, an important physical property of collagen is the three-dimensional organization of the collagen fibers.

The collagen-rich tissues can be thought of as a composite polymeric material in which the highly oriented crystalline collagen fibrils are embedded in the amorphous ground substance of noncollagenous **polysaccharides**, **glycoproteins**, and elastin. When the tissue is heated, its specific volume increases, exhibiting a glass transition at about 40°C and a melting of the crystalline collagen fibrils at about 56°C. The melting temperature of crystalline collagen fibrils is referred to as the *denaturation temperature* of collagenous tissues.

The stress–strain curves of a collagenous tissue such as tendon exhibit nonlinear behavior (Figure 7.6). This nonlinear behavior of stress–strain of tendon collagen is similar to that observed in synthetic fibers. The initial toe region represents alignment of fibers in the direction of stress. The steep rise in slope represents the majority of fibers stretched along their long axes. The decrease in slope following the steep rise may represent the breaking of individual fibers prior to the final catastrophic failure. Table 7.3 summarizes some mechanical properties of collagen and elastic fibers. The difference in biomechanical

TABLE 7.2 Composition of Collagen-Rich Soft Tissues

Component	Composition, %
Collagen	75 (dry), 30 (wet)
Proteoglycans and polysaccharides	20 (dry)
Elastin and glycoproteins	<5 (dry)
Water	60–70

Source: Park and Lakes (1992).

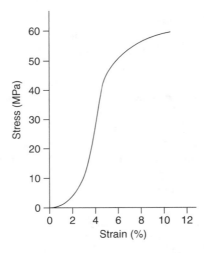

FIGURE 7.6   A typical stress-strain curve for tendon [Rigby et al., 1959].

TABLE 7.3   Elastic Properties of Collagen and Elastic Fibers

Fibers	Modulus of elasticity, MPa	Tensile strength, MPa	Ultimate elongation, %
Collagen	1000	50–100	10
Elastin	0.6	1	100

Source: Park, J.B. and Lakes, R.S. 1992. Structure-property relationships of biological materials. In *Biomaterials: An Introduction*, 2nd ed., pp. 185–222, Plenum Press, New York.

properties between collagen and elastin is a good example of the requirements for these proteins to serve their specific functions in the body.

Unlike tendon or ligament, skin consists of collagen fibers randomly arranged in layers or lamellae. Thus skin tissues show mechanics anisotropy (Figure 7.7). Another feature of the stress-strain curve of the skin is its extensibility under small load as compared to tendon. At small load the fibers are straightened and aligned rather than stretched. Upon further stretching the fibrous lamellae align with respect to each other and resist further extension. When the skin is highly stretched the modulus of elasticity approaches that of tendon as expected of the aligned collagen fibers.

Cartilage is another collagen-rich tissue which has two main physiological functions. One is the maintenance of shape (ear, tip of nose, and rings around the trachea), and the other is to provide bearing surfaces at joints. It contains very large and diffuse proteoglycan (protein-polysaccharide) molecules which form a gel in which the collagen-rich molecules entangled. They can affect the mechanical properties of the collagen by hindering the movements through the interstices of the collagenous matrix network.

The joint cartilage has a very low coefficient of friction (<0.01). This is largely attributed to the squeeze-film effect between cartilage and synovial fluid. The synovial fluid can be squeezed out through highly fenestrated cartilage upon compressive loading, and the reverse action will take place in tension. The lubricating function is carried out in conjunction with **glycosaminoglycans** (GAG), especially **chondroitin sulfates.** The modulus of elasticity (10.3 to 20.7 MPa) and tensile strength (3.4 MPa) are quite low. However, wherever high stress is required the cartilage is replaced by purely collagenous tissue. Mechanical properties of some collagen-rich tissues are given in Table 7.4 as a reference.

7.1.2.2 Physiochemical Properties

Electrostatic properties: A collagen molecule has a total of approximately 240 ε-amino and guanidino groups of lysines, hydroxylysines, and arginines and 230 carboxyl groups of aspartic and glutamic acids.

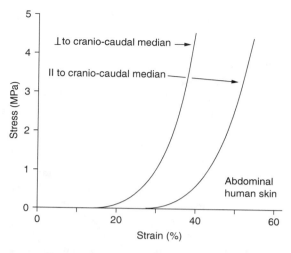

FIGURE 7.7   Stress–strain curves of human abdominal skin [Daly, 1966].

TABLE 7.4   Mechanical Properties of Some Nonmineralized Human Tissues

Tissues	Tensile strength, MPa	Ultimate elongation, %
Skin	7.6	78.0
Tendon	53.0	9.4
Elastic cartilage	3.0	30.0
Heart valves (aortic)		
Radial	0.45	15.3
Circumferential	2.6	10.0
Aorta		
Transverse	1.1	77.0
Longitudinal	0.07	81.0

Source: Park, J.B. and Lakes, R.S. 1992. Structure-property relationships of biological materials. In *Biomaterials: An Introduction*, 2nd ed., pp. 185–222, Plenum Press, New York.

These groups are charged under physiological conditions. In a native fibril, most of these groups interact either intra- or intermolecularly forming salt-linkages providing significant stabilization energy to the collagen fibril [Li et al., 1975]. Only a small number of charged groups are free. However, the electrostatic state within a collagen fibril can be altered by changing the pH of the environment. Since the pK for an amino group is about 10 and about 4 for a carboxyl group, the electrostatic interactions are significantly perturbed at a pH below 4 and above 10. The net result of the pH change is a weakening of the intra- and intermolecular electrostatic interactions, resulting in a swelling of the fibrils. The fibril swelling can be prevented by chemically introducing covalent intermolecular crosslinks. Any bifunctional reagent which reacts with amino, carboxyl, and hydroxyl groups can serve as a crosslinking agent. The introduction of covalent intermolecular crosslinks fixes the physical state of the fibrillar structure and balances the swelling pressures obtained from any pH changes.

Another way of altering the electrostatic state of a collagen fibril is by chemically modifying the electrostatic side groups. For example, the positively charged ε-amino groups of lysine and hydroxylysine can be chemically modified with acetic anhydride, which converts the ε-amino groups to a neutral acetyl group [Green et al., 1954]. The result of this modification increases the number of the net negative charges of the fibril. Conversely, the negatively charged carboxyl groups of aspartic and glutamic acid can be chemically modified to a neutral group by methylation [Fraenkel-Conrat, 1944]. Thus, by adjusting the pH of the solution and applying chemical modification methods, a range of electrostatic properties of collagen can be obtained.

Ion and macromolecular binding properties: In the native state and under physiological conditions, a collagen molecule has only about 60 free carboxyl groups [Li et al., 1975]. These groups have the capability of binding cations such as calcium with a free energy of formation for the protein-COO–Ca^{++} of about 1.2 kcal/mol. This energy is not large enough to compete for the hydrogen bonded salt-linkage interactions, which have a free energy of formation of about −1.6 kcal/mol. The extent of ion binding, however, can be enhanced in the presence of lyotropic salts such as KCNS, which breaks the salt-linkages, or by shifting the pH away from the *isoelectric point* of collagen. Macromolecules can bind to collagen via covalent bonding, cooperative ionic binding, entrapment, entanglement, and a combination of the above. In addition, binding of charged ions and macromolecules can be significantly increased by modifying the charge profile of collagen as described previously. For example, a complete *N*-acetylation of collagen will eliminate all the positively charged *ε*-amino groups and, thus, will increase the free negatively charged groups. The resulting acetylated collagen enhances the binding of positively charged ions and macromolecules. On the other hand, the methylation of collagen will eliminate the negatively charged carboxyl groups and, thus, will increase the free positively charge moieties. The methylated collagen, therefore, enhances the binding of negatively charged ions and macromolecules [Li and Katz, 1976].

Fiber-forming properties: Native collagen molecules are organized in tissues in specific orders. *Polymorphic* forms of collagen can be reconstituted from the collagen molecules, obtained either from enzymatic digestion of collagenous tissues or by extracting the tissues with salt solutions. The formation of polymorphic aggregates of collagen depends on the environment for reconstitution [Piez, 1984]. Native arrangement of the collagen molecules is formed under physiological conditions. Various polymorphic molecular aggregates may be formed by changing the state of intermolecular interactions. For example, when collagen molecules are aggregated under high concentrations of a neutral salt or under nonaqueous conditions, the collagen molecules associate into random arrays having no specific regularities detectable by electron microscopy. The collagen molecules can be induced to aggregate into other polymorphic forms such as the **segment-long-spacing** (SLS) form where all heads are aligned in parallel and the **fibrous-long-spacing** (FLS) form where the molecules are randomly aligned in either a head-to-tail, tail-to-tail, or head-to-head orientation.

7.1.2.3 Biologic Properties

Hemostatic properties: Native collagen aggregates are intrinsically hemostatic. The mechanism of collagen-induced hemostasis has been the subject of numerous investigations [Wilner et al., 1968; Jaffe and Deykin, 1974; Wang et al., 1978]. The general conclusion from these studies is that **platelets** first adhere to a collagen surface. This induces the release of platelet contents, followed by platelet aggregation, leading to the eventual hemostatic plug. The hemostatic activity of collagen is dependent on the size of the collagen aggregate and the native organization of the molecules [Wang et al., 1978]. Denatured collagen (**gelatin**) is not effective in inducing hemostasis [Jonas et al., 1988].

Cell interaction properties: Collagen forms the essential framework of the tissues and organs. Many cells, such as epithelial and endothelial cells, are found resting on the collagenous surfaces or within a collagenous matrix such as that of many connective tissue cells. Collagen-cell interactions are essential features during the development stage and during wound healing and tissue remodeling in adults [Kleinman et al., 1981; Linbold and Kormos, 1991]. Studying collagen-cell interactions is useful in developing simulated tissue and organ structures and in investigating cell behavior in the *in vivo* simulated systems. Numerous studies have aimed at developing viable tissues and organs *in vitro* for transplantation applications [Bell et al., 1981; Montesano et al., 1983; Silbermann, 1990; Bellamkonda and Aebischer, 1994; Hubbel, 1995; Moghe et al., 1996; Sittinger et al., 1996].

Immunologic properties: **Soluble collagen** has long been known to be a poor immunogen [Timpl, 1982]. A significant level of antibodies cannot be raised without the use of Freund's complete adjuvant (a mixture of mineral oil and heat-killed *mycobacteria*) which augments antibody response. It is known that insoluble collagen is even less immunogenic [Stenzel et al., 1974]. Thus, xenogeneic collagenous tissue devices such as porcine and bovine pericardial heart valves are acceptable for long-term implantation in humans. The reasons for the low antibody response against collagen are not known. It may be related to the homology of the collagen structure from different species (low level of foreignness) or to certain structural features associated with collagen [Timpl, 1982].

7.2 Biotechnology of Collagen

7.2.1 Isolation and Purification of Collagen

There are two distinct ways of isolating and purifying collagen material. One is the molecular technology and the other is the fibrillar technology. These two technologies are briefly reviewed here.

7.2.1.1 Isolation and Purification of Soluble Collagen Molecules

The isolation and purification of soluble collagen molecules from a collagenous tissue is achieved by using a proteolytic enzyme such as pepsin to cleave the telopeptides [Miller and Rhodes, 1982]. Since telopeptides are the natural crosslinking sites of collagen, the removal of telopeptides renders the collagen molecules and small collagen aggregates soluble in an aqueous solution. The pepsin-solubilized collagen can be purified by repetitive precipitation with a neutral salt. Pepsin-solubilized collagen in monomeric form is generally soluble in a buffer solution at low temperature. The collagen molecules may be reconstituted into fibrils of various **polymorphisms**. However, the reconstitution of the pepsin-solubilized collagen into fibrils of native molecular packing is not as efficient as the intact molecules, since the telopeptides facilitate fibril formation [Comper and Veis, 1977].

7.2.1.2 Isolation and Purification of Fibrillar Collagen

The isolation and purification of collagen fibers relies on the removal of noncollagenous materials from the collagenous tissue. Salt extraction removes the newly synthesized collagen molecules that have not been covalently incorporated into the collagen fibrils. Salt also removes the noncollagenous materials that are soluble in aqueous conditions and are bound to collagen fibrils by nonspecific interactions. **Lipids** are removed by low-molecular-weight organic solvents such as low-molecular-weight ethers and alcohols. Acid extraction facilities the removal of acidic proteins and glycosaminoglycans due to weakening of the interactions between the acidic proteins and collagen fibrils. Alkaline extraction weakens the interaction between the basic proteins and collagen fibrils and thus facilitates the removal of basic proteins. In addition, various enzymes other than **collagenase** can be used to facilitate the removal of the small amounts of glycoproteins, proteoglycans, and elastins from the tissue. Purified collagen fibers can be obtained through these sequential extractions and enzymatic digestions from the collagen-rich tissues.

7.2.2 Matrix Fabrication Technology

The purified collagen materials obtained from either the molecular technology or from the fibrillar technology are subjected to additional processing to fabricate the materials into useful devices for specific medical applications. The different matrices and their medical applications are summarized in Table 7.5. The technology in fabricating these matrices are briefly outlined below.

7.2.2.1 Membranous Matrix

Collagen membranes can be produced by drying a collagen solution or a fibrillar collagen dispersion cast on a nonadhesive surface. The thickness of the membrane is governed by the concentration and the initial thickness of the cast solution or dispersion. In general, membrane thickness of up to 0.5 mm can be easily obtained by air drying a cast collagen material. Additional chemical crosslinking is required to stabilize the membrane from dissolution or dissociation. The membrane produced by casting and air drying does not permit manipulation of the pore structure. Generally, the structure of a cast membrane is dense and amorphous with minimal **permeability** to macromolecules [Li et al., 1991]. Porous membranes

TABLE 7.5 Summary of Different Collagen Matrices and Their Medical Applications

Matrix form	Medical application
Membrane (film, sheet)	Oral tissue repair; wound dressings; dura repair; patches
Porous (sponge, felt, fibers)	Hemostats; wound dressings; cartilage repair; soft-tissue augmentation
Gel	Drug and biologically active macromolecule delivery; soft- and hard-tissue augmentation
Solution	Soft-tissue augmentation; drug delivery
Filament	Tendon and ligament repair; sutures
Tubular (membrane, sponge) Composite	Nerve repair; vascular repair
Collagen/synthetic polymer	Vascular repair; skin repair; wound dressings
Collagen/biological polymer	Soft-tissue augmentation; skin repair
Collagen/ceramic	Hard-tissue repair

may be obtained by freeze-drying a cast solution or dispersion of a predetermined density or by partially compressing a preformed porous matrix to a predetermined density and pore structure.

7.2.2.2 Porous Matrix

Porous collagen matrices are generally obtained by freeze-drying an aqueous volume of collagen solution or dispersion. The freeze-dried porous matrix requires chemical crosslinking to stabilize the structure. A convenient way to stabilize the porous matrix is to crosslink the matrix by vapor using a volatile crosslinking agent such as formaldehyde or glutaraldehyde. The pore structure of the matrix depends, to a large extent, on the concentration of the collagen in the solution or dispersion. Other factors that contribute to the pore structure include the rate of freezing, the size of fibers in the dispersion, and the presence and absence of other macromolecules. **Apparent densities** from 0.05 to 0.3 g matrix per cubic centimeter matrix volume can be obtained. These porous matrices generally have pores from about 50 μm to as large as 1500 μm.

7.2.2.3 Gel Matrix

A *gel matrix* may be defined as a homogeneous phase between a liquid and a solid. As such, a gel may vary from a simple viscous fluid to a highly concentrated putty-like material. Collagen gels may be formed by shifting the pH of a dispersion away from its isoelectric point. Alternatively, the collagen material may be subjected to a chemical modification procedure to change its charge profile to a net positively charged or a net negatively charged protein before hydrating the material to form a gel matrix. For example, native fibers dispersed in water at pH 7 will be in the form of two phases. The dispersed fibers become gel when the pH changes from 7 to 3. Succinylating the primary amino groups of collagen, which converts the positively charged amino groups to negatively charged carboxyl groups, changes the isoelectric point of collagen from about 7 to about 4.5. Such a collagen material swells to a gel at a pH of 7.

7.2.2.4 Solution Matrix

A collagen solution is obtained by dissolving the collagen molecules in an aqueous solution. Collagen molecules are obtained by digesting the insoluble tissue with pepsin to cleave the crosslinking sites of collagen (telopeptides) as previously described. The solubility of collagen depends on the pH, the temperature, the ionic strength of the solution, and the molecular weight. Generally, collagen is more soluble in the cold. Collagen molecules aggregate into fibrils when the temperature of the solution increases to the body temperature. pH plays an important role in solubilizing collagen. Collagen is more soluble at a pH away from the isoelectric point of the protein. Collagen is less soluble at higher ionic strength of a solution. The solubility of collagen decreases with increasing the size of molecular aggregates. Thus, collagen becomes increasingly less soluble with increasing the extent of crosslinking [Bailey et al., 1970].

7.2.2.5 Filamentous Matrix

Collagen filaments can be produced by extrusion techniques [Kemp et al., 1995; Li and Stone, 1993; Schimpf and Rodriquez, 1976]. A collagen solution or dispersion having a concentration in the range of 0.5 to 1.5% (w/v) is first prepared. Collagen is extruded into a coacervation bath containing a high concentration of a salt or into an aqueous solution at a pH of the isoelectric point of the collagen. Tensile strength of 30 MPa has been obtained for the reconstituted filaments.

7.2.2.6 Tubular Matrix

Tubular matrices may be formed by extrusion through a coaxial cylinder [Stenzl et al., 1974], or by coating collagen onto a mandrel [Li, 1990]. Different properties of the tubular membranes can be obtained by controlling the drying properties.

7.2.2.7 Composite Matrix

Collagen can form a variety of homogeneous composites with other water-soluble materials. Ions, peptides, proteins, and polysaccharides can all be uniformly incorporated into a collagen matrix. The methods of homogeneous composite formation include ionic and covalent bonding, entrapment, entanglement, and coprecipitation. A heterogeneous composite can be formed between collagen, ceramics, and synthetic polymers that have distinct properties for medical applications [Li, 1988].

7.3 Design of a Resorbable Collagen-Based Medical Implant

Designing a medical implant for tissue or organ repair requires a thorough understanding of the structure and function of the tissue and organ to be repaired, the structure and properties of the materials used for repair, and the design requirements. There are at present two schools of thought regarding the design of an implant, namely the permanent implant and the *resorbable implant*. The permanent implants are intended to permanently replace the damaged tissues or organs are fabricated from various materials including metals and natural or synthetic polymers. For example, most of the weight-bearing orthopedic and oral implants are made of metals or alloys. Non-weight-bearing tissues and organs are generally replaced with implants that are fabricated either from synthetic or natural materials. Implants for blood vessel, heart valve, and most soft tissue repair fall into this class. Permanent implants, particularly those made of synthetic and biological materials, frequently suffer from the long-term effects of material degradation. Material degradation can result from biological processes such as enzymatic degradation or environmentally induced degradation from mechanical, metal-catalyzed oxidation, and from the permeation of body fluids into the polymeric devices [Bruck, 1991]. The material degradation is particularly manifested in applications where there is repetitive stress-strain on the implant, such as artificial blood vessels and heart valves.

As a result of the lack of suitable materials for long-term implantation, the concept of using a resorbable template to guide host tissue regeneration (guided tissue regeneration) has received vigorous attention in recent years. This area of research can be categorized into synthetic and biological templates. **Polyglycolic acid** (PGA), **polylactic acid** (PLA), polyglycolic-polylactic acid copolymers, and **polydioxanone** are among the polymers most selected for resorbable medical implant development. Among the biological materials used for resorbable medical implant development, *collagen* has been one of the most popular materials in this category. Collagen-based templates have been developed for skin [Yannas and Burke, 1981], peripheral nerve [Li et al., 1990; Yannas et al., 1985], oral tissue [Altman and Li, 1990; Blumenthal, 1988], and meniscal regeneration [Li et al., 1994; Stone et al., 1997]. A variety of other collagen based templates are being developed for tissue repair and regeneration applications [Goldstein et al., 1989; Ma et al., 1990; Li, et al., 1997].

The following discussion is useful in designing a template for tissue repair and regeneration applications. By way of an example, the design parameters listed below are specifically applied to the development of a **resorbable collagen** based template for guiding meniscal tissue repair and regeneration in the knee joint.

Menisci are semilunar fibrocartilages that are anatomically located between the femoral condyles and tibial plateau, providing stability, weight bearing, shock absorption and assisting in lubrication of the knee joint. A major portion of the meniscal tissue is avascular except the peripheral rim, which comprises about 10 to 30% of the total width of the structure and which is nourished by the peripheral vasculature [Arnoczky and Warren, 1982]. Collagen is the major matrix material of the **meniscus**, and the fibers are oriented primarily in the circumferential direction in the line of stress for mechanical function. Repair of damaged meniscal tissue in the peripheral vascular rim can be accomplished with sutures. However, in cases where the injured site is in the avascular region, partial or total removal of the meniscal tissue is often indicated. This is primarily due to the inadequacy of the *fibrochondrocytes* alone to self-repair the damaged meniscal tissue. Studies in animals and humans have shown that removal of the meniscus is a prelude to degenerative knees manifested by the development of **osteoarthritis** [Hede and Sarberg, 1992; Shapiro and Glimcher, 1980]. At present there is no suitable permanent substitute for meniscal tissue.

7.3.1 Biocompatibility

Biocompatibility of the materials and their degraded products is a prerequisite for resorbable implant development. Purified collagen materials have been used either as implants or have been extensively tested in clinical studies as implants without adverse effects. The meniscus template can be fabricated from purified type I collagen fibers that are further crosslinked chemically to increase the stability and reduce the immunogenicity *in vivo*. In addition, small amounts of noncollagenous materials such as glycosaminoglycans and growth factors can be incorporated into the collagen matrix to improve the osmotic properties as well as the rate of tissue ingrowth.

Since the primary structure of a collagen molecule from bovine is homologous to human collagen [Miller, 1984], the *in vivo* degradation of bovine collagen implant should be similar to the normal host tissue remodeling process during wound healing. For a resorbable collagen template, the matrix is slowly degraded by the host over time. It is known that a number of cell types such as **polymorphonuclear leukocytes**, **fibroblasts**, and **macrophages**, during the wound healing period, are capable of secreting enzyme collagenases which cleave a collagen molecule at 1/4 position from the C-terminal end of the molecule [Woolley, 1984]. The enzyme first reduces a collagen molecule to two smaller triple helices which are not stable at body temperature and are subsequently denatured to random coiled polypeptides. These polypeptides are further degraded by proteases into amino acids and short peptides that are metabolized through normal metabolic pathways [Nimni and Harkness, 1988].

Despite the safety record of collagen materials for implantation, during the process of preparing the collagen template, small amounts of unwanted noncollagenous materials could be incorporated into the device such as salts and crosslinking agents. Therefore, a series of biocompatibility testing must be conducted to ensure the residuals of these materials do not cause any safety issues. The FDA has published a new guideline for biocompatibility testing of implantable devices (Biological Evaluation of Medical Devices, 1995).

7.3.2 Physical Dimension

The physical dimension of a template defines the boundary of regeneration. Thus, the size of the collagen template should match the tissue defect to be repaired. A properly sized meniscal substitute has been found to function better than a substitute which mismatches the physical dimension of the host meniscus [Rodkey et al., 1998; Sommerlath et al., 1991]. For a porous, elastic matrix such as the one designed from collagen for meniscal tissue repair, the shape of the meniscus is further defined *in vivo* by the space available between the femoral condyles and tibial plateau within the synovial joint.

7.3.3 Apparent Density

The apparent density as defined as the weight of the dry matrix in a unit volume of matrix. Thus, the apparent density is a direct measure of the empty space which is not occupied by the matrix material per se in the dry state. For example, for a collagen matrix of an apparent density 0.2 g/cm^3, the empty space would be 0.86 cm^3 for a 1 cm^3 total space occupied by the matrix, taking the density of collagen to be 1.41 g/cm^3 [Noda, 1992]. The apparent density is also directly related to the mechanical strength of a matrix. In weight-bearing applications, the apparent density has to be optimized such that the mechanical properties are not compromised for the intended function of the resorbable implant as described in the mechanical properties section.

7.3.4 Pore Structure

The dimension of a mammalian fibrogenic cell body is on the order of 10 to 50 μm, depending on the substrate to which the cell adheres [Folkman et al., 1978]. In order for cells to infiltrate into the interstitial space of a matrix, the majority of the pores must be significantly larger than the dimension of a cell such that both the cell and its cellular processes can easily enter the interstitial space. In a number of studies using collagen-based matrices for tissue regeneration, it has been found that pore size plays an important role in the effectiveness of the collagen matrix to induce host tissue regeneration [Chvapil, 1982; Dagalailis et al., 1980; Yannas, 1996]. It was suggested that pore size in the range of 100 to 400 μm was optimal for tissue regeneration. Similar observations were also found to be true for porous metal implants in total hip replacement [Cook et al., 1991]. The question of interconnecting pores may not be a critical issue in a collagen template as collagenases are synthesized by most **inflammatory cells** during wound healing and remodeling processes. The interporous membranes which exist in the noninterconnecting pores should be digested as part of resorption and wound healing processes.

7.3.5 Mechanical Property

In designing a resorbable collagen implant for weight-bearing applications, not only the initial mechanical strength is important, but the gradual strength reduction of the partially resorbed template has to be compensated by the strength increase from the regenerated tissue such that at any given time point, the total mechanical properties of the template are maintained. In order to accomplish this goal, one must first be certain that the initial mechanical properties are adequate for supporting the weight-bearing application. For example, compressing the implant with multiple body weights should not cause fraying of the collagen matrix material. It is also of particular importance to design an implant having an adequate and consistent suture pullout strength in order to reduce the incidence of detachment of the implant from the host tissue. The suture pullout strength is also important during surgical procedures as the lack of suture pull strength may result in retrieval and reimplantation of the template. In meniscal tissue repair the suture pullout strength of 1 kg has been found to be adequate for arthroscopically assisted surgery in simulated placement procedures in human cadaver knees, and this suture pullout strength should be maintained as the minimal strength required for this particular application.

7.3.6 Hydrophilicity

Hydration of an implant facilitates nutrient diffusion. The extent of hydration would also provide information on the space available for tissue ingrowth. The porous collagen matrix is highly hydrophilic and therefore facilitates cellular ingrowth. The biomechanical properties of the hydrophilic collagen matrix such as fluid outflow under stress, fluid inflow in the absence of stress, and the resiliency for shock absorption are the properties also found in the weight-bearing cartilagenous tissues.

7.3.7 Permeability

The permeability of ions and macromolecules is of primary importance in tissues that do not rely on vascular transport of nutrients to the end organs. The diffusion of nutrients into the interstitial space ensures the survival of the cells and their continued ability of growth and synthesis of tissue specific extracellular matrix. Generally, the permeability of a macromolecule the size of the bovine serum albumin

(MW 67,000) can be used as a guideline for probing accessibility of the interstitial space of a collagen template [Li et al., 1994].

7.3.7.1 *In Vivo* Stability

As stated above, the rate of template resorption and the rate of new tissue regeneration have to be balanced so that the adequate mechanical properties are maintained at all times. The rate of *in vivo* resorption of a collagen-based implant can be controlled by controlling the density of the implant and the extent of intermolecular crosslinking. The lower the density, the greater the interstitial space and generally the larger the pores for cell infiltration, leading to a higher rate of matrix degradation. The control of the extent of intermolecular crosslinking can be accomplished by using bifunctional crosslinking agents under conditions that do not denature the collagen. Glutaraldehyde, formaldehyde, adipyl chloride, hexamethylene diisocyanate, and carbodiimides are among the many agents used in crosslinking the collagen-based implants. Crosslinking can also be achieved through vapor phase of a crosslinking agent. The vapor phase crosslinking is effective using crosslinking agents of high vapor pressures such as formaldehyde and glutaraldehyde. The vapor crosslinking is particularly useful for thick implants of vapor permeable dense fibers where crosslinking in solution produces nonuniform crosslinking. In addition, intermolecular crosslinking can be achieved by heat treatment under high vacuum. This treatment causes the formation of an amide bond between an amino group of one molecule and the carboxyl group of an adjacent molecule and has often been referred to in the literature as dehydrothermal crosslinking.

The shrinkage temperature of the crosslinked matrix has been used as a guide for *in vivo* stability of a collagen implant [Li, 1988]. The temperature of shrinkage of collagen fibers measures the transition of the collagen molecules from the triple helix to a random coil conformation. This temperature depends on the number of intermolecular crosslinks formed by chemical means. Generally, the higher the number of intermolecular crosslinks, the higher the thermal shrinkage temperature and more stable the material *in vivo*.

A second method of assessing the *in vivo* stability is to determine the crosslinking density by applying the theory of rubber elasticity to denatured collagen [Wiederhorn and Beardon, 1952]. Thus, the *in vivo* stability can be directly correlated with the number of intermolecular crosslinks introduced by a given crosslinking agent.

Another method that has been frequently used in assessing the *in vivo* stability of a collagen-based implant is to conduct an *in vitro* collagenase digestion of a collagen implant. Bacterial collagenase is generally used in this application. The action of bacterial collagenase on collagen is different from that of mammalian collagenase [Woolley, 1984]. In addition, the enzymatic activity used in *in vitro* studies is arbitrarily defined. Thus, the data generated from the bacterial collagenase should be viewed with caution. The bacterial collagenase digestion studies, however, are useful in comparing a prototype with a collagen material of known rate of *in vivo* resorption.

Each of the above parameters should be considered in designing a resorbable implant. The interdependency of the parameters must also be balanced for maximal efficacy of the implant.

7.4   Tissue Engineering for Tissue and Organ Regeneration

Biomedical applications of collagen have entered a new era in the past decade. The potential use of collagen materials in medicine has increasingly been appreciated as the science and technology advances.

One major emerging field of biomedical research which has received rigorous attention in recent years is tissue engineering. Tissue engineering is an interdisciplinary science of biochemistry, cell and molecular biology, genetics, materials science, biomedical engineering, and medicine to produce innovative three-dimensional composites having structure/function properties that can be used either to replace or correct poorly functioning components in humans and animals or to introduce better functional components into these living systems. Thus, the field of tissue engineering requires a close collaboration among various disciplines for success.

TABLE 7.6 Survey of Collagen-Based Medical Products, and Research and Development Activities

Applications	Comments
Hemostasis	Commercial Products: Sponge, fiber, and felt forms are used in cardiovascular [Abbott and Austin, 1975]; neurosurgical [Rybock and Long, 1977]; dermatological [Larson, 1988]; ob/gyn [Cornell et al., 1985]; orthopaedic [Blanche and Chaux, 1988]; oral surgical applications [Stein et al., 1985]
Dermatology	Commercial Products: Injectable collagen for soft tissue augmentation [Webster et al., 1984]; collagen based artificial skins [Bell et al., 1981; Yannas and Burke, 1981]. Research and Development: Collagen based wound dressings [Armstrong et al., 1986]
Cardiovascular Surgery and Cardiology	Commercial Products: Collagen coated and gelatin coated vascular grafts [Jonas et al., 1988, Li, 1988]; chemically processed human vein graft [Dardik et al., 1974]; bovine arterial grafts [Sawyer et al., 1977]; porcine heart valves [Angell et al., 1982]; bovine pericardial heart valves [Walker et al., 1983]; vascular puncture hole seal device [Merino et al., 1992]
Neurosurgery	Research and Development: Guiding peripheral nerve regeneration [Archibald et al., 1991; Yannas et al., 1985]; dura replacement material [Collins et al., 1991]
Periodontal and Oral Surgery	Research and Development: Collagen membranes for periodontal ligament regeneration [Blumenthal, 1988]; resorbable oral tissue wound dressings [Ceravalo and Li, 1988]; collagen/hydroxyapatite for augmentation of alveolar ridge [Gongloff and Montgomery, 1985]
Ophthalmology	Commercial Products: Collagen corneal shield to facilitate epithelial healing [Ruffini et al., 1989]. Research and Development: Collagen shield for drug delivery to the eye [Reidy et al., 1990]
Orthopaedic Surgery	Commercial Products: Collagen with hydroxyapatite and autogenous bone marrow for bone repair [Hollinger et al., 1989]. Research and Development: Collagen matrix for meniscus regeneration [Li et al., 1994]; collagenous material for replacement and regeneration of Achilles tendon [Kato et al., 1991]; reconstituted collagen template for ACL reconstruction [Li et al., 1997]
Other Applications	Research and Development: Drug delivery support [Sorensen et al., 1990]; delivery vehicles for growth factors and bioactive macromolecules [Deatherage and Miller, 1987; Li et al., 1996]; collagenous matrix for delivery of cells for tissue and organ regeneration [Bell et al., 1981]

Tissue engineering consists primarily of three components: (1) extracellular matrix, (2) cells, and (3) regulatory signals (e.g., tissue specific growth factors). One of the key elements in tissue engineering is the extracellular matrix which either provides a scaffolding for cells or acts as a delivery vehicle for regulatory signals such as growth factors.

Type I collagen is the major component of the extracellular matrix and is intimately associated with development, wound healing, and regeneration. The development of the type I collagen based matrices described in this review article will greatly facilitate the future development of tissue engineering products for tissue and organ repair and regeneration applications.

To date, collagen-based implants have been attempted for many tissue and organ repair and regeneration applications. A complete historical survey of all potential medical applications of collagen is a formidable task but a selected survey of collagen-based medical products and the research and development activities are summarized in Table 7.6 as a reference.

Defining Terms

Alanine (Ala): One of the amino acids in collagen molecules.

Allysine: The ε-amino group of lysine has been enzymatically modified to an aldehyde group.

Apparent density: Calculated as the weight of the dry collagen matrix per unit volume of matrix.

Arginine (Arg): One of the amino acids in collagen molecules.

Aspartic acid (Asp): One of the amino acids in collagen molecules.

Atelocollagen: A collagen molecule without the telopeptides.

Chondroitin sulfate: Sulfated polysaccharide commonly found in cartilages,bone, corea, tendon, and skin.

Collagen: A family fibrous insoluble proteins having a triple helical conformation extending over a major part of the molecule. Glycine is present at every third amino acid in the triple helix and proline and hydroxyproline are required in the triple helix.

Collagenase: A proteolytic enzyme that specifically catalyzes the degradation of collagen molecules.

Dehydrohydroxylysinonorleucine (deH-HLNL):   A covalently crosslinked product between an allysine and a hydroxylysine residues in collagen fibrils.

D spacing: The repeat distance observed in collagen fibrils by electron microscopic and x-ray diffraction methods.

Elastin: One of the proteins in connective tissue. It is highly stable at high temperatures and in chemicals. It also has rubberlike properties.

Fiber: A bundled group of collagen fibrils.

Fibril: A self-assembled group of collagen molecules.

Fibroblast: Any cell from which connective tissue is developed.

Fibrochondrocyte: Type of cells that are associated with special types of cartilage tissues such as meniscus of the knee and intervertebral disc of the spine.

Fibrous long spacing (FLS):   One of the polymorphic forms of collagen where the collagen molecules are randomly aligned in either head-to-tail, tail-to-tail, or head-to-head orientation.

Gelatin: A random coiled form (denatured form) of collagen molecules.

Glutamic acid (Glu):   One of the amino acids in collagen molecules.

Glycine (Gly):   One of the amino acids in collagen molecules having the simplest structure.

Glycoprotein: A compound consisting of a carbohydrate protein. The carbohydrate is generally hexosamine, an amino sugar.

Glycosaminoglycan (GAG):   A polymerized sugar (see polysaccharide) commonly found in various connective tissues.

Helical pitch: Repeating distance within a single polypeptide chain in a collagen molecule.

Hemostat: Device or medicine which arrests the flow of blood.

Hydrophilicity: The tendency to attract and hold water.

Hydrophobicity: The tendency to repel or avoid contact with water. Substances generally are nonpolar in character, such as lipids and nonpolar amino acids.

Hydroxylysine (Hyl):   One of the amino acids in collagen molecules.

Hydroxyproline (Hyp):   One of the amino acids uniquely present in collagen molecules.

Inflammatory cell: Cells associated with the succession of changes which occur in living tissue when it is injured. These include macrophages, polymorphonuclear leukocytes, and lymphocytes.

Intermolecular crosslink: Covalent bonds formed *in vivo* between a side group of one moiecule and a side group of another molecule; covalent bonds formed between a side group of one molecule and one end of a bifunctional agent and between a side group of a second molecule and the other end of a bifunctional agent.

Intrafibrillar volume: The volume of a fibril excluding the volume occupied by the collagen molecule.

In vitro: In glass, as in a test tube. An *in vitro* test is one done in the laboratory, usually involving isolated tissues, organs, or cells.

In vivo: In the living body or organism. A test performed in a living organism.

Isoelectric point: Generally used to refer to a particular pH of a protein solution. At this pH, there is no net electric charge on the molecule.

Isoleucine (Ile):   One of the amino acids in collagen molecules.

Leucine (Leu):   One of the amino acids in collagen molecules.

Lipid: Any one of a group of fats or fat-like substances, characterized by their insolubility in water and solubility in fat solvents such as alcohol, ether, and chloroform.

Lysine (Lys):   One of the amino acids in collagen molecules.

Meniscus:　A C-shaped fibrocartilage anatomically located between the femoral condyles and tibial plateau providing stability and shock absorption and assisting in lubrication of the knee joint.

Macrophage:　Cells of the reticuloendothelial system having the ability to phagocytose particulate substances and to store vital dyes and other colloidal substances. They are found in loose connective tissues and various organs of the body.

Mycobacterium:　A genus of acid-fast organisms belonging to the Mycobacteriaceae which includes the causative organisms of tuberculosis and leprosy. They are slender, nonmotile, gram-positive rods and do not produce spores or capsules.

Osteoarthritis:　A chronic disease involving the joint, especially those bearing the weight, characterized by destruction of articular cartilage, overgrown of bone with impaired function.

Permeability:　The space within a collagen matrix, excluding the space occupied by collagen molecules, which is accessible to a given size of molecule.

Pepsin:　A proteolytic enzyme commonly found in the gastric juice. It is formed by the chief cells of gastric glands and produces maximum activity at a pH of 1.5 to 2.0.

Phenolalanine (Phe):　One of the amino acids in collagen molecules.

Platelet:　A round or oval disk, 2 to 4 μm in diameter, found in the blood of vertebrates. Platelets contain no hemoglobin.

Polydioxanone:　A synthetic polymer formed from dioxanone monomers which degrades by hydrolysis.

Polyglycolic acid (PGA):　A synthetic polymer formed from glycolic acid monomers which degrades by hydrolysis.

Polylactic acid (PLA):　A synthetic polymer formed from lactic acid monomers which degrades by hydrolysis.

Polymorphism:　Different types of aggregated states of the collagen molecules.

Polymorphonuclear leukocyte:　A white blood cell which possesses a nucleus composed of two or more lobes or parts; a granulocyte (neutrophil, eosinophil, basophil).

Polypeptide:　Polymerized amino acid molecules formed by enzymatically regulated stepwise polymerization *in vivo* between the carboxyl group of one amino acid and the amino group of a second amino acid.

Polysaccharide:　Polymerized sugar molecules found in tissues as lubricant (synovial fluid) or cement (between osteons, tooth root attachment) or complexed with proteins such as glycoproteins or proteoglycans.

Proline (Pro):　One of the amino acids commonly occurring in collagen molecules.

Proteolytic enzyme:　Enzymes which catalyze the breakdown of native proteins.

Resorbable collagen:　Collagen which can be biodegraded *in vivo*.

Salt-linkage:　An electrostatic bond formed between a negative charge group and a positive charge group in collagen molecules and fibrils.

Segment-long-spacing (SLS):　One of the polymorphic forms of collagen where all heads of collagen molecules are aligned in parallel.

Soluble collagen:　Collagen molecules that can be extracted with salts and dilute acids. Soluble collagen molecules contain the telopeptides.

Telopeptide:　The two short nontriple helical peptide segments located at the ends of collagen molecules.

Valine (Val):　One of the amino acids in collagen molecules.

References

Abbott, W.M. and Austin, W.G. 1975. The effectiveness of mechanism of collagen-induced topical hemostasis. *Surgery* 78: 723–729.

Altman, R. and Li, S.T. 1990. Collagen matrix for oral surgical applications. *Int. J. Oral Implantol.* 7: 75.

Angell, W.W., Angell, J.D., and Kosek, J.C. 1982. Twelve year experience with gluteraldehyde preserved porcine xenografts. *J. Thorac Cardiovasc. Surg.* 83: 493–502.

Archibald, S.J., Krarup, C., Shefner, J., Li, S.T., and Madison, R. 1991. Collagen-based nerve conduits are as effective as nerve grafts to repair transected peripheral nerves in rodents and non-human primates. *J. Comp. Neurol.* 306: 685–696.

Armstrong, R.B., Nichols, J., and Pachance, J. 1986. Punch biopsy wounds treated with Monsel's solution or a collagen matrix. *Arch. Dermatol.* 122: 546–549.

Arnoczky, S.P. and Warren, R.F. 1982. Microvasculature of the human meniscus. *Am. J. Sport Med.* 10: 90–95.

Baily, A.J. and Rhodes, D.N. 1964. Irradiation-induced crosslinking of collagen. *Radiat. Res.* 22: 606–621.

Bear, R.S. 1952. The structure of collagen fibrils. *Adv. Prot. Chem.* 7: 69–160.

Bell, E., Ehrlich, H.P., Buttle, D.J., and Nakatsuji, T. 1981. Living tissue formed *in vitro* and accepted as skin equivalent tissue of full thickness. *Science* 211: 1042–1054.

Bellamkonda, R. and Aebischer, P. 1994. Review: tissue engineering in the nerve system. *Biotechnol. Bioeng.* 43: 543–554.

Biological evaluation and medical devices. Use of international standard ISO-10993. Blue Book memorandon G95-1, Rockville, MD, FDA, CDRH, Office of Device Evaluation, May 1, 1995.

Blanche, C. and Chaux, A. 1988. The use of absorbable microfibrillation collagen to control sternal bone marrow bleeding. *Int. Surg.* 73: 42–43.

Blumenthal, N.M. 1988. The use of collagen membranes to guide regeneration of new connective tissue attachment in dogs. *J. Periodontol* 59: 830–836.

Brodsky, B. and Ramshaw, J.A. 1997. The collagen triple-helix structure. *Matrix Biol.* 15: 545–554.

Bruck, S.D. 1991. Biostability of materials and implants. *J. Long-Term Effects Med. Implants* 1: 89–106.

Ceravolo, F. and Li, S.T. 1988. Alveolar ridge augmentation utilizing collagen wound dressing. *Int. J. Oral Implantol.* 4: 15–18.

Chvapil, M. 1982. Considerations on manufacturing principles of a synthetic burn dressing: a review. *J. Biomed. Mater. Res.* 16: 245–263.

Collins, R.L., Christiansen, D., Zazanis, G.A., and Silver, F.H. 1991. Use of collagen film as a dural substitute: Preliminary animal studies. *J. Biomed. Mater. Res.* 25: 267–276.

Comper, W.D. and Veis, A. 1977. Characterization of nuclei in vitro collagen fibril formation. *Biopolymers* 16: 2133–2142.

Cook, S.D., Thomas, K.A., Dalton, J.E., Volkman, T., and Kay, J.F. 1991. Enhancement of bone ingrowth and fixation strength by hydroxyapatite coating porous implants. *Trans. Orthop. Res. Soc.* 16: 550.

Correll, J.T., Prentice, H.R., and Wise, R.C. 1985. Biological investigations of a new absorbable sponge. *Surg. Gynecol. Obstet.* 81: 585–589.

Dagalailis, N., Flink, J., Stasikalis, P., Burke, J.F., and Yannas, I.V. 1980. Design of an artificial skin. III. Control of pore structure. *J. Biomed. Mater. Res.* 14: 511–528.

Daly, C.H. 1966. The Biomechanical Characteristics of Human Skin. Ph.D. thesis, University of Strathclyde, Scotland.

Dardik, H., Veith, F.J., Spreyregen, S., and Dardik I. 1974. Arterial reconstruction with a modified collagen tube. *Ann. Surg.* 180: 144–146.

Deatherage, J.R. and Miller, E.J. 1987. Packaging and delivery of bone induction factors in a collagen implant. *Collagen Rel. Res.* 7: 225–231.

Eastoe, J.E. 1967. Composition of collagen and allied proteins. In *Treatise on Collagen*, G.N. Ramachandran (Ed.), pp. 1–72, Academic Press, New York.

Ellis, D.L. and Yannas, I.V. 1996. Recent advances in tissue synthesis *in vivo* by use of collagen-glycosaminoglycans copolymers. *Biomaterials* 17: 291–299.

Folkman, J. and Moscona, A. 1978. Role of cell shape in growth control. *Nature* 273: 345–349.

Fraenkel-Conrat, H. and Olcott, H.S. 1945. Esterification of proteins with alcohols of low molecular weight. *J. Biol. Chem.* 161: 259–268.

Fraser, R.D.B., MacRae, T.P., Miller, A., and Suzuki, E. 1983. Molecular conformation and packing in collagen fibrils. *J. Mol. Biol.* 167: 497–510.

Fukai, N., Apte, S.S., and Olsen, B.R. 1994. Nonfibrillar collagens. *Meth. Enzymol.* 245: 3–28.

Fung, Y.C. 1993. Bioviscoelastic solids. In *Biomechanics, Mechanical Properties of Living Tissues*, 2nd ed., p. 255, Springer-Verlag, New York.

Goldstein, J.D., Tria, A.J., Zawadsky, J.P., Kato, Y.P., Christiansen, D., and Silver, F.H. 1989. Development of a reconstituted collagen tendon prosthesis. A preliminary implantation study. *J. Bone Joint Surg.* 71-A: 1183–1191.

Gongloff, R.K., Whitlow, W., and Montgomery, C.K. 1985. Use of collagen tubes for implantation of hydroxylapatite. *J. Oral Maxillofac. Surg.* 43: 570–573.

Green, R.W., Ang K.P., and Lam, L.C. 1953. Acetylation of collagen. *Biochem. J.* 54: 181–187.

Guidon, R., Marcean, D., Rao, T.J., Merhi, Y., Roy, P., Martin, L., and Duval, M. 1987. *In vitro* and *in vivo* characterization of an impervious polyester arterial prosthesis: the Gelseal Triaxial graft. *Biomaterials* 8: 433–441.

Hede, A., Larson, E., and Sanberg, H. 1992. The long term outcome of open total and partial meniscectomy related to the quantity and site of the meniscus removed. *Int. Orthop.* 16: 122–125.

Hodge, A.J. and Petruska, J.A. 1963. Recent studies with the electron microscope on the ordered aggregates of the tropocollagen molecule. In *Aspects of Proteins Structure*, G.N. Ramachandran (Ed.), pp. 289–300, Academic Press, New York.

Hofmann, H. and Kuhn, K. 1981. Statistical analysis of collagen sequences with regard to fibril assembly and evolution. In *Structural Aspects of Recognition and Assembly in Biological Macromolecules*, M. Balaban, J.L. Sussman, W. Traub, and A. Yonath (Eds.), pp. 403–425, Balabann ISS, Rehovot and Philadelphia.

Hollinger, J., Mark, D.E., Bach, D.E., Reddi, A.H., and Seyfer, A.E. 1989. Calvarial bone regeneration using osteogenin. *J. Oral Maxillofac. Surg.* 47: 1182–1186.

Hubbell, J.A. 1995. Biomaterials in tissue engineering. *Bio/technology* 13: 565–576.

Jaffe, R. and Deykin, D.J. 1974. Evidence for a structural requirement for the aggregation of platelet by collagen. *Clin. Invest.* 53: 875–883.

Kato, Y.P., Dunn, M.G., Zawadsky, J.P., Tria, A.J., and Silver, F.H. 1991. Regeneration of Achilles tendon with a collagen tendon prosthesis. *J. Bone Joint Surg.* 73-A: 561–574.

Katz, E.P. and Li, S.T. 1972. The molecular organization of collagen in mineralized and nonmineralized tissues. *Biochem. Biophys. Res. Commun.* 3: 1368–1373.

Katz, E.P. and Li, S.T. 1973a. The intermolecular space of reconstituted collagen fibrils. *J. Mol. Biol.* 73: 351–369.

Katz, E.P. and Li, S.T. 1973b. Structure and function of bone collagen fibrils. *J. Mol. Biol.* 80: 1–15.

Katz, E.P. and Li, S.T. 1981. The molecular packing of type I collagen fibrils. In *The Chemistry and Biology of Mineralized Connective Tissues*, A. Veis (Ed.), pp. 101–105, Elsevier, North Holland.

Kemp, P.D., Cavallaro, J.F., and Hastings, D.N. 1995. Effects of carbodiimide crosslinking and load environment on the remodeling of collagen scaffolds. *Tissue Eng.* 1: 71–79.

Kleinman, H.K., Klebe, R.J., and Martin, G.R. 1981. Role of collagenous matrices in the adhesion and growth of cells. *J. Cell Biol.* 88: 473–485.

Larson, P.O. 1988. Topical hemostatic agents for dermatologic surgery. *J. Dermatol. Surg. Oncol.* (14): 623–632.

Li, S.T. 1990. A multi-layered, semipermeable conduit for nerve regeneration comprised of type I collagen, its method of manufacture and a method of nerve regeneration using said conduit. U.S. Patent 4,963,146.

Li, S.T. and Stone, K.R. 1993. Prosthetic ligament. U.S. Patent 5,263,984.

Li, S.T. 1988. Collagen and vascular prosthesis. In *Collagen*, Vol. III., M.E. Nimni (Ed.), pp. 253–271, CRC Press, Boca Raton, FL.

Li, S.T. and Katz, E.P. 1976. An electrostatic model for collagen fibrils: the interaction of reconstituted collagen with Ca^{++}, Na^{+}, and Cl^{-}. *Biopolymers* 15: 1439–1460.

Li, S.T., Archibald, S.J., Krarup, C., and Madison, R.D. 1991. The development of collagen nerve guiding conduits that promote peripheral nerve regeneration. In *Biotechnology and Polymers*, C.G. Gebelein (Ed.), pp. 282–293, Plenum Press, New York.

Li, S.T., Archibald, S.J., Krarup, C., and Madison, R. 1990. Semipermeable collagen nerve conduits for peripheral nerve regeneration. *Polym. Mater. Sci. Eng.* 62: 575–582.

Li, S.T., Golub, E., and Katz, E.P. 1975. On electrostatic side chain complimentarity in collagen fibrils. *J. Mol. Biol.* 98: 835–839.

Li, S.T., Sullman, S., and Katz, E.P. 1981. Hydrogen bonded salt linkages in collagen. In *The Chemistry and Biology of Mineralized Tissues*, A. Veis (ed.), pp. 123–127, Elsevier, North Holland.

Li, S.T., Yuen, D., Li, P.C., Rodkey, W.G., and Stone, K.R. 1994. Collagen as a biomaterial: an application in knee meniscal fibrocartilage regeneration. *Mater. Res. Soc. Symp. Proc.* 331: 25–32.

Li, S.T., Yuen, D., Charoenkul, W., Ulreich, J.B., and Speer, D.P. 1997. A type I collagen ligament for ACL reconstruction. *Trans. Soc. Biomater.* 407.

Li, S.T., Bolton, W., Helm, G., Gillies, G., and Frenkel, S. 1996. Collagen as a delivery vehicle for bone morphogenetic protein (BMP). *Trans. Orthop. Res. Soc.* p 647.

Lindblad, W.J. and Kormos, A.I. 1991. Collagen: a multifunctional family of proteins. *J. Reconstruct. Microsurg.* 7: 37–43.

Ma, S., Chen, G., and Reddi, A.H. 1990. Collaboration between collagenous matrix and osteoginin is required for bone induction. *Ann. NY Acad. Sci.* 580: 524–525.

Merino, A., Faulkner, C., Corvalan, A., and Sanborn, T.A. 1992. Percutaneous vascular hemostasis device for interventional procedures. *Catheterizat. Cardiovasc. Diagn.* 26: 319–322.

Miller A. 1976. Molecular packing in collagen fibrils. In *Biochemistry of Collagen*, G.N. Ramachandran and H. Reddi (Eds.), pp. 85–136, Plenum Press, New York.

Miller, E.J. 1984. Chemistry of the collagens and their distribution. In *Extracellular Matrix Biochemistry*, K.A. Piez and A.H. Reddi (Eds.), pp. 41–82, Elsevier, New York.

Miller, E.J. and Rhodes, R.K. 1982. Preparation and characterization of the different types of collagen. *Meth. Enzymol.* 82: 33–63.

Moghe, P.V., Berthiaume, F., Ezzell, R.M., Toner, M., Tompkins, R.C., and Yarmush, M.L. 1996. Culture matrix configuration and composition in the maintenance of hepatocyte polarity and function. *Biomaterials* 17: 373–385.

Montesano, R., Mouron, P., Amherdt, M., and Orci, L. 1983. Collagen matrix promotes reorganization of pancreatic endocrine cell monolayers into islet-like organoids. *J. Cell Biol.* 97: 935–939.

Nimni M.E., (Ed.) 1988. *Collagen*, Vols. I, II, and III. CRC Press, Boca Raton, FL.

Nimni, M.E. and Harkness, R.D. 1988. Molecular structures and functions of collagen. In *Collagen*, Vol. I, M.E. Nimni (Ed.), pp. 1–78, CRC Press, Boca Raton, FL.

Noda, H. 1972. Partial specific volume of collagen. *J. Biochem.* 71: 699–703.

Park, J.B. and Lakes, R.S. 1992. Structure-property relationships of biological materials. In *Biomaterials: An Introduction*, 2nd ed., pp. 185–222, Plenum Press, New York.

Piez, K.A. 1984. Molecular and aggregate structures of the collagens. In *Extracellular Matrix Biochemistry*, K.A. Piez and A.H. Reddi (Eds.), p. 5, Elsevier, New York

Ramachandran, G.N. 1967. Structure of collagen at the molecular level. In *Treatise on Collagen*, Vol. I, G.N. Ramachandran (Ed.), pp. 103–183, Academic Press, New York, London.

Reidy, J.J., Limberg, M., and Kaufman, H.E. 1990. Delivery of fluorescein to the anterior chamber using the corneal collagen shield. *Ophthalmology* 97: 1201–1203.

Rich, A. and Crick, F.H.C. 1961. The molecular structure of collagen. *J. Mol. Biol.* 3: 483–505.

Rigby, B.J., Hiraci, N., Spikes, J.D., and Eyring H. 1959. The mechanical properties of rat tail tendon. *J. Gen. Physiol.* 43: 265–283.

Rodkey, W.G., Li, S.T., Arnoczky, S.P., McDevitt, C.A., Woo, SL-Y., and Steadman, J.R. 1998. Type I collagen based template for meniscal tissue regeneration: III. In vivo evaluations in humans. (In preparation.)

Ruffini, J.J., Aquavella, J.V., and LoCascio, J.A. 1989. Effect of collagen shields on corneal epithelialization following penetrating keratoplasty. *Ophthal. Surg.* 20: 21–25.

Rybock, J.D. and Long, D.M. 1977. Use of microfibrillar collagen as a topical hemostatic agent in brain tissue. *J. Neurosurg.* 46: 501–505.

Sawyer, P.N., Stanczewski, B., and Kirschenbaum, D. 1977. The development of polymeric cardiovascular collagen prosthesis. *Artif. Organs* 1: 83–91.

Schimpf, W.C. and Rodriquez F. 1976. Fibers from regenerated collagen. *Ind. Eng. Chem. Prod. Res. Rev.* 16: 90–92.

Shapiro, F. and Glimcher, M.J. 1980. Induction of osteoarthrosis in the rabbit knee joint: histologic changes following meniscectomy and meniscal lesions. *Clin. Orthop.* 147: 287–295.

Silbermann, M. 1990. *In vitro* systems for inducers of cartilage and bone development. *Biomaterials* 11: 47–49.

Sittinger, J.B., Bugia, J., Rotter, N., Reitzel, D., Minuth, W.W., and Burmester, G.R. 1996. Tissue engineering and autologous transplant formation: prectical approaches with resorbable biomaterials and new cell culture techniques. *Biomaterials* 17: 237–242.

Sommerlath, K., Gallino, M., and Gillquist, J. 1991. Biomechanical characteristics of different artificial substitutes for the rabbit medial meniscus and the effect of prosthesis size on cartilage. *Trans. Orthop. Res. Soc.* 16: 375.

Sorensen, T.S., Sorensen, A.I., and Merser, S. 1990. Rapid release of gentamicin from collagen sponge. *Acta Orthop. Scand.* 61: 353–356.

Stein, M.D., Salkin, L.M., Freedman, A.L., and Glushko, V. 1985. Collagen sponge as a topical hemostatic agent in mucogingival surgery. *J. Periodontol.* 56: 35–38.

Stenzl, K.H., Miyata, T., and Rubin, A.L. 1974. Collagen as a biomaterial. *Ann. Rev. Biophys. Bioeng.* 3: 231–253.

Stone, K.R., Steadman, J.R., Rodkey, W.R., and Li, S.T. 1997. Regeneration of meniscal cartilage with the use of a collagen scaffold. *J. Bone Joint Surg.* 79-A: 1770–1777.

Tiffit, J.T. 1980. The organic matrix of bone tissue. In *Fundamental and Clinical Bone Physiology*, M.R. Urist (Ed.), p. 51, JB Lippincott Co., PA.

Timpl, R. 1982. Antibodies to collagen and procollagen. *Meth. Enzymol.* 82: 472–498.

van der Rest, M., Dublet, B., and Champliaud, M.F. 1990. Fibril-associated collagens. *Biomaterials* 11: 28–31.

Walker, W.E., Duncan, J.M., Frazier, O.H., Liversay, J.J., Ott, D.A., Reul, G.J., and Cooly, D.A. 1983. Early experience with the Ionescu-Shiley pericardial xenograft valve. *J. Thorac Cardiovasc. Surg.* 86: 570–575.

Wang, C-L, Miyata, T., Weksler, B., Rubin, A., and Stenzel, K.H. 1978. Collagen-induced platelet aggregation and release: critical size and structural requirements of collagen. *Biochim. Biophys. Acta* 544: 568–577.

Webster, R.C., Kattner, M.D., and Smith, R.C. 1984. Injectable collagen for augmentation of facial areas. *Arch. Otolaryngol.* 110: 652–656.

Wiederhorn, N. and Beardon, G.V. 1952. Studies concerned with the structure of collagen: II. Stress-strain behavior of thermally controlled collagen. *J. Polym. Sci.* 9: 315–325.

Wilner, G.D., Nossel, H.L., and Leroy, E.C. 1968. Activation of Hageman factor by collagen. *J. Clin. Invest.* 47: 2608–2615.

Woolley, D.E. 1984. Mammalian Collagenases. In *Extracellular Matrix Biochemistry*, K.A. Piez and A.H. Reddi (Eds.), pp. 119–151, Elsevier, New York.

Yamuchi, M., Katz, E.P., and Mechanic, G.L. 1986. Intermolecular cross-linking and stereospecific molecular packing in type I collagen fibrils of the periodontal ligament. *Biochemistry* 25: 4907–4913.

Yannas, I.V. and Burke, J.F. 1981. Design of an artificial skin. I. Basic design principles. *J. Biomed. Mater. Res.* 14: 65–80.

Yannas, I.V., Orgill, D.P., Silver, J., Norregaad, T., Ervas, N.N., and Schoene, W.C. 1985. Polymeric template facilitates regeneration of sciatic nerve across a 15 mm gap. *Polym. Mater. Sci. Eng.* 53: 216–218.

8

Soft Tissue Replacements

K.B. Chandran
University of Iowa

K.J.L. Burg
Carolinas Medical Center

S.W. Shalaby
Poly-Med, Inc.

8.1 Blood Interfacing Implants

K.B. Chandran

8.1.1 Introduction

Blood comes in contact with foreign materials for a short term in extracorporeal devices such as **dialysers**, **blood oxygenators**, ventricular assist devices, and **catheters**. Long-term vascular implants include heart valve prostheses, **vascular grafts**, and **cardiac pacemakers** among others. In this section, we will be concerned with development of biomaterials for long-term implants, specifically for heart valve prostheses, total artificial heart (**TAH**), and vascular grafts. The primary requirements for biomaterials for long-term implants are biocompatibility, nontoxicity, and durability. Furthermore, the material should be nonirritating to the tissue, resistant to **platelet** and **thrombus** deposition, nondegradable in the physiological environment, and neither absorb blood constituents nor release foreign substances into the blood stream [Shim and Lenker, 1988]. In addition, design considerations include that the implant should mimic the function of the organ that it replaces without interfering with the surrounding anatomical structures and must be of suitable size and weight. The biomaterials chosen must be easily available, inexpensive, easily machinable, sterilizable, and have a long storage life. The selection of material will also be dictated by the strength requirement for the implant being made. As an example, an artificial heart valve prosthesis is required to open and close on an average once every second. The biomaterial chosen must be such that

TABLE 8.1 Heart Valve Prostheses Developed and Currently Available in the U.S.

Type	Name	Manufacturer
Caged ball	Starr-Edwards	Baxter Health Care, Irvine, CA
Tilting disc	Medtronic-Hall	Medtronic Blood Systems, Minneapolis, MN
	Lillehei-Kaster	Medical Inc., Inner Grove Heights, MN
	Omni-Science	
Bileaflet	St. Jude Medical	St. Jude Medical, Inc., St. Paul, MN
	Carbomedics	Carbomedics, Austin, TX
	ATS Valve[a]	ATS Medical, St. Paul, MN
	On-X Valve[a]	Medical Carbon Research Inst., Austin, TX
Porcine bioprostheses	Carpentier-Edwards Standard	Baxter Health Care, Irvine, CA
	Hancock Standard	Medtronic Blood Systems, Santa Ana, CA
	Hancock modified orifice	
	Hancock II	
Pericardial bioprostheses	Carpentier-Edwards	Edwards Laboratories, Santa Ana, CA

[a] FDA approval pending.

the valve is durable and will not fail under **fatigue stress** after implantation in a patient. As sophisticated measurement techniques and detailed computational analyses become available with the advent of super computers, our knowledge on the complex dynamics of the functioning of the implants is increasing. Improvements in design based on such knowledge and improvements in selection and manufacture of biomaterials will minimize problems associated with blood interfacing implants and significantly improve the quality of life for patients with implants. We will discuss the development of biomaterials for the blood interfacing implants, problems associated with the same, and future directions in the development of such implants.

8.1.2 Heart Valve Prostheses

Attempts at replacing diseased natural human valves with prostheses began about four decades ago. The details of the development of heart valve prostheses, design considerations, *in vitro* functional testing, and durability testing of valve prototypes can be found in several monographs [Shim and Lenker, 1988; Chandran, 1992]. The heart valve prostheses can be broadly classified into **mechanical prostheses** (made of non-biological material) and **bioprostheses** (made of biological tissue). Currently available mechanical and tissue heart valve prostheses in the United States are listed in Table 8.1.

8.1.2.1 Mechanical Heart Valves

Lefrak and Starr [1970] describe the early history of mechanical valve development. The initial designs of mechanical valves were of centrally occluding caged ball or caged disc type. The Starr–Edwards caged ball prostheses, commercially available at the present time, was successfully implanted in the mitral position in 1961. The caged ball prostheses is made of a polished Co–Cr alloy (**Stellite 21®**) cage and a silicone rubber ball (Silastic®) which contains 2% by weight barium sulfate for **radiopacity** (Figure 8.1). The valve **sewing rings** use a silicone rubber insert under a knitted composite polytetrafluorethylene (PTFE-**Teflon®**) and **polypropylene** cloth. Even though these valves have proven to be durable, the centrally occluding design of the valve results in a larger pressure drop in flow across the valve and higher **turbulent stresses** distal to the valve compared to other designs of mechanical valve prostheses [Yoganathan et al., 1979a, b; 1986; Chandran et al., 1983]. The relatively large profile design of caged ball or disc construction also increases the possibility of interference with anatomical structures after implantation. The **tilting disc valves**, with improved hemodynamic characteristics, were introduced in the late 1960s. The initial design consisted of a polyacetal (**Delrin®**) disc with a Teflon® sewing ring. Delrin acetal resins are thermoplastic

FIGURE 8.1   A caged-ball heart valve prosthesis. (Courtesy of Baxter Health Care, Irvine, CA.)

polymers manufactured by the polymerization of **formaldehyde** [Shim and Lenker, 1988]. Even though Delrin exhibited excellent wear resistance and mechanical strength with satisfactory performance after more than 20 years of implantation, it was also found to swell when exposed to humid environments such as **autoclaving** and blood contact. To avoid design and manufacturing difficulties due to the swelling phenomenon, the Delrin disc was soon replaced by the **pyrolytic carbon** disc and has become the preferred material for mechanical valve prostheses occluders to date. Pyrolytic carbons are formed in a fluidized bed by pyrolysis of a gaseous hydrocarbon in the range of 1000 to 2400°C. For biomedical applications, carbon is deposited onto a preformed polycrystalline graphite substrate at temperatures below 1500°C (low temperature isotropic pyrolytic carbon, **LTI** Pyrolite®). Increase in strength and wear resistance is obtained by codepositing silicone (up to 10% by weight) with carbon in applications for heart valve prostheses. The pyrolytic carbon discs exhibit excellent blood compatibility, as well as wear and fatigue resistance. The guiding **struts** of tilting disc valves are made of **titanium** or Co–Cr alloys (**Haynes 25®** and Stellite 21®). The Co–Cr based alloys, along with pure titanium and its alloy (Ti6A14V) exhibit excellent mechanical properties as well as resistance to corrosion and thrombus deposition. A typical commercially available tilting disc valve with a pyrolytic carbon disk is shown in Figure 8.2a. A tilting disc valve with the leaflet made of **ultra high molecular weight polyethylene** (Chitra valve — Figure 8.2b) is currently marketed in India. The advantages of **leaflets** with relatively more flexibility compared to pyrolytic carbon leaflets are discussed in Chandran et al. [1994a]. Another new concept in a tilting disc valve design introduced by Reul et al. [1995] has an S-shaped leaflet with leading and trailing edges being parallel to the direction of blood flow. The housing for the valve is nozzle-shaped to minimize flow separation at the inlet and energy loss in flow across the valve. Results from *in vitro* evaluation and animal implantation have been encouraging.

In the late 1970s, a bileaflet design was introduced for mechanical valve prostheses and several different bileaflet models are being introduced into the market today. The leaflets as well as the housing of the bileaflet valves are made of pyrolytic carbon and the bileaflet valves show improved hemodynamic characteristics especially in smaller sizes compared to tilting disc valves. A typical bileaflet valve is shown in Figure 8.3. Design features to improve the hydrodynamic characteristics of the mechanical valves include the opening angle of the leaflets [Baldwin et al., 1997] as well as having an open-pivot design in which the pivot area protrudes into the orifice and is exposed to the washing action of flowing blood [Drogue and Villafana, 1997]. Other design modifications to improve the mechanical valve function include: the use of double polyester (Dacron®) velour material for the suture ring to encourage rapid and controlled tissue ingrowth, and mounting the cuff on a rotation ring which surrounds the orifice ring to protect the cuff mounting mechanism from deeply placed annulus sutures. A PTFE (Teflon®) insert in the cuff provides pliability without excessive drag on the sutures. Tungsten (20% by weight) is incorporated into the leaflet substrate in order to visualize the leaflet motion *in vivo*.

(a)

(b)

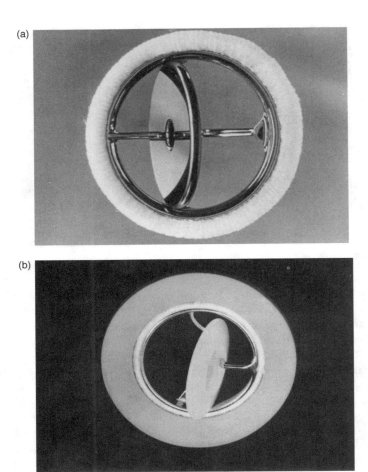

FIGURE 8.2 (a) Photograph of a typical tilting disc valve prosthesis. (Courtesy of Medtronic Heart Valves, Minneapolis, MN.) (b) Chitra tilting disc valve prosthesis with the occluder made of ultra high molecular weight polyethylene. (Courtesy of Sree Chitra Tirunal Institute for Medical Sciences and Technology, India.)

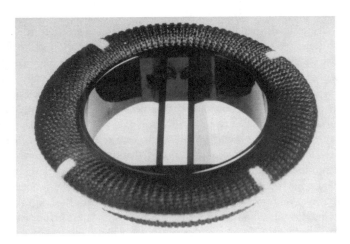

FIGURE 8.3 A CarboMedics bileaflet valve with pyrolytic carbon leaflets and housing. (Courtesy of Sulzer-CarboMedics, Austin, TX.)

(a)

(b)

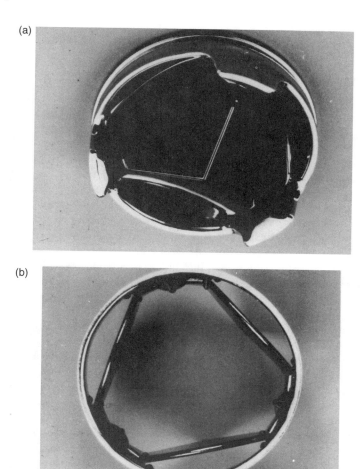

FIGURE 8.4 A tri-leaflet heart valve prosthesis under development. (Courtesy of Triflo Medical, Inc., Costa Mesa, CA.)

Another attempt to design a mechanical valve which mimics the geometry and function of the tri-leaflet aortic valve is that of Lapeyre et al. (1994) (Figure 8.4a, b). The geometry of the valve affords true central flow characteristics with reduced backflow. Accelerated fatigue tests have also shown good wear characteristics for this design and the valve is undergoing further evaluation including animal studies. Other improvements in the mechanical valves which augment performance include: machining of the valve housing to fit a disk so as to produce optimal washing and minimal regurgitation [McKenna, 1997]; a supra-annular design so that a larger-sized valve can be inserted in the aortic position in the case of patients with small aortic annulus [Bell, 1997]; and coating of a titanium alloy ring with a thin, uniform, and strongly adherent film of high-density turbostratic carbon (Carbofilm™) [Bona et al., 1997] in order to integrate the structural stability of the metal alloy to the non-thrombogenecity of pyrolytic carbon. Details of contemporary design efforts in mechanical valve design and potential future biomaterials such as Boralyn® (boron carbide) are discussed in Wieting [1997].

In spite of the desirable characteristics of the biomaterials used in the heart valve prostheses, problems with **thrombo-embolic complications** are significant with implanted valves and patients with mechanical valves are under long-term anticoagulant therapy. The mechanical stresses induced by the flow of blood across the valve prostheses have been linked to the lysis and activation of **formed elements of blood** (red blood cells, white blood cells, and platelets) resulting in the deposition of thrombi in regions with relative stasis in the vicinity of the prostheses. Numerous *in vitro* studies with mechanical valves in pulse duplicators simulating physiological flow have been reported in the literature and have been reviewed by Chandran [1988] and Dellsperger and Chandran [1991]. Such studies have included measurement of velocity profiles and turbulent stresses distal to the valve due to flow across the valve. The aim of these studies has been the correlation of regions prone to thrombus deposition and tissue overgrowth with explanted valves and the experimentally measured bulk turbulent shear stresses as well as regions of relative stasis. In spite of improvements in design of the prostheses to afford a centralized flow with minimal flow disturbances and fluid mechanical stresses, the problems with thrombus deposition remain significant.

Reports of strut failure, material **erosion** and leaflet escapes, as well as **pitting** and erosion of valve leaflets and housing, have resulted in numerous investigations of the **closing dynamics** of mechanical valves. The dynamics of the leaflet motion and its impact with the valve housing or seat stop is very complex and a number of experimental and numerical studies have appeared recently in the literature. As the leaflet impacts against the seat stop and comes to rest instantaneously, high positive and **negative pressure** transients are present on the outflow and inflow side of the occluder, respectively, at the instant when the leaflet impacts against the seat stop or the guiding strut [Leuer, 1986; Chandran et al., 1994a]. The *negative pressure transients* have been shown to reach magnitudes below the **liquid vapor pressure** and have been demonstrated to be a function of the loading rate on the leaflet inducing the valve closure. As the magnitudes of negative pressure transients go below the liquid vapor pressure, **cavitation bubbles** are initiated and the subsequent collapse of the cavitation bubbles may also be a factor in the lysis of red blood cells, platelets, and **valvular structures** [Chandran et al., 1994a; Lee et al., 1994]. Typical cavitation bubbles visualized in an *in vitro* study with tilting disc and bileaflet valves are shown in Figure 8.5. A correlation is also observed between the region where cavitation bubbles are present, even though for a period of time less than a millisecond after valve closure, and sites of pitting and erosion reported in the pyrolytic carbon material in the valve housing and on the leaflets with explanted valves [Kafesjian, 1994] as well as those used in total artificial hearts [Leuer, 1987]. An electron micrograph of pitting and erosion observed in the pyrolytic carbon valve housing of an explanted bileaflet mechanical valve is shown in Figure 8.6. The pressure transients at valve closure are substantially smaller in mechanical valves with a flexible occluder and leaflets made of ultra high molecular weight polyethylene (Figure 8.2b) may prove to be advantageous based on the closing dynamic analysis [Chandran et al., 1994a]. A correlation between the average velocity of the leaflet edge and the negative pressure transients in the same region at the instant of valve closure, as well as the presence of cavitation bubbles has been reported recently [Chandran et al., 1997]. This study demonstrated that for the valves of the same geometry (e.g., tilting disk) and size, the leaflet edge velocity as well as the negative pressure transients were similar. However, the presence of cavitation bubbles depended on the local interaction between the leaflet and the seat stop. Hence, it was pointed out that magnitudes of leaflet velocity or presence of pressure transients below the liquid vapor pressure might not necessarily indicate cavitation inception with mechanical valve closure. Chandran et al. [1998] have also demonstrated the presence of negative pressure transients in the atrial chamber with implanted mechanical valves in the mitral position in animals, demonstrating that potential for cavitation exists with implanted mechanical valves. Similar to the *in vitro* results, the transients were of smaller magnitudes with the Chitra valve made of flexible leaflets, and no pressure transients were observed with tissue valve implanted in the mitral position *in vivo*. The demonstration of the negative pressure transients with mechanical valve closure also shows that this phenomenon is localized and the flow chamber or valve holder rigidity with the *in vitro* experiments will not affect the valve closing dynamics.

The pressure distribution on the leaflets and impact forces between the leaflets and guiding struts have also been experimentally measured in order to understand the causes for strut failure [Chandran et al., 1994b]. The flow through the clearance between the leaflet and the housing at the instant of valve closure [Lee and Chandran, 1994a, b] and in the fully closed position [Reif, 1991] and the resulting wall shear stresses within the clearance are also being suggested as responsible for clinically significant hemolysis and thrombus initiation. Detailed analysis of the complex closing dynamics of the leaflets may also be exploited in improving the design of the mechanical valves to minimize problems with structural failure

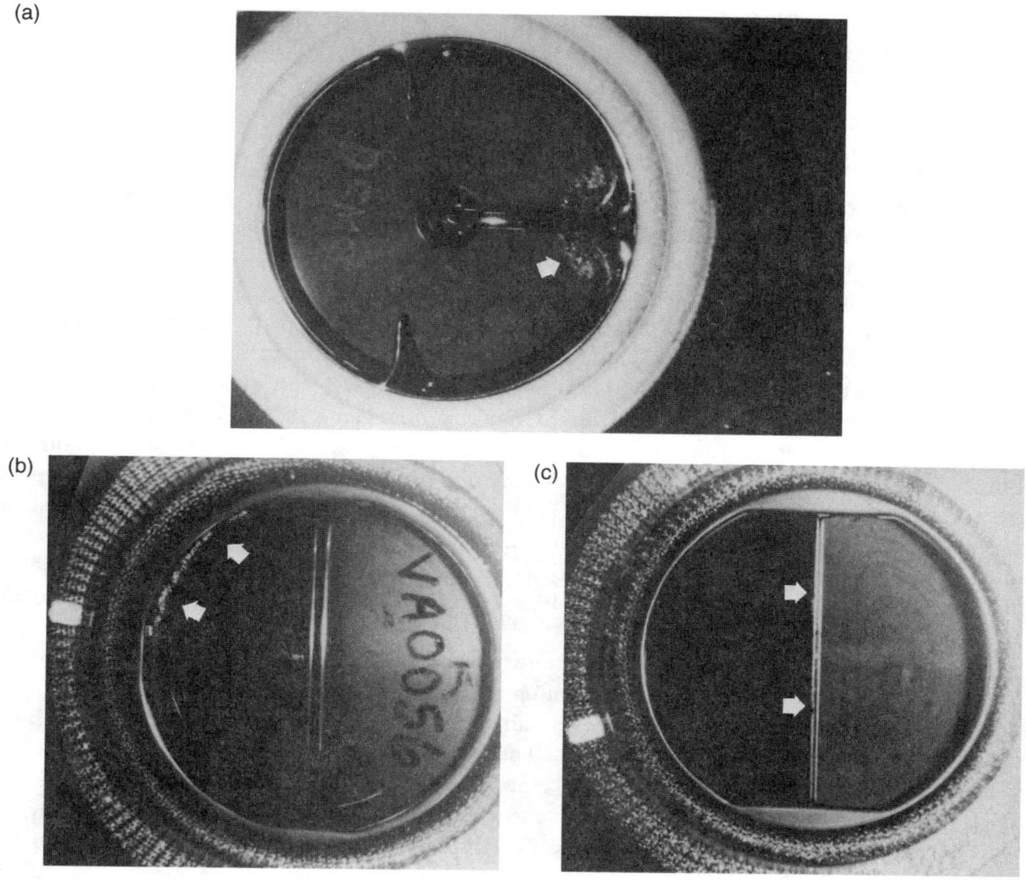

(a)

(b)   (c)

FIGURE 8.5 Cavitation bubbles visualized on the inflow side of the valves *in vitro* [Chandran et al., 1994a]: (a) Medtronic-Hall tilting disc valve; (b) Edwards-Duromedics bileaflet valve; (c) CarboMedics bileaflet valve.

FIGURE 8.6 Photographs showing pitting on pyrolytic carbon surface of a mechanical heart valve. (Courtesy of Baxter Health Care, Irvine, CA.)

[Cheon and Chandran, 1994]. Further improvements in the design of the valves based on the closing dynamics as well as improvements in material may result in minimizing thrombo-embolic complications with implanted mechanical valves.

8.1.2.2 Biological Heart Valves

The first biological valves implanted were **homografts** with valves explanted from cadavers within 48 h after death. Preservation of the valves included various techniques of sterilization, freeze drying, and immersing in antibiotic solution. The use of homografts is not popular due to problems with long term durability and due to limited availability except in a few centers [Shim and Lenker, 1988; Lee and Boughner, 1991]. Attempts were also made in the early 1960s in the use of **xenografts** (valves made from animal tissue) and porcine bioprostheses became commercially available after the introduction of the gluteraldehyde (rather than formaldehyde which was initially used) fixation technique. Gluteraldehyde reacts with tissue proteins to form crosslinks and results in improved durability [Carpentier et al., 1969]. The valves are harvested from 7 to 12 month old pigs and attached to supporting **stents** and preserved. The stent provided support to preserve the valve in the natural shape and to achieve normal opening and closing. Initial supports were made of metal and subsequently flexible polypropylene stents were introduced. The flexible stents provided the advantage of ease of assembling the valve and **finite element analyses** have demonstrated reduction in stresses at the juncture between the stent and tissue leaflets resulting in increased durability and increased leaflet coaptation area [Reis et al., 1971; Hamid et al., 1985]. A typical porcine bioprosthesis is included in Figure 8.7a.

Fixed bovine pericardial tissue is also used to construct heart valves in which design characteristics such as orifice area, valve height, and degree of coaptation can be specified and controlled. Thus, the geometry and flow dynamics past **pericardial prostheses** mimic those of the natural human aortic valves more closely. Due to the low profile design of pericardial prostheses and increased orifice area, these valves are less stenotic compared to porcine bioprostheses, especially in smaller sizes [Chandran et al., 1984]. In the currently available bioprostheses, the stents are constructed from polypropylene, **Acetol®** homopolymer or copolymer, Elgiloy wire, or titanium. A stainless steel radiopaque marker is also introduced to visualize the valve *in vivo*. Other biomaterials, which have been employed in making the bioprostheses, include **fascia lata** tissue as well as human **duramater** tissue. The former was prone to deterioration and hence unsuitable for bioprosthetic application, while the latter lacked commercial availability.

The advantage with bioprostheses is the freedom from thrombo-embolism and hence not requiring long term anticoagulant therapy in general. These prostheses are preferable in patients who do not tolerate anticoagulants. On the other hand, bioprosthetic valves are prone to **calcification** and leaflet tear with an average lifetime of about 10 years before replacement is necessary, and is generally attributed to the tissue fixation process. Numerous attempts are being made to improve the design as well as fixation in bioprostheses in order to minimize problems with calcification and increase duration of the function of the implant. As an example, a bovine pericardial trileaflet valve (Figure 8.7[b]) treated with a non-aldehyde fixation resulting in collagen crosslink formation without a new "foreign" chemical process [Phillips and Printz, 1997] has been introduced in the European market. A non-aldehyde iodine-based sterilization process also sterilizes the valve.

Numerous studies linking the mechanical stresses on the leaflets with calcification, focal thinning, and leaflet failure [Thubrikar et al., 1982a; Sabbah et al., 1985], and design improvements to minimize the stresses on the leaflets [Thubrikar et al., 1982b] have been reported in the literature. Further details on the effects of tissue fixation and mechanical effects of fixation on the leaflets are reported elsewhere [Lee and Boughner, 1991]. Improvements in fixation techniques as well as in design of the bioprostheses are continually being made in order to minimize problems with calcification of the leaflets and improve the durability and functional characteristics of bioprosthetic heart valves [Piwnica and Westaby, 1998]. The biomaterials used in commercially available mechanical and bioprosthetic heart valves are included in Table 8.2. Table 8.3 includes a summary of the problems associated with implanted artificial heart valves.

(a)

(b)

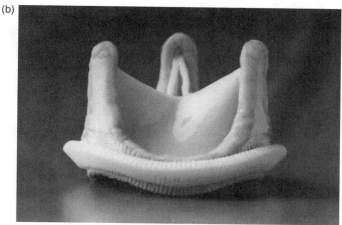

FIGURE 8.7 Typical bioprostheses: (a) Hancock porcine bioprosthesis (courtesy of Medtronic Heart Valves, Minneapolis, MN); (b) PhotoFix™ α pericardial prosthesis. (Courtesy of Sulzer-CarboMedics, Inc., Austin, TX.)

8.1.2.3 Synthetic Heart Valves

Concurrently, efforts have also been made in the development of valve prostheses made of synthetic material. Several attempts to make bileaflet [Braunwald et al., 1960] and trileaflet valves [Roe et al., 1958; Hufnagel, 1977; Gerring et al., 1974; Ghista and Reul, 1977] made of polyurethanes, polyester fabrics, and silicone rubber were not successful due to problems with durability of relatively thin leaflets made of synthetic material. With the advent of the total artificial hearts (TAH) and **left ventricular assist devices** (LVAD) in the 1980s, an additional impetus on the development of synthetic valves is present. Due to problems with thrombus deposition in the vicinity of the mechanical valves used in the TAH and subsequent stroke episodes in patients with permanent implants, the use of the device is currently restricted as a bridge to transplantation. In such temporary use before a donor heart becomes available (on an average of several weeks), the four mechanical prostheses used in the TAH results in substantial cost. Hence, efforts are being made to replace the mechanical valves with those made with synthetic material. With **vacuum forming** or **solution casting** techniques, synthetic valves can be made at a fraction of the cost of mechanical valves, provided their function in a TAH environment for several weeks will be satisfactory. Implantation of synthetic trileaflet valves [Russel et al., 1980; Harold et al., 1987], even more recently, have resulted in limited success due to leaflet failure and calcification. Hemodynamic comparison

TABLE 8.2 Biomaterial Used in Heart Valve Prostheses

Type	Component	Biomaterial
Caged ball	Ball/occluder	Silastic
	Cage	Stellite 21®/Titanium
	Suture ring	Silicone rubber insert under knitted composite Teflon®/polypropylene cloth
Tilting disc	Leaflet	Delrin®; Pyrolytic carbon (carbon deposited on graphite substrate); ultra high molecular polyethylene (UHMPE)
	Housing/strut	Haynes 25®/Titanium
	Suture ring	Teflon®/Dacron®
Bileaflet	Leaflets	Pyrolytic carbon
	Housing	Pyrolytic carbon
	Suture ring	Double velour Dacron® tricot knit polyester
Porcine bioprostheses	Leaflets	Porcine aortic valve fixed by stabilized gluteraldehyde
	Stents	Polypropylene stent covered with Dacron®; lightweight Elgiloy wire covered with porous knitted Teflon® cloth
	Suture ring	Dacron®; soft silicone rubber insert covered with porous, seamless Teflon® cloth
Pericardial bioprostheses	Leaflets	Porcine pericardial tissue fixed by stabilized gluteraldehyde before leaflets are sewn to the valve stents
	Stents	Polypropylene stent covered with Dacron®; Elgiloy wire and nylon support band covered with polyester and Teflon® cloth
	Suture ring	PTFE fabric over silicone rubber filter

Source: Shim and Lenker, 1988; Dellsperger and Chandran, 1988.

TABLE 8.3 Common Problems with Implanted Prosthetic Heart Valves

I. *Mechanical valves*
(a) Thromboembolism
(b) Structural failure
(c) Red blood cell and platelet destruction
(d) Tissue overgrowth
(e) Damage to endothelial lining
(f) Paravalvular/perivalvular leakage
(g) Tearing of sutures
(h) Infection

II. *Bioprosthetic valves*
(a) Tissue calcification
(b) Leaflet rupture
(c) Paravalvular/perivalvular leakage
(d) Infection

Source: Yoganathan et al., 1979a; Shim and Lenker, 1988; Chandran, 1992.

of vacuum formed and solution cast trileaflet valves to currently available bioprostheses have produced satisfactory results [Chandran et al., 1989a, b]. *Finite element analysis* of synthetic valves can be exploited in design improvements similar to those reported for bioprostheses [Chandran et al., 1991a].

8.1.3 Total Artificial Hearts (TAH) or Ventricular Assist Devices (VAD)

Artificial circulatory support can be broadly classified into two categories. The first category is for those patients who undergo open heart surgery to correct **valvular disorders**, ventricular *aneurysm*, or coronary **artery** disease. In several cases, the heart may not recover sufficiently after surgery to take over the pumping action. In such patients ventricular assist devices are used as extracorporeal devices to maintain circulation until the heart recovers. Other ventricular assist devices include **intra-aortic balloon pumps** as well as

cardiopulmonary bypass. Within several days or weeks, when the natural heart recovers, these devices will be removed. In the second category are patients with advanced stages of cardiomyopathy and are subjects for heart transplantation. Due to problems in the availability of suitable donor hearts, not all patients with a failed heart are candidates for heart transplantation. For those patients not selected for transplantation, the concept of replacing the natural heart with a total artificial heart has gained attention in recent years [Akutsu and Kolff, 1958; Jarvick, 1981; DeVries and Joyce, 1983; Unger, 1989; Kambic and Nose, 1991]. A number of attempts in the permanent implantation of TAH with pneumatically powered units were made in the 1980s. However, due to neurological complications as a result of thrombo-embolism, infection, and hematological and renal complications, permanent implantations are currently suspended. If a suitable donor heart is not readily available, TAHs can be used as "bridge to transplantation" for several weeks until a donor heart becomes available. Until recently, most of the circulatory assist devices were pneumatically driven and a typical pneumatic heart is shown in Figure 8.8a. It has two chambers for the left and right ventricle with inlet and outlet valves for each of the chambers. A line coming from the external pneumatic driver passes through the skin and is attached to the diaphragm housing through the connector shown in the photograph. Thus, the patient is tethered to an external pneumatic drive. He can move around for a short period of time by attaching the pneumatic line to a portable driver that he can carry.

Electrically driven blood pumps, which can afford tether-free operation within the body, unlike those of the pneumatically powered pumps, are currently at various stages of development for long-term use (of more than 2 years). The components of such devices include the blood pump in direct contact with blood, energy converter (from electrical to mechanical energy), variable column compensator, implantable batteries, transcutaneous energy transmission system, and external batteries. The blood pump configuration in these devices includes sac, diaphragm, and **pusher plate devices**. Materials used in blood contacting surfaces in these devices are synthetic polymers (polyurethanes, segmented polyurethanes, **Biomer®**, and others). Segmented polyurethane elastomer used in prosthetic ventricles with a thromboresistant additive modifying the polymeric surface have resulted in improved blood compatibility and reduced thrombo-embolic risk in animal trials [Farrar et al., 1988]. Design considerations include reduction of regions of stagnation of blood within the blood chamber and minimizing the mechanical stresses induced on the formed elements in blood. Apart from the characteristics of these materials to withstand repetitive high mechanical stresses and minimize failure due to fatigue, surface interaction with blood is also another crucial factor. An electrically powered total artificial heart intended for long term implantation is shown in Figure 8.8b. The details of the design considerations for the circulatory assist devices are included in Rosenberg [1995a] and details of the evaluation of the electrically powered heart is included in Rosenberg et al. [1995b].

Due to significant problems with thrombo-embolic complications and subsequent neurological problems with long-term implantation of TAH in humans, attention has been focused on minimizing factors responsible for thrombus deposition. In order to eliminate crevices formed with the quick connect system, valves sutured in place at the inflow and outflow orifices were offered as an alternative in the Philadelphia Heart [Wurzel et al., 1988]. An alternative quick connect system using precision machined components has been demonstrated to reduce valve- and connector-associated thrombus formation substantially [Holfert et al., 1987]. Several *in vitro* studies have been reported in the literature in order to assess the effect of fluid dynamic stresses on thrombus deposition [Phillips et al., 1979; Tarbell et al., 1986; Baldwin et al., 1990; Jarvis et al., 1991]. These have included flow visualization and **laser Doppler anemometry** velocity and turbulence measurements within the ventricular chamber as well as in the vicinity of the inflow and outflow orifices. The results of such studies indicate that the flow within the chamber generally has a smooth washout of blood in each pulsatile flow cycle with relatively large turbulent stresses and regions of stasis found near the valves. The thrombus deposition found with implanted TAH in the vicinity of the inflow valves also indicates that the major problem with the working of these devices are still with the flow dynamics across the mechanical valves. Computational flow dynamic analysis within the ventricular chamber may also be exploited to improve the design of the valve chambers and the mechanical valves in order to reduce the turbulent stresses near the vicinity of the inflow and outflow orifices [Kim et al., 1992]. Structural failure of the mechanical valves, initially reported with the TAH may have been the result of

(a)

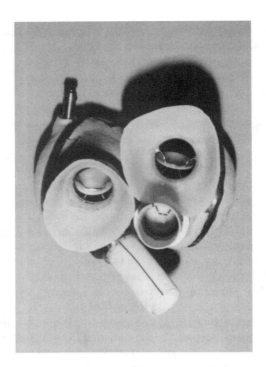

(b)

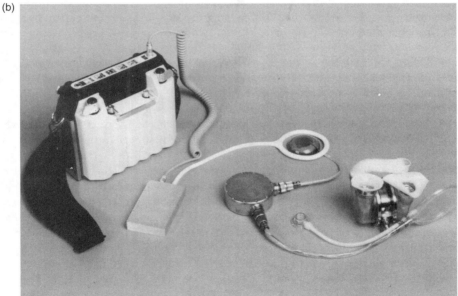

FIGURE 8.8 Typical prototype designs of total artificial hearts: (a) pneumatically powered TAH. The right and left ventricular chambers, inflow and outflow valves, as well as the connector for the pneumatic line are visible in the photograph; (b) electrically powered TAH. Shown are the external battery pack, transcutaneous energy transmission system (TETS) primary and secondary coils, implanted electronics, energy converter and the blood pumps, compliance chamber and the subcutaneous access port. (Courtesy of G. Rosenberg, Pennsylvania State University.)

TABLE 8.4 Classification of Vascular Prostheses

Prosthesis	Comments
Surgically-implanted biological grafts	
Autograft	Graft transplanted from part of a patient's body to another
	Example: saphenous vein graft for peripheral bypass
Allograft	Homograft. Transplanted vascular graft tissues derived from the
	same species as recipient. Example: glutaraldehyde treated umbilical cord vein graft
Xenograft	Heterograft. Surgical graft of vascular tissues derived from one
	species to a recipient of another species. Example: modified bovine heterograft
Surgically-implanted synthetic grafts	
Dacron (polyethylene terephthalate)	Woven, knitted
PTFE (polytetrafluoroethylene)	Expanded, knitted
Other	Nylon, polyurethane

increased load on the valves during closure due to the relatively large dp/dt (p is pressure, t is time) at which the TAH was operated. Attempts at reducing the dp/dt during closure of the inflow valves have also been reported with modified designs of the artificial heart driver [Wurzel et al., 1988]. Due to the relatively large dp/dt at which TAHs are operated, there is increased possibility of cavitation bubble initiation and subsequent collapse of the bubbles may also be another important reason for thrombus deposition near the mechanical valve at the inflow orifice. Introducing synthetic valves to replace the mechanical valves [Chandran et al., 1991b] may prove to be advantageous with respect to cavitation initiation and may minimize thrombus formation.

8.1.4 Vascular Prostheses

In advanced stages of vascular diseases such as obstructive **atherosclerosis** and aneurysmal dilatation, when other treatment modalities fail, replacement of diseased segments with vascular prostheses is a common practice. Vascular prostheses can be classified as given in Table 8.4.

8.1.4.1 Surgically Implanted Biological Grafts

Arterial homografts, even though initially used in large scale, resulted in aneurysm formation especially in the proximal suture line [Strandness and Sumner, 1975]. Still, a viable alternative is to use the saphenous vein graft from the same patient. Vein grafts have a failure rate of about 20% in one year and up to 30% in five years after implantation. Vein grafts from the same patients are also unavailable or unsuitable in about 10 to 30% of the patients [Abbott and Bouchier-Hayes, 1978]. Modified **bovine heterograft** and gluteraldehyde treated **umbilical cord vein grafts** have also been employed as vascular prostheses with less success compared to autologous vein grafts.

8.1.4.2 Surgically Implanted Synthetic Grafts

Prostheses made of synthetic material for vascular replacement have been used for over 40 years. Polymeric material currently used as implants include **nylon**, polyester, **polytetrafluoroethylene (PTFE)**, polypropylene, polyacrylonitrile, and silicone rubber [Park and Lakes, 1992]. However, Dacron® (polyethylene terephthalate) and PTFE are the more common vascular prostheses materials currently available. These materials exhibit the essential qualities for implants — they are biocompatible, resilient, flexible, durable, and resistant to sterilization and biodegradation. Detailed discussion on the properties, manufacturing techniques, and testing of Dacron® prostheses is included in Guidoin and Couture [1992]. Figure 8.9a depicts a Dacron vascular graft having a bifurcated configuration. Figure 8.9b shows expanded PTFE vascular grafts having a variety of configurations and sizes: straight, straight with external reinforcement rings (to resist external compression), and bifurcated.

(a)

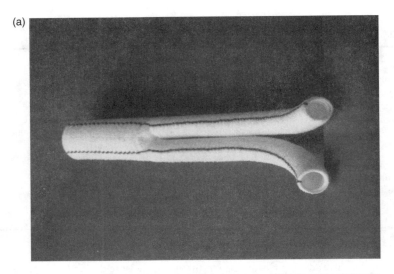

(b)

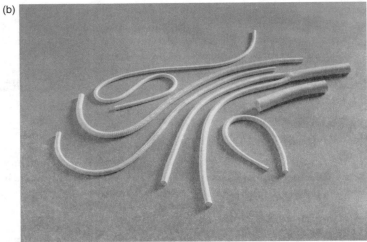

FIGURE 8.9 (a) Photograph of a Dacron vascular graft having a bifurcated configuration. (Courtesy of W.L. Gore and Associates, Inc., Flagstaff, AZ.) (b) Photographs of expanded PTFE vascular grafts with straight, straight with external reinforcement rings to resist compression, and bifurcated configurations. (Courtesy of W.L. Gore and Associates, Inc., Flagstaff, AZ.)

Synthetic vascular grafts implanted as large-vessel replacements have resulted in reasonable degrees of success. However, in medium- and small-diameter prostheses (less than 6 mm in diameter), loss of **patency** within several months after implantation is more acute. Graft failure due to thrombosis or intimal hyperplasia with thrombosis is primarily responsible in failures within 30 days after implantation, and intimal hyperplasia formation is the reason for failure within 6 months after surgery. Soon after implantation, a layer of **fibrin** and fibrous tissue covers the intimal and outer surface of the prosthesis, respectively. A layer of **fibroblasts** replaces the fibrin and is referred to as **neointima**. In the later stages, **neointimal hyperplasia** formation occurs and ultimately results in the occlusion of the vessels in small-diameter vascular grafts. Attempts are being made currently in suitably modifying the surface characteristics of the prostheses in order to reduce the problems with loss of patency. Studies are also being performed in order to understand the mechanical stresses induced at the anastomotic region, which may result in deposits on the intimal surface and occlusion of the vessels [Chandran and Kim, 1994]. The alterations in mechanical stresses with the implantation of vascular prostheses in the arterial circulation may include changes in the deformation and stress concentrations at the anastomotic site. Altered fluid shear stresses at the intimal surface in the vicinity of the anastomosis has also been suggested as important particularly since the loss of patency is present more often at the distal anastomosis. The vascular prostheses should have the same dynamic response after implantation as the host artery in order to reduce the effect of abnormal mechanical stresses at the junction. For a replacement graft of the same size as the host artery, mismatch in **compliance** may be the most important factor resulting in graft failure [Abbott and Bouchier-Hayes, 1978]. In implanting the prostheses, **end-to-end configuration** is common in the reconstruction of peripheral arteries. **End-to-side configuration** is common in coronary artery bypass where blood will flow from the host artery (aorta) to the prosthesis branching out at the anastomotic site. At the other end, the graft is attached distal to the occlusion in the host (coronary) vessel to enable perfusion of the vascular bed downstream from the occlusion. Numerous studies analyzing the abnormal flow dynamics within the anastomotic geometry and stress distribution within the vascular material at the junction to the prostheses have been reported in delineating the causes for intimal hyperplasia formation and loss of patency [Kim and Chandran, 1993; Kim et al., 1993; Ojha et al., 1990; Keynton et al., 1991; Chandran et al., 1992; Rodgers et al., 1987; Rhee and Tarbell, 1994] and a detailed discussion on the mechanical aspects of vascular prostheses can be found in Chandran and Kim [1994]. Improvements in the blood–surface interactions are also being attempted in order to improve the functioning capability of vascular grafts. Attempts at seeding the grafts with **endothelial cells** [Hunter et al., 1983], and modifying the graft material properties by removing the **crimping** and heat fusing a coil of bendable and dimensionally stable polypropylene at the outer surface to make it kink resistant [Guidoin et al., 1983], and employing a compliant and biodegradable graft which will promote regeneration of arterial wall in small caliber vessels [Van der Lei et al., 1985, 1986] are a few examples of such improvements.

8.1.4.3 Transluminally Placed Endovascular Prostheses (Stent-Grafts)

Endoluminal approaches to treating vascular disease involve the insertion of a prosthetic device into the vasculature through a small, often percutaneous, access site created in a remote vessel, followed by the intraluminal delivery and deployment of a prosthesis via transcatheter techniques [Veith et al., 1995]. In contrast to conventional surgical therapies for vascular disease, the use of transluminally placed endovascular prostheses are distinguished by their "minimally invasive" nature. Because these techniques do not require extensive surgical intervention, they have the potential to simplify the delivery of vascular therapy, improve procedural outcomes, decrease procedural costs, reduce morbidity, and broaden the patient population that may benefit from treatment. Not surprisingly, endoluminal therapies have generated intense interest within the vascular surgery, interventional radiology, and cardiology communities over recent years.

The feasibility of using transluminally placed endovascular prostheses, or stent-grafts, for the treatment of traumatic vascular injury [Marin et al., 1994], atherosclerotic obstructions [Cragg and Dake, 1993], and aneurysmal vascular disease [Parodi et al., 1991; Yusuf et al., 1994; Dake et al., 1994] has been demonstrated in human beings. Endoluminal stent-grafts continue to evolve to address a number of cardiovascular pathologies at all levels of the arterial tree. Figure 8.10a depicts endoluminal stent-grafts having a variety of configurations (straight, bifurcated) and functional diameters (peripheral, aortic) that are currently under clinical investigation.

Endoluminal stent-grafts are catheter-deliverable endoluminal prostheses comprised of an intravascular stent component and a biocompatible graft component. The function of these devices is to provide an intraluminal conduit that enables blood flow through pathologic vascular segments without the need for open surgery. The stent component functions as an arterial attachment mechanism and provides structural support to both the graft and the treated vascular segment. By design, stents are delivered to the vasculature in a low profile, small diameter delivery configuration, and can be elastically or plastically expanded to a secondary, large diameter configuration upon deployment. Vascular attachment is achieved by the interference fit created when a stent is deployed within the lumen of a vessel having a diameter smaller than that of the stent. The graft component, on the other hand, is generally constructed from a biocompatible material such as expanded polytetrafluoroethylene (ePTFE), woven polyester (Dacron), or polyurethane. The

(a)

(b)

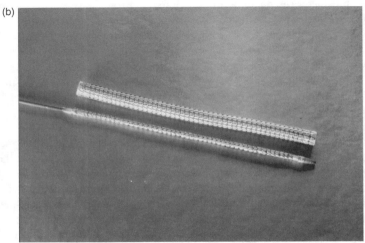

FIGURE 8.10   (a) Endoluminal stent-grafts of straight and bifurcated configurations and sizes currently under clinical investigation. (Courtesy of W.L. Gore and Associates, Inc., Flagstaff, AZ.) (b) A stent-graft implant consisting of an expanded PTFE graft that is externally supported by a self-expanding nitinol stent. (Courtesy of W.L. Gore and Associates, Inc., Flagstaff, AZ.)

graft component has a number of real and theoretical functions, including: segregating potential thromboemboli or atheroemboli from the bloodstream, presenting a physical barrier to mass transport between the bloodstream and arterial wall, and mitigating cellular infiltration and the host inflammatory response. Figure 8.10b shows a stent-graft implant consisting of an expanded polytetrafluoroethylene (ePTFE) graft that is externally supported along its entire length by a self-expanding nitinol stent. The implant is radially constrained and attached to the leading end of a dual lumen polyethylene delivery catheter that allows transluminal delivery and deployment. Following introduction into the vascular system, the implant is positioned fluoroscopically within the diseased segment and released from the delivery system.

Mechanical properties play an important role in determining the *in vivo* performance of an endoluminal stent-graft. Since the graft component typically lacks significant structural integrity, the mechanical behavior of the stent-graft predominantly depends upon the mechanical properties of its stent component. The type of mechanism required to induce dilatation from the delivery (small diameter) configuration, to the deployed (large diameter) configuration typically classifies stents. Self-expanding stents are designed to spontaneously dilate (i.e., elastically recover) from the delivery diameter up to a maximal, pre-determined deployed diameter; whereas balloon-expandable stents are designed to be plastically enlarged over a range of values with the use of appropriately sized and pressurized dilatation balloons. Consequently, self-expanding stents exert a continuous, radially outward directed force on periluminal tissues, while balloon-expandable stents assume a fixed diameter that resists recoil of the surrounding periluminal tissues. Both types of stents exhibit utilitarian features. For example, in comparison to balloon-expandable devices, self-expanding stents can be rapidly deployed without the use of dilatation balloons, are elastic and therefore less prone to external compression, can radially adapt to post-deployment vascular remodeling, and retain some of the natural compliance of the vascular tissues. In contrast, balloon-expandable stents are much more versatile when it comes to conforming to irregular vascular morphologies because their diameter can be radially adjusted via balloon dilatation. Since the luminal diameter of self-expanding stents cannot be adjusted (i.e., enlarged) to any appreciable degree, accurate sizing of the host vessel is critical. A sizing mismatch resulting in oversizing can cause overcompression of the self-expanding stent and obstructive invagination of the stent into the lumen. Undersizing, in turn, can result in a poor interference fit, inadequate anchoring, device migration, and/or leakage of blood into the abluminal compartment. In either case, the stent provides a scaffold that structurally supports the graft material. Ongoing work in the field of biomedical engineering is directed at optimizing the biomechanical and biological performance of these devices.

8.1.5 Conclusions

In the last four decades, we have observed significant advances in the development of biocompatible materials to be used in blood interfacing implants. In the case of mechanical heart valve prostheses, pyrolytic carbon has become the material of choice for the occluder and the housing. The pyrolytic carbon is chemically inert and exhibits very little wear even after more than 20 years of use. However, thrombo-embolic complications still remain significant with mechanical valve implantation. The complex dynamics of valve function and the resulting mechanical stresses on the formed elements of blood appear to be the main cause for initiation of thrombus. More recent reports of structural failure with implanted mechanical valves and pitting and erosion observed on the pyrolytic carbon surfaces have resulted in investigations on cavitation bubble formation during valve closure. Along with further improvements in biomaterials for heart valves, detailed analysis of the closing dynamics and design improvements to minimize the adverse effects of mechanical stresses may be the key to reducing thrombus deposition. Improvements on mechanical heart valves or further developments in durable synthetic leaflet valves may also be vital for the development of TAHs for long-term implantation without neurologic complications.

In the case of vascular grafts, the mismatch of material properties (compliance) between the host artery and the graft, as well as geometric considerations in end-to-side anastomoses, appear to be important for the loss of patency within several months after implantation particularly with medium and small diameter arterial replacement. Most of the vascular grafts are stiffer compared to the host artery and it has been suggested that the mechanical stresses resulting from the discontinuity at the junction is the major cause for neointimal hyperplasia formation and subsequent occlusion of the conduit. Developments with more compliant grafts and in modifying the surface interaction of the graft with blood (endothelialization or other treatment of the graft material) may result in reducing the problems with loss of patency. Recent advances in the use of minimally invasive stent-grafts also show promise in improving the quality of life of patients with vascular disease.

Defining Terms

Acetol: Product of the addition of two moles of alcohol to one of an aldehyde.
Aneurysms: Abnormal bulging or dilatation of a segment of a blood vessel or myocardium.
Artery: Blood vessel transporting blood in a direction away from the heart.
Atherosclerosis: Lipid deposits in the intima of arteries.

ATS valve: A bileaflet mechanical valve made by ATS (Advancing The Standard) Inc.

Autoclaving: Sterilizing by steam under pressure.

Biomer®:   Segmented polyurethane elastomer.

Bioprostheses: Prosthetic heart valves made of biological tissue.

Blood oxygenators: Extracorporeal devices to oxygenate blood during heart bypass surgery.

Bovine heterograft: Graft material (arterial) transplanted from bovine species.

Calcification: Deposition of insoluble salts of calcium.

Cardiac pacemakers: Prosthesis implanted to stimulate cardiac muscles to contract.

Cardiopulmonary bypass: Connectors bypassing circulation to the heart and the lungs.

Catheters: Hollow cylindrical tubing to be passed through the blood vessels or other canals.

Cavitation bubbles (vapor cavitation):   Formation of vapor bubbles due to transient reduction in pressure to below the liquid vapor pressure.

Closing dynamics: Dynamics during the closing phase of heart valves.

Compliance: A measure of ease with which a structure can be deformed; ratio of volumetric strain to increase in unit pressure.

Crimping: Creasing of the synthetic vascular grafts in the longitudinal direction to accommodate the large intermittent flow of blood.

Delrin®:   Polyacetal made by Union Carbide.

Dialysers: Devices to filter the blood of waste products taking over the function of the kidney.

dp/dt: Slope of the pressure vs. time curve of the ventricles.

Duramater: A tough fibrous membrane forming the outer cover of the brain and the spinal cord.

Electrohydraulic blood pump: Blood pumps energized by the conversion of electrical to hydraulic energy.

End-to-end configuration: End of the vascular graft anastamosed to the end of the host artery.

End-to-side configuration: End of the vascular graft anastamosed to the side of the host.

Endothelial cells: A layer of flat cells lining the intimal surface of blood vessels.

Erosion: A state of being worn away.

Fascia lata: A sheet of fibrous tissue enveloping the muscles of the thigh.

Fatigue stress: Level of stress below which the material would not undergo fatigue failure (107 cycles is used as the normal limit).

Fibrin: An elastic filamentous protein derived from fibrinogen in coagulation of the blood.

Fibroblasts: An elongated cell with cytoplasmic processes present in connective tissue capable of forming collagen fibers.

Finite element analysis: Structural analysis with the aid of a computer which divides the structure into finite elements and applies the laws of mechanics on each element.

Formaldehyde: Formic aldehyde, methyl aldehyde, a pungent gas used as antiseptic.

Formed elements in blood: Red blood cells, white blood cells, platelets, and other cells in whole blood.

Haynes 25®:   Co–Cr alloy.

Homografts: Transplants (heart valves, arterial segments, etc.) from the same species.

Intra-aortic balloon pumps: A balloon catheter inserted in the descending aorta and alternately inflated and deflated timed to the EKG in order to assist the ventricular pumping.

Laser Doppler anemometry: A velocity measurement device using the principle of Doppler shifted frequency of laser light by particles moving with the fluid.

Leaflets: Occluders on valves which open and close to aid blood flow in one direction.

Left ventricular assist devices: Prosthetic devices to assist the left ventricle in pumping blood.

Liquid vapor pressure: Pressure at which liquid vaporizes.

LTI: Low temperature (below 1500°C) isotropic pyrolytic carbon.

Mechanical prostheses: Prostheses made of non-biological material.

Negative pressure transients: Reduction in pressure for a short duration.

Neointima: Newly formed intimal surface.

Neointimal hyperplasia: Growth of new intimal surface formed by fibroblasts.

Nylon: Synthetic polymer with condensation polymerization.

Patency: State of being freely open.

Pericardial prostheses: Heart valve prosthesis made with fixed bovine pericardial tissue.

Pitting: Depression or indent on a surface.

Platelet: One of the formed elements of blood responsible for blood coagulation.

Polypropylene: One of the vinyl polymers with good flex life and good environmental stress crack resistance.

Polytetrafluoroethylene (PTFE):   A fluorocarbon polymer known as Teflon®.

Pusher plate devices: Artificial heart devices working with pusher plates moving the blood.

Pyrolytic carbon: Carbon deposited onto preformed polycrystalline graphite substrate.

Radiopacity: Being opaque to x-ray.

Sewing rings: Rings surrounding the housing of artificial heart valves used to sew the valve to the tissue orifice with suture.

Solution casting: Casting by pouring molten material on dyes to form a structure.

Stellite 21®:   Co–Cr alloy.

Stent: A device used to maintain the bodily orifice or cavity.

Strut: A projection in the structure such as guiding struts in heart valves used to guide the leaflets during opening and closing.

TAH: Total artificial heart replacing a failed natural heart.

Teflon®:   See PTFE.

Thrombo-embolic complications: Complications due to breaking away (emboli) of thrombus blocking the distal blood vessels.

Thrombus: A clot in the blood vessels or in the cavities of the heart formed from the constituents of blood.

Tilting disc valves: Valves with a single leaflet tilting open and shut.

Titanium: Highly reactive metal having low density, good mechanical properties, and biocompatibility due to tenacious oxide layer formation.

Turbulent stresses: Stresses generated in the fluid due to agitated random motion of particles.

Ultra high molecular weight polyethylene: Linear thermoplastics with very high molecular weight ($>2 \times 10^6$ g/mol) used for orthopedic devices such as acetabular cup for hip joint replacement.

Umbilical cord vein grafts: Vascular graft made from umbilical cord veins.

Vacuum forming: A manufacturing technique for thermoplastic polymer in which a sheet is heated and formed over a mold while a vacuum is present under the sheet.

Valvular disorders: Diseased states of valves such as stenosis.

Valvular structures: Components of valves such as leaflets, struts, etc.

Vascular grafts: Grafts to replace segments of diseased vessels.

Xenografts: Grafts obtained from species other than that of the recipient.

References

Abbott, W.M. and Bouchier-Hayes, D.J. 1978. The role of mechanical properties in graft design. In *Graft Materials in Vascular Surgery*, Dardick, H., Ed. Year Book Medical Publishers, Chicago, IL, pp. 59–78.

Akutsu, T. and Kolff, W.J. 1958. Permanent substitutes for valves and hearts. *Trans. Am. Soc. Art. Intern. Organs (ASAIO)* 4: 230–235.

Baldwin, J.T., Tarbell, J.M., Deutsch, S., Geselowitz, D.B., and Rosenberg, G. 1988. Hot-film wall shear probe measurements inside a ventricular assist device. *Am. Soc. Mech. Eng. (ASME) J. Biomech. Eng.* 110: 326–333.

Baldwin, J.T., Campbell, A., Luck, C., Ogilvie, W., and Sauter, J. 1997. Hydrodynamics of the CarboMedics® aortic kinetic™ prosthetic heart valve. In *Surgery for Acquired Aortic Valve Disease*. Piwnica, A. and Westaby, S., Eds., ISIS Medical Media, Oxford, pp. 365–370.

Bell, R.S. 1997. CarboMedics® supra-annular Top Hat™ aortic valve. In *Surgery for Acquired Aortic Valve Disease*. Piwnica, A. and Westaby, S., Eds., ISIS Medical Media, Oxford, pp. 371–375.

Bona, G., Rinaldi, S., and Vallana, F. 1997. Design characteristics of the BICARBON™ bileaflet heart valve prosthesis. In *Surgery for Acquired Aortic Valve Disease*. Piwnica, A. and Westaby, S. Eds., ISIS Medical Media, Oxford, pp. 392–396.

Braunwald, N.S., Cooper, T., and Morrow, A.G. 1960. Complete replacement of the mitral valve: successful application of a flexible polyurethane prosthesis. *J. Thorac. Cardiovasc. Surg.* 40: 1–11.

Carpentier, A., Lamaigre, C.G., Robert, L., Carpentier, S., and Dubost, C. 1969. Biological factors affecting long-term results of valvular heterografts. *J. Thorac. Cardiovasc. Surg.* 58: 467–483.

Chandran, K.B., Cabell, G.N., Khalighi, B., and Chen, C.J. 1983. Laser anemometry measurements of pulsatile flow past aortic valve prostheses. *J. Biomech.* 16: 865–873.

Chandran, K.B., Cabell, G.N., Khalighi, B., and Chen, C.J. 1984. Pulsatile flow past aortic valve bioprostheses in a model human aorta. *J. Biomech.* 17: 609–619.

Chandran, K.B. 1988. Heart valve prostheses: *in vitro* flow dynamics. In *Encyclopedia of Medical Devices and Instrumentation*, Vol. 3. Webster, J.G., Ed., Wiley Interscience, New York, pp. 1475–1483.

Chandran, K.B., Fatemi, R., Schoephoerster, R., Wurzel, D., Hansen, G., Pantalos, G., Yu, L.-S., and Kolff, W.J. 1989a. *In vitro* comparison of velocity profiles and turbulent shear distal to polyurethane trileaflet and pericardial prosthetic valves. *Artif. Organs* 13: 148–154.

Chandran, K.B., Schoephoerster, R.T., Wurzel, D., Hansen, G., Yu, L.-S., Pantalos, G., and Kolff, W.J. 1989b. Hemodynamic comparison of polyurethane trileaflet and bioprosthetic heart valves. *Trans. Am. Soc. Artif. Intern. Organs (ASAIO)* 35: 132–138.

Chandran, K.B., Kim, S.-H., and Han, G. 1991a. Stress distribution on the cusps of a polyurethane trileaflet heart valve prosthesis in the closed position. *J. Biomech.* 24: 385–395.

Chandran, K.B., Lee, C.S., Shipkowitz, T., Chen, L.D., Yu, L.S. and Wurzel, D. 1991b. *In vitro* hemodynamic analysis of flexible artificial ventricle. *Artif. Organs* 15: 420–426.

Chandran, K.B., Gao, D., Han, G., Baraniewski, H., and Corson, J.D. 1992. Finite element analysis of arterial anastomosis with vein, Dacron® and PTFE grafts. *Med. Biol. Eng. Comp.* 30: 413–418.

Chandran, K.B. 1992. *Cardiovascular Biomechanics*. New York University Press, New York.

Chandran, K.B., Lee, C.S., and Chen, L.D. 1994a. Pressure field in the vicinity of mechanical valve occluders at the instant of valve closure: correlation with cavitation initiation. *J. Heart Valve Dis.* 3 (Suppl. 1): S 65–S 76.

Chandran, K.B., Lee, C.S., Aluri, S., Dellsperger, K.C., Schreck, S., and Wieting, D.W. 1994b. Pressure distribution near the occluders and impact forces on the outlet struts of Björk–Shiley convexo-concave valves during closing. *J. Heart Valve Dis.* 5: 199–206.

Chandran, K.B. and Kim, Y.H. 1994. Mechanical aspects of vascular graft-host artery anastomoses. *IEEE Eng. Med. Biol. Mag.* 13: 517–524.

Chandran, K.B. and Aluri, S. 1997. Mechanical valve closing dynamics. Relationship between velocity of closing, pressure transients, and cavitation initiation. *Ann. Biomed. Eng.* 25: 926–938.

Chandran, K. B., Dexter, E. U., Aluri, S., and Richenbacher, W.E. 1998. Negative pressure transients with mechanical heart-valve closure: Correlation between *in vitro* and *in vivo* results. *Ann. Biomed. Eng.* 26: 546–556.

Cheon, G.J. and Chandran, K.B. 1994. Transient behavior analysis of a mechanical monoleaflet heart valve prosthesis in the closing phase. *Am. Soc. Mech. Eng. J. Biomech. Eng.* 116: 452–459.

Cragg A.H. and Dake M.D. 1993. Percutaneous femoropopliteal graft placement. *Radiology* 187: 643–648.

Dake M.D., Miller D.C., Semba C.P. et al. 1994. Transluminal placement of endovascular stent-grafts for the treatment of descending thoracic aortic aneurysms. *N. Engl. J. Med.* 331: 1729–34.

Dellsperger, K.C. and Chandran, K.B. 1991. Prosthetic heart valves. In *Blood Compatible Materials and Devices. Perspectives towards the 21st Century*. Sharma, C.P. and Szycher, M., Eds., Technomic Publishing Company Inc., Lancaster, PA, pp. 153–165.

DeVries, W.C. and Joyce, L.D. 1983. The artificial heart. *CIBA Clin. Symp.*, 35.

Drogue, J., and Villafana, M. 1997. ATS Medical open pivot™ valve. In *Surgery for Acquired Aortic Valve Disease*. Piwnica, A. and Westaby, S., Eds., ISIS Medical Media, Oxford, pp. 410–416.

Farrar, D.J., Litwak, P., Lawson, J.H., Ward, R.S., White, K.A., Robinson, A.J., Rodvein, R., and Hill, J.D. 1988. *In vivo* evaluations of a new thromboresistant polyurethane for artificial heart blood pumps. *J. Thorac. Cardiovasc. Surg.* 95: 191–200.

Gerring, E.L., Bellhouse, B.J., Bellhouse, F.H., and Haworth, F.H. 1974. Long term animal trials of the Oxford aortic/pulmonary valve prosthesis without anticoagulants. *Trans. ASAIO* 20: 703–708.

Ghista, D.N. and Reul, H. 1977. Optimal prosthetic aortic leaflet valve: design, parametric and longevity analysis: Development of the avcothane-51 leaflet valve based on the optimal design analysis. *J. Biomech.* 10: 313–324.

Guidoin, R., Gosselin, C., Martin, L., Marios, M., Laroche, F., King, M., Gunasekara, K., Domurado, D., and Sigot-Luizard, M.F. 1983. Polyester prostheses as substitutes in the thoracic aorta of dogs. I. Evaluation of commercial prostheses. *J. Biomed. Mater. Res.* 17: 1049–1077.

Guidoin, R. and Couture, J. 1991. Polyester prostheses: The outlook for the future. In *Blood Compatible Materials and Devices. Perspectives towards the 21st Century*. Sharma, C.P. and Szycher, M., Eds., Technomic Publishing Company Inc., Lancaster, PA, pp. 153–165.

Hamid, M.S., Sabbah, H.N., and Stein, P.D. 1985. Finite element evaluation of stresses on closed leaflets of bioprosthetic heart valves with flexible stents. *Finite Elem. Anal. Des.* 1: 213–225.

Harold, M., Lo, H.B., Reul, H., Muchter, H., Taguchi, K., Gierspien, M., Birkle, G., Hollweg, G., Rau, G., and Messmer, B.J. 1987. The Helmholtz Institute tri-leaflet polyurethane heart valve prosthesis: Design, manufacturing, and first *in vitro* and *in vivo* results. In *Polyurethanes in Biomedical Engineering II*. Planck, H., Syre, I., and Dauner, M., Eds., Elsevier Publishing Co., Amsterdam, pp. 321–356.

Holfert, J.W., Reibman, J.B., Dew, P.A., De Paulis, R., Burns, G.L., and Olsen, D.B. 1987. A new connector system for total artificial hearts: preliminary results. *Trans. ASAIO* 10: 151–156.

Hunter, G.C., Schmidt, S.P., Sharp, W.V., and Malindzak, G.S. 1983. Controlled flow studies in 4 mm endothelialized Dacron® grafts. *Trans. ASAIO* 29: 177–182.

Hufnagel, C.A. 1977. Reflections on the development of valvular prostheses. *Med. Instrum.* 11: 74–76.

Jarvick, R.K. 1981. The total artificial heart. *Sci. Am.* 244: 66–72.

Jarvis, P., Tarbell, J.M., and Frangos, J.A. 1991. An *in vitro* evaluation of an artificial heart. *Trans. ASAIO* 37: 27–32.

Kafesjian, R., Howanec, M., Ward, G.D., Diep, L., Wagstaff, L.S., and Rhee, R. 1994. Cavitation damage of pyrolytic carbon in mechanical heart valves. *J. Heart Valve Dis.* 3 (Suppl. 1): S 2–S 7.

Kambic, H.E. and Nose, Y. 1991. Biomaterials for blood pumps. In *Blood Compatible Materials and Devices. Perspectives Towards the 21st Century*. Sharma, C.P. and Szycher, M., Eds., Technomic Publishing Company Inc., Lancaster, PA, pp. 141–151.

Keynton, R.S., Rittgers, S.E., and Shu, M.C.S. 1991. The effect of angle and flow rate upon hemodynamics in distal vascular graft anastomoses: An *in vitro* model study. *ASME J. Biomech. Eng.* 113: 458–463.

Kim, S.H., Chandran, K.B., and Chen, C.J. 1992. Numerical simulation of steady flow in a two-dimensional total artificial heart model. *ASME J. Biomech. Eng.* 114: 497–503.

Kim, Y.H., Chandran, K.B., Bower, T.J., and Corson, J.D. 1993. Flow dynamics across end-to-end vascular bypass graft anastomoses. *Ann. Biomed. Eng.* 21: 311–320.

Kim, Y.H. and Chandran, K.B. 1993. Steady flow analysis in the vicinity of an end-to-end anastomosis. *Biorheol.* 30: 117–130.

Lapeyre, D.M., Frazier, O.H., and Conger, J.L. 1994. *In vivo* evaluation of a trileaflet mechanical heart valve. *ASAIO J.* 40: M707–M713.

Lee, C.S., Chandran, K.B., and Chen, L.D. 1994. Cavitation dynamics of mechanical heart valve prostheses. *Artif. Organs* 18: 758–767.

Lee, C.S. and Chandran, K.B. 1994. Instantaneous backflow through peripheral clearance of Medtronic Hall valve at the moment of closure. *Ann. Biomed. Eng.* 22: 371–380.

Lee, C.S. and Chandran, K.B. 1995. Numerical simulation of instantaneous backflow through central clearance of bileaflet mechanical heart valves at the moment of closure: shear stress and pressure fields within the clearance. *Med. Biol. Eng. Comp.* 33: 257–263.

Lee, J.M. and Boughner, D.R. 1991. Bioprosthetic heart valves: Tissue mechanics and implications for design. In *Blood Compatible Materials and Devices. Perspectives Towards the 21st Century*. Sharma, C.P. and Szycher, M., Eds., Technomic Publishing Company Inc., Lancaster, PA, pp. 167–188.

Lefrak, E.A. and Starr, A., Eds. 1970. *Cardiac Valve Prostheses*. Appleton-Century-Crofts, New York.

Leuer, L. 1987. Dynamics of mechanical valves in the artificial heart. *Proc. 40th Ann. Conf. Eng. Med. Biol. (ACEMB)*, p. 82.

Marin, M.L., Veith, F.J., Panetta, T.F. et al. 1994. Transluminally placed endovascular stented graft repair for arterial trauma. *J. Vasc. Surg.* 20: 466–73.

McKenna, J. 1997. The Ultracor™ prosthetic heart valve. In *Surgery for Acquired Aortic Valve Disease*. Piwnica, A. and Westaby, S., Eds., ISIS Medical Media, Oxford, pp. 337–340.

Ojha, M., Ethier, C.R., Johnston, K.W., and Cobbold, R.S.C. 1990. Steady and pulsatile flow fields in an end-to-side arterial anastomosis model. *J. Vasc. Surg.* 12: 747–753.

Park, J.B. and Lakes, R.S. 1992. *Biomaterials: An Introduction*, 2nd ed., Plenum Press, New York.

Parodi, J.C., Palmaz, J.C., and Barone, H.D. 1991. Transfemoral intraluminal graft implantation for abdominal aortic aneurysms. *Ann. Vasc. Surg.* 5: 491–9.

Phillips, W.M., Brighton, J.A., and Pierce, W.S. 1979. Laser Doppler anemometer studies in unsteady ventricular flows. *Trans. ASAIO* 25: 56–60.

Phillips, R.E., and Printz, L.K. 1997. PhotoFix™ α: a pericardial aortic prosthesis. In *Surgery for Acquired Aortic Valve Disease*. Piwnica, A. and Westaby, S., Eds., ISIS Medical Media, Oxford, pp. 376–381.

Piwnica, A. and Westaby, S., Eds. 1997. *Surgery for Acquired Aortic Valve Disease*. ISIS Medical Media, Oxford.

Reif, T.H. 1991. A numerical analysis of the back flow between the leaflets of a St. Jude Medical cardiac valve prosthesis. *J. Biomech.* 24: 733–741.

Reis, R.L., Hancock, W.D., Yarbrough, J.W., Glancy, D.L., and Morrow, A.G. 1971. The flexible stent. *J. Thorac. Cardiovasc. Surg.* 62: 683–691.

Reul, H., Steinseifer, U., Knoch, M., and Rau, G. 1995. Development, manufacturing and validation of a single leaflet mechanical heart valve prosthesis. *J. Heart Valve Dis.* 4: 513–519.

Rhee, K. and Tarbell, J.M. 1994. A study of wall shear rate distribution near the end-to-end anastomosis of a rigid graft and a compliant artery. *J. Biomech.* 27: 329–338.

Rodgers, V.G.J., Teodori, M.F., and Borovetz, H.S. 1987. Experimental determination of mechanical shear stress about an anastomotic junction. *J. Biomech.* 20: 795–803.

Roe, B.B., Owsley, J.W., and Boudoures, P.C. 1958. Experimental results with a prosthetic aortic valve. *J. Thorac. Cardiovasc. Surg.* 36: 563–570.

Rosenberg, G. 1995a. Artificial heart and circulatory assist devices. In *The Biomedical Engineering Handbook*. Bronzino, J.D., Ed., CRC Press, Boca Raton, FL, pp. 1839–1846.

Rosenberg, G., Snyder, A.J., Weiss, W.J., Sapirstein, J.S., and Pierce, W.S. 1995b. *In vivo* testing of a clinical-size totally implantable artificial heart. In *Assisted Circulation 4*. F. Unger, Ed., Springer-Verlag, Berlin, pp. 235–248.

Russel, F.B., Lederman, D.M., Singh, P.I., Cumming, R.D., Levine, F.H., Austen, W.G., and Buckley, M.J. 1980. Development of seamless trileaflet valves. *Trans. ASAIO* 26: 66–70.

Sabbah, H.N., Hamid, M.S., and Stein, P.D. 1985. Estimation of mechanical stresses on closed cusps of porcine bioprosthetic valves: effect of stiffening, focal calcium and focal thinning. *Am. J. Cardiol.* 55: 1091–1097.

Shim, H.S. and Lenker, J.A. 1988. Heart valve prostheses. In *Encyclopedia of Medical Devices and Instrumentation*, Vol. 3. Webster, J.G., Ed., Wiley Interscience, New York, pp. 1457–1474.

Strandness, D.E. and Sumner, D.S. 1975. Grafts and grafting. In *Hemodynamics for Surgeons*, Grune and Stratton, New York, pp. 342–395.

Tarbell, J.M., Gunishan, J.P., Geselowitz, D.B., Rosenberg, G., Shung, K.K., and Pierce, W.S. 1986. Pulsed ultrasonic Doppler velocity measurements inside a left ventricular assist device. *ASME J. Biomech. Eng.* 108: 232–238.

Thubrikar, M.J., Skinner, J.R., and Nolan, S.P. 1982a. Design and stress analysis of bioprosthetic valves *in vivo*. In *Cardiac Bioprostheses.* Cohn, L.H. and Gallucci, V., Eds., Yorke Medical Books, New York, pp. 445–455.

Thubrikar, M.J., Skinner, J.R., Eppink, T.R., and Nolan, S.P. 1982b. Stress analysis of porcine bioprosthetic heart valves *in vivo. J. Biomed. Mater. Res.* 16: 811–826.

Unger, F. 1989. *Assisted Circulation*, Vol. 3. Springer-Verlag, Berlin.

Van der Lei, B., Wildevuur, C.R.H., Niewenhuis, P., Blaauw, E.H., Dijk, F., Hulstaert, C.E., and Molenaar, I. 1985. Regeneration of the arterial wall in microporous, compliant, biodegradable vascular grafts after implantation into the rat abdominal aorta. *Cell Tissue Res.* 242: 569–578.

Van der Lei, B., Wildevuur, C.R.H., and Nieuwenhuis, P. 1986. Compliance and biodegradation of vascular grafts stimulate the regeneration of elastic laminae in neoarterial tissue: an experimental study in rats. *Surgery* 99: 45–51.

Veith, F.J., Abbott, W.M., Yao, J.S.T. et al. 1995. Guidelines for development and use of transluminally placed endovascular prosthetic grafts in the arterial system. *J. Vasc. Surg.* 21: 670–85.

Wieting, D.W. 1997. Prosthetic heart valves in the future. In *Surgery for Acquired Aortic Valve Disease.* Piwnica, A. and Westaby, S., Eds., ISIS Medical Media, Oxford, pp. 460–478.

Wurzel, D., Kolff, J., Missfeldt, W., Wildevuur, W., Hansen, G., Brownstein, L., Reibman, J., De Paulis, R., and Kolff, W.J. 1988. Development of the Philadelphia heart system. *Artif. Organs* 12: 410–422.

Yoganathan, A.P., Corcoran, W.H., and Harrison, E.C. 1979a. *In vitro* velocity measurements in the vicinity of aortic prostheses. *J. Biomech.* 12: 135–152.

Yoganathan, A.P., Corcoran, W.H., and Harrison, E.C. 1979b. Pressure drops across prosthetic aortic heart valves under steady and pulsatile flow — *in vitro* measurements. *J. Biomech.* 12: 153–164.

Yoganathan, A.P., Woo, Y.R., and Sung, H.W. 1986. Turbulent shear stress measurements in the vicinity of aortic heart valve prostheses. *J. Biomech.* 19: 433–442.

Yusef, S.W., Baker, D.M., Chuter, T.A.M. et al. 1994. Transfemoral endoluminal repair of abdominal aortic aneurysm with bifurcated graft. *Lancet* 344: 650–1.

8.2 Non-Blood-Interfacing Implants for Soft Tissues

K.J.L. Burg and S.W. Shalaby

Most tissues other than bone and cartilage are of the soft category. Implants do not generally interface directly with blood; the exceptions are located primarily in the cardiovascular systems. Non-blood-interfacing soft tissue implants are used to augment or replace natural tissues or to redirect specific biological functions. The implants can be transient; that is, of short-term function and thus made of absorbable materials, or they can be long-term implants which are expected to have prolonged functions and are made of nonabsorbable materials.

Toward the successful development of a new biomedical device or implant, including those used for soft tissues, the following milestones must be achieved: (1) acquire certain biologic and biomechanic data about the implant site and its function to aid in the selection of materials and engineering design of such an implant, to meet carefully developed product requirements; (2) construct a prototype and evaluate its physical and biologic properties both *in vitro* and *in vivo*, using the appropriate animal model; and (3) conduct a clinical study following a successful battery of animal safety studies depending on intended application and availability of historical safety and clinical data on the material or design. Extent of the studies associated with any specific milestone can vary considerably. Although different applications require different materials with specific properties, minimum requirements for soft-tissue implants should be met. The implant must (1) exhibit physical properties (e.g., flexibility and texture) which are equivalent or comparable to those called for in the product profile; (2) maintain the expected physical properties after implantation for a specific period; (3) elicit no adverse tissue reaction; (4) display no carcinogenic, toxic, allergenic, and/or immunogenic effect; and (5) achieve assured sterility without compromising the physicochemical properties. In addition to these criteria, a product of potentially broad applications is expected to (1) be easily mass produced at a reasonable cost; (2) have acceptable aesthetic quality; (3) be enclosed in durable, properly labeled, easy-access packaging; and (4) have adequate shelf stability.

The most common types of soft-tissue implants are (1) sutures and allied augmentation devices; (2) percutaneous and cutaneous systems; (3) maxillofacial devices; (4) ear and eye prostheses; (5) space-filling articles; and (6) fluid transfer devices.

8.2.1 Sutures and Allied Augmentation Devices

Sutures and staples are the most common types of augmentation devices. In recent years, interest in using tapes and adhesives has increased and may continue to do so, should new efficacious systems be developed.

8.2.1.1 Sutures and Suture Anchors

Sutures are usually packaged as a thread attached to a metallic needle. Although most needles are made of stainless steel alloys, the thread component can be made of various materials, and the type used determines the class of the entire suture. In fact, it is common to refer to the thread as the suture. Presently, most needles are drilled (mechanically or by laser) at one end for thread insertion. Securing the thread in the needle hole can be achieved by crimping or adhesive attachment. Among the critical physical properties of sutures are their diameter, *in vitro* knot strength, needle-holding strength, needle penetration force, ease of knotting, knot security, and *in vitro* strength retention profile.

Two types of threads are used in suture manufacturing and are distinguished according to the retention of their properties in the biologic environment, namely, absorbable and nonabsorbable. These may also be classified according to their source of raw materials, that is, natural (catgut, silk, and cotton), synthetic (nylon, polypropylene, polyethylene terephthalate, and polyglycolide and its copolymers), and metallic sutures (stainless steel and tantalum). Sutures may also be classified according to their physical form, that is, monofilament and twisted or braided multifilament (or simply braids).

The first known suture, the absorbable catgut, is made primarily of collagen derived from sheep intestinal submucosa. It is usually treated with a chromic salt to increase its *in vivo* strength retention and through imparted crosslinking that retards absorption. Such treatment extends the functional performance of catgut suture from 1 to 2 weeks up to about 3 weeks. The catgut sutures are packaged in a specially formulated fluid to prevent drying and maintain necessary compliance for surgical handling and knot formation.

The use of synthetic absorbable sutures exceeded that of catgut over the past two decades. This is attributed to many factors including (1) higher initial breaking strength and superior handling characteristics; (2) availability of sutures with a broad range of *in vivo* strength retention profiles; (3) considerably milder tissue reactions and no immunogenic response; and (4) reproducible properties and highly predictable *in vivo* performance. Polyglycolide (PG) was the first synthetic absorbable suture to be introduced, about three decades ago. Because of the high modulus of oriented fibers, PG is made mostly in the braided form. A typical PG suture braid absorbs in about 4 months and retains partial *in vivo* strength after 3 weeks. However, braids made of the 90/10 glycolide/l-lactide copolymer have a comparable or improved strength retention profile and faster absorption rate relative to PG. The copolymeric sutures absorb in about 3 months and have gained wide acceptance by the surgical community.

As with other types of braided sutures, an absorbable coating which improves suture handling and knot formation has been added to the absorbable braids. To minimize the risk of infection and tissue drag that are sometimes associated with braided sutures, four types of monofilament sutures have been commercialized. The absorbable monofilaments were designed specifically to approach the engineering compliance of braided sutures, by combining appropriate materials to achieve low moduli, for example, polydioxanone and copolymers of glycolide with caprolactone or trimethylene carbonate.

Members of the nonabsorbable family of sutures include braided silk (a natural protein), nylon, and polyethylene terephthalate (PET). These braids are used as coated sutures. Although silk sutures have retained wide acceptance by surgeons, nylon and particularly PET sutures are used for critical procedures where high strength and predictable long-term performance are required. Meanwhile, the use of cotton sutures is decreasing constantly because of their low strength and occasional tissue reactivity due to contaminants. Monofilaments are important forms of nonabsorbable sutures and are made primarily of polypropylene, nylon, and stainless steel. An interesting application of monofilament sutures is illustrated in the use of polypropylene loops (or haptics) for intraocular lenses. The polypropylene sutures exhibit not only the desirable properties of monofilaments but also the biologic inertness reflected in the minimal tissue reactions associated with their use in almost all surgical sites. With the exception of its natural tendency to undergo hydrolytic degradation and, hence, continued loss of mechanical strength postoperatively, nylon monofilament has similar attributes to those of polypropylene. Because of their exceptionally high modulus, stainless steel sutures are not used in soft-tissue repair because they can tear these tissues. All sutures can be sterilized by gamma radiation except those made of synthetic absorbable polymers, polypropylene, or cotton, which are sterilized by ethylene oxide.

Related to the suture is the tissue suture anchor, used to attach soft tissue to bone. The anchor is embedded into bone and the suture can be used to reattach the soft tissue. The most common anchor is polylactide-based and is used in shoulder repair.

8.2.1.2 Nonsuture Fibrous and Microporous Implants

Woven PET and polypropylene fabrics are commonly used as surgical meshes for abdominal wall repair and similar surgical procedures where surgical "patching" is required. Braid forms and similar construction made of multifilament PET yarns have been used for repairing tendons and ligaments. Microporous foams of polytetrafluoroethylene (PTFE) are used as pledgets (to aid in anchoring sutures to soft tissues) and in repair of tendons and ligaments. Microporous collagen-based foams are used in wound repair to accelerate healing.

8.2.1.3 Clips, Staples, and Pins

Ligating clips are most commonly used for temporary or long-term management of the flow in tubular tissues. Titanium clips are among the oldest and still-versatile types of clips. Thermoplastic polymers such as nylon can be injection-molded into different forms of ligating clips. These are normally designed to have a latch and living hinge. Absorbable polymers made of lactide/glycolide copolymers and polydioxanone have been successfully converted to ligating clips with different design features for a broad range of applications.

Metallic staples were introduced about three decades ago as strong competitors to sutures for wound augmentation; their use has grown considerably over the past 10 years for everything from skin closure procedures to a multiplicity of internal surgical applications. Major advantages associated with the use of staples are ease of application and minimized tissue trauma. Metallic staples can be made of tantalum, stainless steel, or titanium–nickel alloys. Staples are widely used to facilitate closure of large incisions produced in procedures such as Caesarean sections and intestinal surgery. Many interesting applications of small staples have been discovered for ophthalmic and endoscopic use, a fast-growing area of minimally invasive surgery.

Thermoplastic materials based on lactide/glycolide copolymers have been used to produce absorbable staples for skin and internal wound closures. These staples consist primarily of two interlocking components, a fastener and receiver. They are advantageous in that they provide a quick means of closure with comparable infection resistance. They are limited to locations which do not have large tensile loads and/or thicker or more sensitive tissue.

A new form of ligating device is the subcutaneous pin. This is designed with a unique applicator to introduce the pin parallel to the axis of the wound. During its application, the linear pin acquires a zig-zag-like configuration for stabilized tissue anchoring. The pins are made of lactide/glycolide polymers.

8.2.1.4 Surgical Tapes

Surgical tapes are intended to minimize necrosis, scar tissue formation, problems of stitch abscesses, and weakened tissues. The problems with surgical tapes are similar to those experienced with traditional skin tapes. These include (1) misaligned wound edges, (2) poor adhesion due to moisture or dirty wounds, and (3) late separation of tapes when hematoma or wound drainage occur.

Wound strength and scar formation in skin may depend on the type of incision made. If the subcutaneous muscles in the fatty tissue are cut and the overlying skin is closed with tape, then the muscles retract. This, in turn, increases the scar area, causing poor cosmetic appearance when compared to a suture closure. Tapes also have been used successfully for assembling scraps of donor skin for skin graft.

8.2.1.5 Tissue Adhesives

The constant call for tissue adhesives is particularly justified when dealing with the repair of exceptionally soft tissues. Such tissues cannot be easily approximated by sutures, because sutures inflict substantial mechanical damage following the traditional knotting scheme and associated shear stresses. However, the variable biological environments of soft tissues and their regenerative capacity make the development of an ideal tissue adhesive a difficult task. Experience indicates that an ideal tissue adhesive should (1) be able to wet and bond to tissues; (2) be capable of onsite formation by the rapid polymerization of a liquid monomer without producing excessive heat or toxic byproducts; (3) be absorbable; (4) not interfere with the normal healing process; and (5) be easily applied during surgery. The two common types of tissue adhesives currently used are based on alkyl-o-cyanoacrylates and fibrin. The latter is a natural adhesive derived from fibrinogen, which is one of the clotting components of blood. Although fibrin is used in Europe, its use in the United States has not been approved because of the risk of its contamination with hepatitis and/or immune disease viruses. Due to its limited mechanical strength (tensile strength and elastic modulus of 0.1 and 0.15 MPa, respectively), fibrin is used mostly as a sealant and for adjoining delicate tissues as in nerve anastomoses. Meanwhile, two members of the cyanoacrylate family of adhesives, n-butyl- and iso-butyl-cyano-acrylates, are used in a number of countries as sealants, adhesives, and blocking agents. They are yet to be approved for use in the United States because of the lack of sufficient safety data. Due to a fast rate of polymerization and some limited manageability in localizing the adhesive to the specific surgical site, the *in vivo* performance of cyanoacrylates can be unpredictable. Because of the low strength of the adhesive joints or sealant films produced on *in vivo* polymerization of these cyanoacrylates, their applications generally are limited to use in traumatized fragile tissues (such as spleen, liver, and kidney) and after extensive surgery on soft lung tissues. A major safety concern of these alkyl cyanoacrylates is related to their nonabsorbable nature. Hence, a number of investigators have directed their attention to certain alkoxy-alkyl cyanoacrylates which can be converted to polymeric adhesives with acceptable absorbable profiles and rheological properties. Methoxypropyl cyanoacrylate, for example, has demonstrated both the absorbability and high compliance that is advantageous to soft tissue repair.

8.2.2 Percutaneous and Skin Implants

The need for percutaneous (*trans* or through the skin) implants has been accelerated by the advent of artificial kidneys and hearts and the need for prolonged injection of drugs and nutrients. Artificial skin is urgently needed to maintain the body temperature and prevent infection in severely burned patients. Actual permanent replacement of skin by biomaterials is still a great clinical challenge.

8.2.2.1 Percutaneous Devices

The problem of obtaining a functional and viable interface between the tissue (skin) and an implant (percutaneous device) is primarily due to the following factors. First, although initial attachment of the tissue into the interstices of the implant surface occurs, attachment cannot be maintained for a sustained time since the dermal tissue cells turn over continuously. Downgrowth of epithelium around the implant or overgrowth on the implant leads to extrusion or invagination, respectively. Second, any opening near the implant that is large enough for bacteria to penetrate may result in infection, even though initially there may be a tight seal between skin and implant. Several factors are involved in the development of percutaneous devices:

1. Type of end use — this may deal with transmission of information (biopotentials, temperature, pressure, blood flow rate), energy (electrical stimulation, power for heart-assist devices), transfer of matter (cannula for blood), and load (attachment of a prosthesis);
2. Engineering factors — these may address materials selection (polymers, ceramics, metals, and composites), design variation (button, tube with and without skirt, porous or smooth surface), and mechanical stresses (soft and hard interface, porous or smooth interface);
3. Biologic factors — these are determined by the implant host (human, dog, hog, rabbit, sheep), and implant location (abdominal, dorsal, forearm);
4. Human factors — these can pertain to postsurgical care, implantation technique, and esthetic look.

No percutaneous devices are completely satisfactory. Nevertheless, some researchers believe that hydroxyapatite may be part of a successful approach. In one experimental trial, a hydroxyapatite-based percutaneous device was associated with less epidermal downgrowth (1 mm after 17 months vs. 4.6 mm after 3 months) when compared with a silicone rubber control specimen in the dorsal skin of canines. Researchers have also investigated coatings such as laminin-5 which has been shown to enhance epithelial attachment.

8.2.2.2 Artificial Skins

Artificial skin is another example of a percutaneous implant, and the problems are similar to those described above. Important for this application is a material which can adhere to a large (burned) surface and thus prevent the loss of fluids, electrolytes, and other biomolecules until the wound has healed.

In one study on wound-covering materials with controlled physicochemical properties, an artificial skin was designed with a crosslinked collagen-polysaccharide (chondroitin 6-sulfate) composite membrane. This was specifically chosen to have controlled porosity (5 to 150 μm in diameter), flexibility (by varying crosslink density), and moisture flux rate.

Several polymeric materials and reconstituted collagen have also been examined as burn dressings. Among the synthetic ones are the copolymers of vinyl chloride and vinyl acetate as well as polymethyl cyanoacrylate (applied as a fast-polymerizing monomer). The latter polymer and/or its monomer were found to be too brittle and histotoxic for use as a burn dressing. The ingrowth of tissue into the pores of polyvinyl alcohol sponges and woven fabric (nylon and silicone rubber velour) was also attempted without much success. Nylon mesh bonded to a silicone rubber membrane, another design attempt, prevented water evaporation but has not been found to induce fibrovascular growth.

Rapid epithelial layer growth by culturing cells *in vitro* from the skin of the burn patient for covering the wound area may offer a practical solution for less severely burned patients. Implantation of an allogenic fibroblast/polymer construct has proven useful for providing long-term skin replacement. Related to this, temporary tissue engineered replacements are possible alternatives for burns requiring larger area coverage. These can be similar to the synthetic dressing, a nylon mesh and silicone rubber component, but incorporates allogeneic fibroblasts. This temporary covering hopefully will stimulate or allow fibrovascular growth into the wound bed by providing the appropriate matrix proteins and growth factors.

8.2.3 Maxillofacial Implants

There are two types of maxillofacial implants: extraoral and intraoral. The former deals with the use of artificial substitutes for reconstructing defective regions in the maxilla, mandible, and face. Useful polymeric materials for extraoral implants require (1) match of color and texture with those of the patient; (2) mechanical and chemical stability (i.e., material should not creep or change color or irritate skin); and (3) ease of fabrication. Copolymers of vinyl chloride and vinyl acetate (with 5 to 20% acetate), polymethyl methacrylate, silicones, and polyurethane rubbers are currently used. Intraoral implants are used for repairing maxilla, mandibular, and facial bone defects. Material requirements for the intraoral implants are similar to those of the extraoral ones. For the latter group of implants, metallic materials such as tantalum, titanium, and CoCr alloys are commonly used. For soft tissues, such as gum and chin, polymers such as silicone rubber and polymethylmethacrylate are used for augmentation.

8.2.4   Ear and Eye Implants

Implants can be used to restore conductive hearing loss from otosclerosis (a hereditary defect which involves a change in the bony tissue of the ear) and chronic otitis media (the inflammation of the middle ear which can cause partial or complete impairment of the ossicular chain). A number of prostheses are available for correcting these defects. The porous polyethylene total ossicular implant is used to achieve a firm fixation by tissue ingrowth. The tilt-top implant is designed to retard tissue ingrowth into the section of the shaft which may diminish sound conduction. Materials used in fabricating these implants include polymethyl methacrylate, polytetrafluoroethylene, polyethylene, silicone rubber, stainless steel, and tantalum. More recently, polytetrafluoroethylene–carbon composites, porous polyethylene, and pyrolytic carbon have been described as suitable materials for cochlear (inner ear) implants.

Artificial ear implants capable of processing speech have been developed with electrodes to stimulate cochlear nerve cells. Cochlear implants also have a speech processor that transforms sound waves into electrical impulses that can be conducted through coupled external and internal coils. The electrical impulses can be transmitted directly by means of a percutaneous device.

Eye implants are used to restore the functionality of damaged or diseased corneas and lenses. Usually the cornea is transplanted from a suitable donor. In cataracts, eye lenses become cloudy and can be removed surgically. Intraocular lenses (IOL) are implanted surgically to replace the original eye lens and to restore function. IOL are made from transparent acrylics, particularly polymethyl methacrylate, which has excellent optical properties. Infection and fixation of the lens to the tissues are frequent concerns, and a number of measures are being used to address them. Transplantation of retinal pigmented epithelium can be used in the treatment of adult onset blindness; the challenge is developing readily detachable or absorbable materials on which to culture sheets of these cells.

8.2.5   Space-Filling Implants

Breast implants are common space-filling implants. At one time, the enlargement of breasts was done with various materials such as paraffin wax and silicone fluids, by direct injection or by enclosure in a rubber balloon. Several problems have been associated with directly injected implants, including progressive instability and ultimate loss of original shape and texture as well as infection and pain. One of the early efforts in breast augmentation was to implant a sponge made of polyvinyl alcohol. However, soft tissues grew into the pores and then calcified with time, and the so-called marble breast resulted. Although the enlargement or replacement of breasts for cosmetic reasons alone is not recommended, prostheses have been developed for patients who have undergone radical mastectomy or who have nonsymmetrical deformities. The development of the tissue-engineered breast is ongoing, where fat or normal breast tissue may be derived from the patient and combined with an absorbable scaffold for transplantation. A silicone rubber bag filled with silicone gel and backed with polyester mesh to permit tissue ingrowth for fixation is a widely used prosthesis, primarily for psychological reasons. The artificial penis, testicles, and vagina fall into the same category as breast implants, in that they make use of silicones and are implanted for psychological reasons rather than to improve physical health.

8.2.6   Fluid Transfer Implants

Fluid transfer implants are required for cases such as hydrocephalus, urinary incontinence, glaucoma-related elevated intraocular pressure, and chronic ear infection. Hydrocephalus, caused by abnormally high pressure of the cerebrospinal fluid in the brain, can be treated by draining the fluid (essentially an ultrafiltrate of blood) through a cannula. Earlier shunts had two one-way valves at either end. However, the more recent Ames shunt has simple slits at the discharging end, which opens when enough fluid pressure is exerted. The Ames shunt empties the fluid in the peritoneum while others drain into the blood stream through the right internal jugular vein or right atrium of the heart. The simpler peritoneal shunt shows less incidence of infection.

The use of implants for correcting the urinary system has not been successful because of the difficulty of adjoining a prosthesis to the living system for achieving fluid tightness. In addition, blockage of the passage by deposits from urine, salt for example, and constant danger of infection are major concerns. Several materials have been used for producing these implants, with limited long-term success; these include Dacron™, glass, polyvinyl alcohol, polyethylene, rubber, silver, tantalum, Teflon™, and Vitallium™. Tissue engineered devices also have application in urinary repair; for example, an alginate-chondrocyte system has been used clinically to treat vesicoureteral reflux. Preliminary results suggest that by transplanting the hydrogel system as a bulking agent below a refluxing ureter, neocartilage gradually develops to correct the reflux. Uroepithelial cells, combined with porous, absorbable polyester matrices show promise in the replacement of urologic tissues.

Drainage tubes, which are impermanent implants for chronic ear infection, can be made from polytetra-fluoroethylene (Teflon™). Glaucoma-related elevated intraocular pressure may be relieved by implanting a tube connecting the anterior eye chamber to the external subconjunctival space in order to direct aqueous humor. Complications can arise with occlusion of the tube due to wound healing processes; however, researchers are investigating combining this device with an absorbable drug delivery plug to regulate flow and deliver drugs to regulate the wound healing process.

8.2.7 Technologies of Emerging Interest

A new process for uniaxial solid-state orientation, using a range of compressive forces and temperatures, has been developed to produce stock sheets of polyether-ether ketone (PEEK) for machining dental implants with substantially increased strength and modulus as compared to their unoriented counterparts.

Surface treatment of materials is another emerging interest, a process potentially influencing both surface charge, topography, and conductivity. The former has obvious effects on cell adhesion and integration of traditional implants with surrounding tissues, and it has a profound effect on the success or failure of a tissue-engineered device.

Pertinent to the effect of surface chemistry and texture on tissue regeneration/ingrowth, recent studies on surface-phosphonylation to create hydroxyapatite-like substrates show that (1) surface-microtextured polypropylene and polyethylene transcortical implants in goat tibia having phosphonate functionalities (with or without immobilized calcium ions) do induce bone ingrowth, and (2) microtexture and surface phosphonylated (with and without immobilized calcium ions) rods made of PEEK and similarly treated rods of carbon fiber-reinforced PEEK induce bone ingrowth when implanted in the toothless region of the lower jaw of goats. The use of surface-phosphonylated and post-treated (with a bridging agent) UHMW-PE fibers and fabric do produce high strength and modulus composites at exceptionally low filler loading. Among these composites are those based on methyl methacrylate matrices, similar to those used in bone cement.

Conductivity manipulation may be extremely useful in such areas as biosensor design. Surface-conducting phosphonylated, ultra-high molecular weight polyethylene may be formed by exposure to aqueous pyrole solution. Preliminary results suggest that this is a stable process, yielding no apparent cytotoxic effects due to the material or leachables. New surface treatments will be instrumental in the development of the two rapidly expanding areas of biosensor design and tissue engineering, both areas which encompass non-blood-interfacing soft tissue implants.

Further Reading

Allan, J. 1993. The molecular binding of inherently conducting polymers to thermoplastic substrates. M.S. thesis, Clemson University, Clemson, SC.

Allan, J.M., Wrana, J.S., Dooley, R.L., Budsberg, S., and Shalaby, S.W. 1999. Bone ingrowth into surface-phosphonylated polyethylene and polypropylene. *Trans. Soc. Biomat.* (submitted).

Allan, J.M., Wrana, J.S., Linden, D.E., Dooley, R.L., Farris, H., Budsberg, S., and Shalaby, S.W. 1999. Osseointegration of morphologically and chemically modified polymeric dental implants. *Trans. Soc. Biomater.* (submitted).

Allan, J.M., Kline, J.D., Wrana, J.S., Flagle, J.A., Corbett, J.T., and Shalaby, S.W. 1999. Absorbable gel-forming sealant/adhesives as a staple adjuvant in wound repair. *Trans. Soc. Biomater.* (submitted).

Chvapil, M. 1982. Considerations on manufacturing principles of a synthetic burn dressing: A review. *J. Biomed. Mater. Res.* 16:245.

Deng, M., Allan, J.M., Lake, R.A., Gerdes, G.A., and Shalaby, S.W. 1999. Effect of phosphonylation on UHMW-PE fabric-reinforced composites. *Trans. Soc. Biomater.* (submitted).

Deng, M., Wrana, J.S., Allan, J.M., and Shalaby, S.W. 1999. Tailoring mechanical properties of polyether-ether ketone for implants using solid-state orientation. *Trans. Soc. Biomater.* (submitted).

El-Ghannam, A., Starr, L., and Jones, J. 1998. Laminin-5 coating enhances epithelial cell attachment, spreading, and hemidesmosome assembly on Ti-6Al-4V implant material *in vitro. J. Biomed. Mater. Res.* 41:30.

Gantz, B.J. 1987. Cochlear implants: An overview. *Acta Otolaryng. Head Neck Surg.* 1:171.

Holder, W.D., Jr., Gruber, H.E., Moore, A.L., Culberson, C.R., Anderson, W., Burg, K.J.L., and Mooney, D.J. 1998. Cellular ingrowth and thickness changes in poly-L-lactide and polyglycolide matrices implanted subcutaneously in the rat. *J. Biomed. Mater. Res.* 41:412–421.

Kablitz, C., Kessler, T., Dew, P.A. et al. 1979. Subcutaneous peritoneal catheter: 1 1/2-years experience. *Artif. Organs* 3:210.

Lanza, R.P., Langer, R., and Chick, W.L., Eds. 1997. *Principles of Tissue Engineering*, Academic Press, San Diego, CA.

Lynch, W. 1982. *Implants: Reconstructing the Human Body.* Van Nostrand Reinhold, New York.

Park, J.B. and Lakes, R.S. 1992. *Biomaterials Science and Engineering*, 2nd ed., Plenum Press, New York.

Postlethwait, R.W., Schaube, J.F., and Dillan, M.L. et al. 1959. An evaluation of surgical suture material. *Surg. Gyn. Obstet.* 108:555.

Shalaby, S.W. 1985. Fibrous materials for biomedical applications. In *High Technology Fibers: Part A.* Lewin, M. and Preston, J., Eds., Dekker, New York.

Shalaby, S.W. 1988. Bioabsorbable polymers. In *Encyclopedia of Pharmaceutical Technology*, Vol. 1. Boylan, J.C. and Swarbrick, J., Eds., Dekker, New York.

Shalaby, S.W., Ed. 1994. *Biomedical Polymers Designed to Degrade Systems*, Hanser, New York.

Shall, L.M. and Cawley, P.W. 1994. Soft tissue reconstruction in the shoulder. Comparison of suture anchors, absorbable staples, and absorbable tacks. *Amer. J. Sports Med.* 22: 715.

VonRecum, A.G. and Park, J.B. 1979. Percutaneous devices. *Crit. Rev. Bioeng.* 5: 37.

Yannas, I.V. and Burke, I.F. 1980. Design of an artificial skin: 1. Basic design principles. *J. Biomed. Mater. Res.* 14: 107.

9

Hard Tissue Replacements

Sang-Hyun Park
Orthopedic Hospital

Adolfo Llinás
Pontificia Universidad Javeriana

Vijay K. Goel
J.C. Keller
University of Iowa

9.1 Bone Repair and Joint Implants

S.-H. Park, A. Llinás, and V.K. Goel

The use of biomaterials to restore the function of traumatized or degenerated connective tissues and thus improve the quality of life of a patient has become widespread. In the past, implants were designed with insufficient cognizance of biomechanics. Accordingly, the clinical results were not very encouraging. An upsurge of research activities into the mechanics of joints and biomaterials has resulted in better designs with better *in vivo* performance. The improving long-term success of total joint replacements for the lower limb is testimony to this. As a result, researchers and surgeons have developed and used fixation devices for the joints, including artificial spine discs. A large number of devices are also available for the repair of the bone tissue. This chapter provides an overview of the contemporary scientific work related to the use of biomaterials for the repair of bone (e.g., fracture) and joint replacements ranging from a hip joint to a spine.

9.1.1 Long Bone Repair

The principal functions of the skeleton are to provide a frame to support the organ-systems, and to determine the direction and range of body movements. Bone provides an anchoring point (insertion), for most skeletal muscles and ligaments. When the muscles contract, long bones act as levers, with the joints functioning as pivots, to cause body movement.

Bone is the only tissue able to undergo spontaneous regeneration and to remodel its micro- and macro-structure. This is accomplished through a delicate balance between an *osteogenic* (bone forming) and *osteoclastic* (bone removing) process [Brighton, 1984]. Bone can adapt to a new mechanical environment by changing the equilibrium between osteogenesis and osteoclasis. These processes will respond to changes in the static and dynamic stress applied to bone; that is, if more stress than the physiological is applied, the equilibrium tilts toward more osteogenic activity. Conversely, if less stress is applied the equilibrium tilts toward osteoclastic activity (this is known as **Wolff's Law** of bone remodeling) [Wolff, 1986].

Nature provides different types of mechanisms to repair fractures in order to be able to cope with different mechanical environments about a fracture [Hulth, 1989; Schenk, 1992]. For example, incomplete fractures (cracks), which only allow micromotion between the fracture fragments, heal with a small amount of fracture-line *callus*, known as **primary healing**. In contrast, complete fractures which are unstable, and therefore generate macromotion, heal with a voluminous callus stemming from the sides of the bone, known as **secondary healing** [Brighton, 1984; Hulth, 1989].

The goals of fracture treatment are obtaining rapid healing, restoring function, and preserving cosmesis without general or local complications. Implicit in the selection of the treatment method is the need to avoid potentially deleterious conditions, for example, the presence of excessive motion between bone fragments which may delay or prevent fracture healing [Brighton, 1984; Brand and Rubin, 1987].

Each fracture pattern and location results in a unique combination of characteristics ("fracture personality") that require specific treatment methods. The treatments can be non-surgical or surgical. Examples of non-surgical treatments are immobilization with casting (plaster or resin) and bracing with a plastic apparatus. The surgical treatments are divided into external fracture fixation, which does not require opening the fracture site, or internal fracture fixation, which requires opening the fracture.

With external fracture fixation, the bone fragments are held in alignment by pins placed through the skin onto the skeleton, structurally supported by external bars. With internal fracture fixation, the bone fragments are held by wires, screws, plates, and/or intramedullary devices. Figure 9.1a,b show radiographs of externally and internally fixed fractures.

All the internal fixation devices should meet the general requirement of biomaterials, that is, biocompatability, sufficient strength within dimensional constraints, and corrosion resistance. In addition, the device should also provide a suitable mechanical environment for fracture healing. From this perspective, stainless steel, cobalt–chrome alloys, and titanium alloys are most suitable for internal fixation. Detailed mechanical properties of the metallic alloys are discussed in the chapter on metallic biomaterials. Most internal fixation devices persist in the body after the fracture has healed, often causing discomfort and requiring removal. Recently, biodegradable polymers, for example, polylactic acid (PLA) and polyglycolic acid (PGA), have been used to treat minimally loaded fractures, thereby eliminating the need for a second surgery for implant removal. A summary of the basic application of biomaterials in internal fixation is presented in Table 9.1. A description of the principal failure modes of internal fixation devices is presented in Table 9.2.

9.1.1.1 Wires

Surgical wires are used to reattach large fragments of bone, like the greater trochanter, which is often detached during total hip replacement. They are also used to provide additional stability in long-oblique or spiral fractures of long bones which have already been stabilized by other means (Figure 9.1b). Similar approaches, based on the use of wires, have been employed to restore stability in the lower cervical spine region and in the lumbar segment as well (Figure 9.1c).

Twisting and knotting is unavoidable when fastening wires to bone; however, by doing so, the strength of the wire can be reduced by 25% or more due to stress concentration [Tencer et al., 1993]. This can be partially overcome by using a thicker wire, since its strength increases directly proportional to its diameter. The deformed regions of the wire are more prone to corrosion than the un-deformed because of the higher strain energy. To decrease this problem and ease handling, most wires are annealed to increase the ductility.

(a)

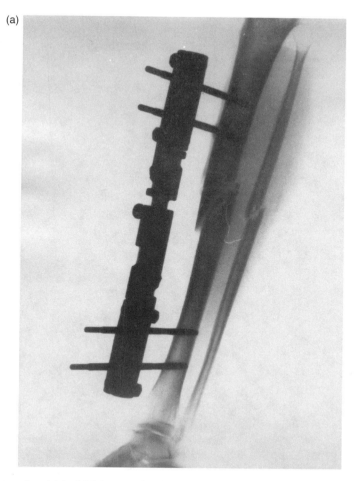

FIGURE 9.1 Radiographs of (a) tibial fracture fixed with four pins and an external bar; (b) a total hip joint replacement in a patient who sustained a femoral fracture and was treated with double bone plates, screws, and surgical wire (arrows); (c) application of wires, screws, and plates in the spine.

Braided multistrain (multifilament) wire is an attractive alternative because it has a similar tensile strength than a monofilament wire of equal diameter, but more flexibility and higher fatigue strength [Taitsman and Saha, 1977]. However, bone often grows into the grooves of the braided multistrain wire, making it exceedingly difficult to remove, since it prevents the wire from sliding when pulled. When a wire is used with other metallic implants, the metal alloys should be matched to prevent galvanic corrosion [Park and Lakes, 1992].

9.1.1.2 Pins

Straight wires are called Steinmann pins; however if the pin diameter is less than 2.38 mm, it is called Kirschner wire. They are widely used primarily to hold fragments of bones together provisionally or permanently and to guide large screws during insertion. To facilitate implantation, the pins have different tip designs which have been optimized for different types of bone (Figure 9.2). The trochar tip is the most efficient in cutting; hence, it is often used for cortical bone.

The holding power of the pin comes from elastic deformation of surrounding bone. In order to increase the holding power to bone, threaded pins are used. Most pins are made of 316L stainless steel; however, recently, biodegradable pins made of polylactic or polyglycolic acid have been employed for the treatment of minimally loaded fractures.

(b)

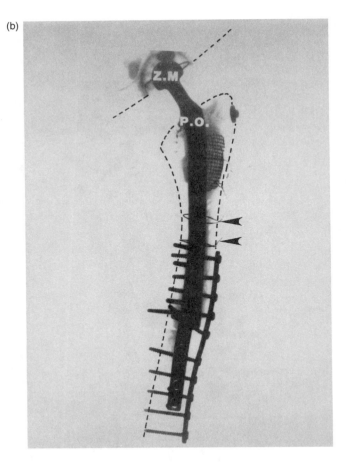

FIGURE 9.1 Continued.

The pins can be used as part of elaborate frames designed for external fracture fixation (Figure 9.1a). In this application, several pins are placed above and below the fracture, but away from it. After the fracture fragments are manually approximated (reduced) to resemble the intact bone, the pins are attached to various bars, which upon assembly will provide stability to the fracture.

9.1.1.3 Screws

Screws are the most widely used devices for fixation of bone fragments. There are two types of bone screws: (1) cortical bone screws, which have small threads, and (2) cancellous screws, which have large threads to get more thread-to-bone contact. They may have either V or buttress threads (Figure 9.3). The cortical screws are subclassified further according to their ability to penetrate, into self-tapping and non-self-tapping (Figure 9.3). The self-tapping screws have cutting flutes which thread the pilot drill-hole during insertion. In contrast, the non-self-tapping screws require a tapped pilot drill-hole for insertion.

The holding power of screws can be affected by the size of the pilot drill-hole, the depth of screw engagement, the outside diameter of the screw, and quality of the bone [Cochran, 1982; DeCoster et al., 1990]. Therefore, the selection of the screw type should be based on the assessment of the quality of the bone at the time of insertion. Under identical conditions, self-tapping screws provide a slightly greater holding power than non-self-tapping screws [Tencer et al., 1993].

Screw pullout strength varies with time after insertion *in vivo*, and it depends on the growth of bone into the screw threads and/or resorption of the surrounding bone [Schatzker et al., 1975]. The bone immediately adjacent to the screw often undergoes *necrosis* initially, but if the screw is firmly fixed, when

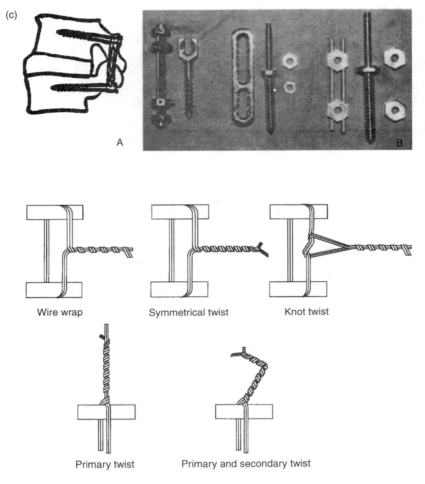

FIGURE 9.1 Continued.

the bone revascularizes, permanent secure fixation may be achieved. This is particularly true for titanium alloy screws or screws with a roughened thread surface, with which bone ongrowth results in an increase in removal torque [Hutzschenreuter and Brümmer, 1980]. When the screw is subject to micro- or macro-movement, the contacting bone is replaced by a membrane of fibrous tissue, the purchase is diminished, and the screw loosens.

The two principal applications of bone screws are (1) as interfragmentary fixation devices to "lag" or fasten bone fragments together, or (2) to attach a metallic plate to bone. Interfragmentary fixation is used in most fractures involving cancellous bone and in those oblique fractures in cortical bone. In order to lag the fracture fragments, the head of the screw must engage the cortex on the side of insertion without gripping the bone, while the threads engage cancellous bone and/or the cortex on the opposing side. When screws are employed for bone plate fixation, the bone screw threads must engage both cortices. Screws are also used for the fixation of spine fractures (for plate fixation or compression of bone fragment; Figure 9.1c).

9.1.1.4 Plates

Plates are available in a wide variety of shapes and are intended to facilitate fixation of bone fragments. They range from the very rigid, intended to produce primary bone healing, to the relatively flexible, intended to facilitate physiological loading of bone.

TABLE 9.1 Biomaterials Applications in Internal Fixation

Materials	Properties	Application
Stainless steel	Low cost, easy fabrication	Surgical wire (annealed) Pin, plate, screw IM nail
Ti alloy	High cost Low density and modulus Excellent bony contact	Surgical wire Plate, screws, IM nails
Co–Cr alloys (wrought)	High cost High density and modulus Difficult fabrication	Surgical wire IM nails
Poly lactic acid Poly glycolic acid	Resorbable Weak strength	Pin, screw
Nylon	Non-resorbable plastic	Cerclage band

TABLE 9.2 Failure Modes of Internal Fixation Devices

Failure mode	Failure location	Reasons for failure
Overload	Bone fracture site Implant screw hole Screw thread	Small size implant Unstable reduction Early weight bearing
Fatigue	Bone fracture site Implant screw hole Screw thread	Early weight bearing Small size implant Unstable reduction Fracture non-union
Corrosion	Screw head-plate hole Bent area	Different alloy implants Over-tightening screw Misalignment of screw Over-bent
Loosening	Screw	Motion Wrong choice of screw type Osteoporotic bone

The rigidity and strength of a plate in bending depend on the cross-sectional shape (mostly thickness) and material of which it is made. Consequently, the weakest region in the plate is the screw hole, especially if the screw hole is left empty, due to a reduction of the cross-sectional area in this region. The effect of the material on the rigidity of the plate is defined by the elastic modulus of the material for bending, and by the shear modulus for twisting [Cochran, 1982]. Thus, given the same dimensions, a titanium alloy plate will be less rigid than a stainless steel plate, since the elastic modulus of each alloy is 110 and 200 GPa, respectively.

Stiff plates often shield the underlying bone from the physiological loads necessary for its healthful existence [Perren et al., 1988; O'Slullivan et al., 1989]. Similarly, flat plates closely applied to the bone prevent blood vessels from nourishing the outer layers of the bone [Perren, 1988]. For these reasons, the current clinical trend is to use more flexible plates (titanium alloy) to allow micromotion, and low-contact plates (only a small surface of the plate contacts the bone, LCP), to allow restoration of vascularity to the bone [Uhthoff and Finnegan, 1984; Claes, 1989]. The underlying goals of this philosophical change are to increase the fracture healing rate, to decrease the loss of bone mass in the region shielded by the plate, and consequently, to decrease the incidence of re-fractures which occur following plate removal.

The interaction between bone and plate is extremely important since the two are combined into a composite structure. The stability of the plate-bone composite and the service life of the plate depends upon accurate fracture reduction. The plate is most resistant in tension; therefore, in fractures of long bones, the plate is placed along the side of the bone which is typically loaded in tension. Having excellent

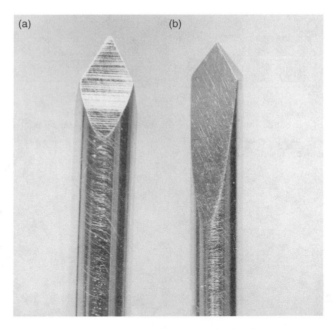

FIGURE 9.2   Types of metallic pin tip: (a) trochar end and (b) diamond end.

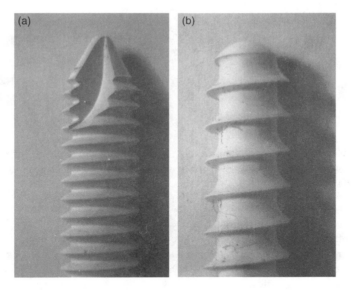

FIGURE 9.3   Bone screws: (a) a self-tapping V-threaded (has a cutting flute), and (b) a non-self-tapping, buttress threaded screw.

apposition of the bone fragments, as well as developing adequate compression between them, is critical in maintaining the stability of the fixation and preventing the plate from repetitive bending and fatigue failure. Interfragmentary compression also creates friction at the fracture surface, increasing resistance to torsional loads [Perren, 1991; Tencer et al., 1993]. On the contrary, too much compression causes micro fractures and necrosis of contacting bone due to the collapse of vesicular canals.

Compression between the fracture fragments can be achieved with a special type of plate called a *dynamic compression plate* (DCP). The dynamic compression plate has elliptic shape screw holes with its

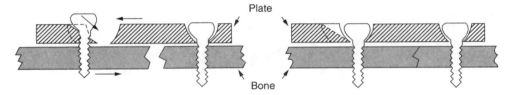

FIGURE 9.4   Principle of a dynamic compression plate for fracture fixation. During tightening a screw, the screw head slides down on a ramp in a plate screw hole which results in pushing the plate away from a fracture end and compressing the bone fragments together.

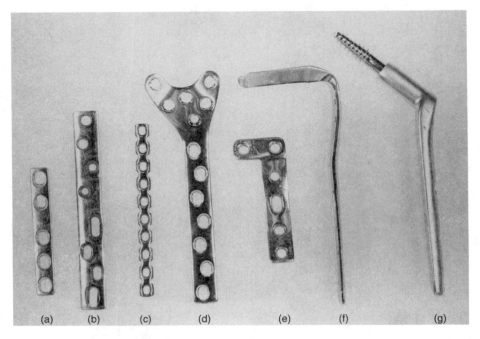

FIGURE 9.5   Bone plates: (a) dynamic compression plate, (b) hybrid compression plate (lower part has dynamic compression screw holes), (c) reconstruction bone plate (easy contouring), (d) buttress bone plate, (e) L shaped buttress plate, (f) nail plate (for condylar fracture), and (g) dynamic compression hip screw.

long axis oriented parallel to that of the plate. The screw hole has a sliding ramp to the long axis of the plate. Figure 9.4 explains the principle of the dynamic compression plate.

Bone plates are often contoured in the operating room in order to conform to an irregular bone shape to achieve maximum contact of the fracture fragments. However, excessive bending decreases the service life of the plate. The most common failure modes of a bone plate-screw fixation are screw loosening and plate failure. The latter typically occurs through a screw hole due to fatigue and/or crevice corrosion [Weinstein et al., 1979].

In the vicinity of the joints, where the diameter of long bones is wider, the cortex thinner, and cancellous bone abundant, plates are often used as a buttress or retaining wall. A buttress plate applies force to the bone perpendicular to the surface of the plate, and prevents shearing or sliding at the fracture site. Buttress plates are designed to fit specific anatomic locations and often incorporate other methods of fixation besides cortical or cancellous screws, for example, a large lag screw or an I-beam. For the fusion of vertebral bodies following diskectomy, spinal plates are used along with bone grafts. These plates are secured to the vertebral bodies using screws (Figure 9.1c). Similar approaches have been employed to restore stability in the thoracolumbar and cervical spine region as well. Figure 9.5 illustrates a variety of types of bone plates.

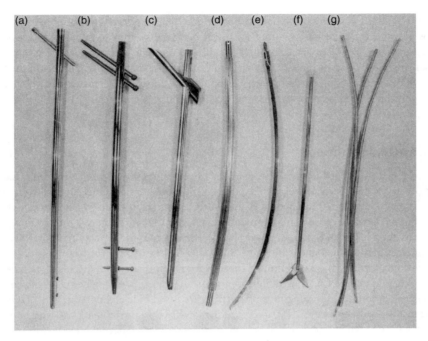

FIGURE 9.6 Intramedullary devices: (a) Gross-Kempf (slotted), (b) Uniflex (Ti alloy, slotted), (c) Kuntscher, (d) Samson, (e) Harris, (f) Brooker-Wills distal locking pin, and (g) Enders pins.

9.1.1.5 Intramedullary Nails

Intramedullary devices (IM nails) are used as internal struts to stabilize long bone fractures. IM nails are also used for fixation of femoral neck or intertrochanteric bone fractures; however, this application requires the addition of long screws. A gamut of designs are available, going from solid to cylindrical, with shapes such as cloverleaf, diamond, and "C" (slotted cylinders). Figure 9.6 shows a variety of intramedullary devices.

Compared to plates, IM nails are better positioned to resist multi-directional bending than a plate or an external fixator, since they are located in the center of the bone. However, their torsional resistance is less than that of the plate [Cochran, 1982]. Therefore, when designing or selecting an IM nail, a high polar moment of inertia is desirable to improve torsional rigidity and strength. The torsional rigidity is proportional to the elastic modulus and the moment of inertia. For nails with a circular cross-section, torsional stiffness is proportional to the fourth power of the nail's radius. The wall thickness of the nail also affects the stiffness. A slotted, open section nail is more flexible in torsion and bending, and it allows easy insertion into a curved medullary canal for example, that of the femur [Tencer, 1993]. However, in bending, a slot is asymmetrical with respect to rigidity and strength. For example, a slotted nail is strongest when bending is applied so that the slot is near the neutral plane; the nail is weakest when oriented so that the slot is under tension.

In addition to the need to resist bending and torsion, it is vital for an IM nail to have a large contact area with the internal cortex of the bone to permit torsional loads to be transmitted and resisted by shear stress. Two different concepts are used to develop shear stress: (1) a three-point, high pressure contact, achieved with the insertion of curved pins, and (2) a positive interlocking between the nail and intramedullary canal, to produce a unified structure. Positive interlocking can be enhanced by reaming the intramedullary canal. Reaming permits a larger, longer, nail-bone contact area and allows the use of a larger nail, with increased rigidity and strength [Kessler et al., 1986].

The addition of screws through the bone and nail, proximal and distal to the fracture, known as interlocking, increases torsional stability and prevents shortening of the bone, especially in unstable fractures

[Perren, 1989]. The IM nail which has not been interlocked allows interfragmentary compressive force due to its low resistance to axial load. Another advantage of IM nails is that they do not require opening the fracture site, since they can be inserted through a small skin incision, typically located in one extreme of the bone. The insertion of an intramedullary nail, especially those that require reaming of the medullary canal, destroys the intramedullary vessels which supply two thirds of the cortex. However, this is not of clinical significance because revascularization occurs rapidly [Kessler et al., 1986; O'Slullivan et al., 1989].

9.1.2 Joint Replacements

Our ability to replace damaged joints with prosthetic implants has brought relief to millions of patients who would otherwise have been severely limited in their most basic activities and doomed to a life in pain. It is estimated that about 16 million people in the United States are affected by osteoarthritis, one of the various conditions that may cause joint degeneration and may lead a patient to a total joint replacement.

Joint degeneration is the end-stage of a process of destruction of the articular cartilage, which results in severe pain, loss of motion, and occasionally, an angular deformity of the extremity [Buckwalter et al., 1993]. Unlike bone, cartilage has a very limited capacity for repair [Salter, 1989]. Therefore, when exposed to a severe mechanical, chemical, or metabolic injury, the damage is permanent and often progressive.

Under normal conditions, the functions of cartilage are to provide a congruent articulation between bones, to transmit load across the joint, and to allow low-friction movements between opposing joint surfaces. The sophisticated way in which these functions are performed becomes evident from some of the mechanical characteristics of normal cartilage. For example, due to the leverage geometry of the muscles and the dynamic nature of human activity, the cartilage of the hip is exposed to about eight times body weight during fast walking [Paul, 1976]. Over a period of 10 years, an active person may subject the cartilage of the hip to more than 17 million weight bearing cycles [Jeffery, 1994]. From the point of view of the optimal lubrication provided by synovial fluid, cartilage's extremely low frictional resistance makes it 15 times easier to move opposing joint surfaces than to move an ice-skate on ice [Mow and Hayes, 1991; Jeffery, 1994].

Cartilage functions as a unit with subchondral bone, which contributes to shock absorption by undergoing viscoelastic deformation of its fine trabecular structure. Although some joints, like the hip, are intrinsically stable by virtue of their shape, the majority require an elaborate combination of ligaments, meniscus, tendons, and muscles for stability. Because of the large multidirectional forces that pass through the joint, its stability is a dynamic process. Receptors within the ligaments fire when stretched during motion, producing an integrated muscular contraction that provides stability for that specific displacement. Therefore, the ligaments are not passive joint restraints as once believed. The extreme complexity and high level of performance of biologic joints determine the standard to be met by artificial implants.

Total joint replacements are permanent implants, unlike those used to treat fractures, and the extensive bone and cartilage removed during implantation makes this procedure irreversible. Therefore, when faced with prosthesis failure and the impossibility to reimplant, the patient will face severe shortening of the extremity, instability or total rigidity of the joint, difficulty in ambulation, and often will be confined to a wheel chair.

The design of an implant for joint replacement should be based on the kinematics and dynamic load transfer characteristic of the joint. The material properties, shape, and methods used for fixation of the implant to the patient determines the load transfer characteristics. This is one of the most important elements that determines long-term survival of the implant, since bone responds to changes in load transfer with a remodeling process, mentioned earlier as Wollff's law. Overloading the implant-bone interface or shielding it from load transfer may result in *bone resorption* and subsequent loosening of the implant [Sarmiento et al., 1990]. The articulating surfaces of the joint should function with minimum friction and produce the least amount of wear products [Charnley, 1973]. The implant should be securely fixed to the body as early as possible (ideally immediately after implantation); however, removal of the implant should not require destruction of a large amount of surrounding tissues. Loss of tissue, especially

TABLE 9.3 Biomaterials for Total Joint Replacements

Materials	Properties	Application
Co–Cr alloy	Stem, head (ball)	Heavy, hard, stiff
(casted or wrought)	Cup, porous coating	High wear resistance
	Metal backing	
Ti alloy	Stem, porous coating	Low stiffness
	Metal backing	Low wear resistance
Pure titanium	Porous coating	Excellent osseousintegration
Tantalum	Porous structure	Excellent osseousintegration
	Good mechanical strength	
Alumina	Ball, cup	Hard, brittle
		High wear resistance
Zirconia	Ball	Heavy and high toughness
		High wear resistance
UHMWPE	Cup	Low friction, wear debris
		Low creep resistance
PMMA	Bone cement fixation	Brittle, weak in tension
		Low fatigue strength

Note: Stem: femoral hip stem/chondylar knee stem; head: femoral head of the hip stem; cup: acetabular cup of the hip.

TABLE 9.4 Types of Total Joint Replacements

Joint	Types
Hip	Ball and socket
Knee	Hinged, semiconstrained, surface replacement
	Unicompartment or bicompartment
Shoulder	Ball and socket
Ankle	Surface replacement
Elbow	Hinged, semiconstrained, surface replacement
Wrist	Ball and socket, space filler
Finger	Hinged, space filler

of bone, makes re-implantation difficult and often shortens the life span of the second joint replacement [Dupont and Charnley, 1972].

Decades of basic and clinical experimentation have resulted in a vast number of prosthetic designs and material combinations (Table 9.3 and Table 9.4) [Griss, 1984]. In the following section, the most relevant achievements in fixation methods and prosthetic design for different joints will be discussed at a conceptual level. Most joints can undergo partial replacement (hemiarthroplasty), that is, reconstruction of only one side of the joint while retaining the other. This is indicated in selected conditions when global joint degeneration has not taken place. This section will focus on total joint replacement, since this allows for a broader discussion of the biomaterials used.

9.1.2.1 Implant Fixation Method

The development of a permanent fixation mechanism of implants to bone has been one of the most formidable challenges in the evolution of joint replacement. There are three types of methods of fixation: (1) mechanical interlock, which is achieved by press-fitting the implant [Cameron, 1994a], by using polymethylmethacrylate, which is called bone cement, as a grouting agent [Charnley, 1979], or by using threaded components [Albrektsson et al., 1994]; (2) biological fixation, which is achieved by using textured or porous surfaces, which allow bone to grow into the interstices [Cameron, 1994b]; and (3) direct chemical bonding between implant and bone, for example, by coating the implant with calcium hydroxyapatite, which has a similar mineral composition to bone [Morscher, 1992]. Recently, direct bonding with bone was observed with Bioglass, a glass-ceramic, through selective dissolution of the surface film [Hench, 1994]; however, the likelihood of its clinical application is still under investigation.

Each of the fixation mechanisms has an idiosyncratic behavior, and their load transfer characteristics as well as the failure mechanisms are different. Further complexity arises from prostheses which combine two or more of the fixation mechanisms in different regions of the implant. Multiple mechanisms of fixation are used in an effort to customize load transfer to requirements of different regions of bone in an effort to preserve bone mass. Loosening, unlocking, or de-bonding between implant and bone constitute some of the most important mechanisms of prosthetic failure.

9.1.2.2 Bone Cement Fixation

Fixation of implants with polymethylmethacrylate (PMMA, bone cement) provides immediate stability, allowing patients to bare all of their weight on the extremity at once. In contrast, implants which depend on bone ingrowth require the patient to wait about 12 weeks to bear full weight.

Bone cement functions as a grouting material; consequently, its anchoring power depends on its ability to penetrate between bone trabeculae during the insertion of the prosthesis [Charnley, 1979]. Being a viscoelastic polymer, it has the ability to function as a shock absorber. It allows loads to be transmitted uniformly between the implant and bone, reducing localized high-contact stress.

Fixation with bone cement creates bone-cement and cement-implant interfaces, and loosening may occur at either one. The mechanisms to enhance the stability of the metal-cement interface constitute an area of controversy in joint replacement. Some investigators have focused their efforts on increasing the bond between metal and cement by roughening the implant, or pre-coating it with PMMA to prevent sinking of the prosthesis within the cement mantle, and circulation of debris within the interface [Park et al., 1978; Barb et al., 1982; Harris, 1988]. In contrast, others polish the implant surfaces and favor wedge-shaped designs which encourage sinking of the prosthesis within the cement, to profit from the viscoelastic deformation of the mantle by loading the cement in compression [Ling, 1992].

The problems with bone-cement interface may arise from intrinsic factors, such as the properties of the PMMA and bone, as well as extrinsic factors such as the cementing technique. Refinements in the cementing technique, such as pulsatile lavage of the medullary canal, optimal hemostasis of the cancellous bone, as well as drying of the medullary canal and pressurized insertion of the prosthesis, can result in a cement-bone interface free of gaps, with maximal interdigitation with cancellous bone [Harris and Davies, 1988]. Despite optimal cementing technique, a thin *fibrous membrane* may appear in various regions of the interface, due to various factors such as the toxic effect of free methylmethacrylate monomer, necrosis of the bone resulting from high polymerization temperatures, or devascularization during preparation of the canal [Goldring et al., 1983]. Although a fibrous membrane in the bone cement interface may be present in a well-functioning implant, it may also increase in width over time (most probably as a result of the accumulation of polyethylene wear debris from the bearing couple), and may result in macromotion, bone loss, and eventual loosening [Ebramzadeh et al., 1994]. Finally, the cement strength itself may be improved by removing air bubbles by mixing monomer and polymer under vacuum and/or centrifuging it [Harris, 1988]. During implantation, various devices are used to guarantee uniform thickness of the mantle to minimize risk of fatigue failure of the cement [Oh et al., 1978].

9.1.2.3 Porous Ingrowth Fixation

Bone ingrowth can occur with inert implants which provide pores larger than 25 μm in diameter, which is the size required to accommodate an osteon. For the best ingrowth in clinical practice, pore size range should be 100 to 350 μm and pores should be interconnected with each other with similar size of opening [Cameron, 1994b]. Implant motion inhibits bony ingrowth and large bone-metal gaps prolong or prevent the **osseointegration** [Curtis et al., 1992]. Therefore, precise surgical implantation and prevention of post-operative weight bearing for about 12 weeks are required for implant fixation.

The porous coated implants require active participation of the bone in the fixation of the implant, in contrast to cementation where the bone has a passive role. Therefore, porous coated implants are best indicated in conditions where the bone mass is near-normal. The implant design should allow ingrown

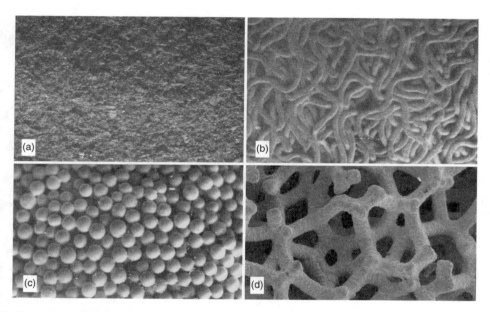

FIGURE 9.7 Scanning electron micrographs of four different types of porous structures: (a) plasma sprayed coating (7×), (b) sintered wire mesh coating (7×), (c) sintered beads coating (20×), and (d) Hedrocell porous tantalum (50×).

bone to be subjected to continuous loading within a physiologic range in order to prevent loss of bone mass due to **stress shielding**. Porous ingrowth prostheses are notoriously difficult to remove, and substantial bone damage often results from the removal process. For this reason, they should be optimized to provide predictable ingrowth with a minimal area of surgically accessible porous coated surface.

Commercially pure titanium, titanium alloy, tantalum, and calcium hydroxyapatite (HA) are currently used as porous coating materials. With pure titanium, three different types of porosity can be achieved: (1) plasma spray coating, (2) sintering of wire mesh, or (3) sintering of beads on an implant surface (Figure 9.7) [Morscher, 1992]. Thermal processing of the porous coating may weaken the underlying metal (implant). Additional problems may result from flaking of the porous coating materials, since loosened metal particles may cause severe wear when they migrate into the articulation (bearing couple) [Agins et al., 1988]. A thin calcium hydroxyapatite coating over the porous titanium surface has been used in an effort to enhance osseointegration; however, it improves only early-stage interfacial strength [Friedman, 1992; Capello and Bauer, 1994]. The long-term degradation and/or resorption of hydroxyapatite is still under investigation.

Recently, a cellular, structural biomaterial comprised of 15 to 25% tantalum (75 to 85% porous) has been developed. The average pore size is about 550 μm, and the pores are fully interconnected. The porous tantalum is a bulk material (i.e., not a coating) and is fabricated via a proprietary chemical vapor infiltration process in which pure tantalum is uniformly precipitated onto a reticulated vitreous carbon skeleton. The porous tantalum possesses sufficient compressive strength for most physiological loads, and tantalum exhibits excellent biocompatibility [Black, 1994]. This porous tantalum can be mechanically attached or diffusion bonded to substrate materials such as Ti alloy. Current commercial applications included polyethylene-porous tantalum acetabular components for total hip joint replacement and repair of defects in the acetabulum.

9.1.3 Total Joint Replacements

9.1.3.1 Hip Joint Replacement

The prosthesis for total hip replacement consists of a femoral component and an acetabular component (Figure 9.8a). The femoral stem is divided into head, neck, and shaft. The femoral stem is made of

(a)

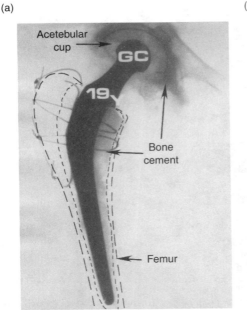

(b)

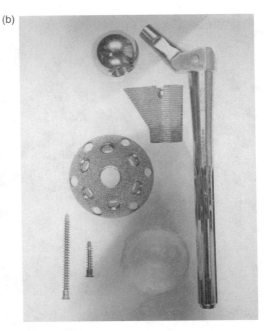

FIGURE 9.8   (a) Radiograph of bone cement fixed Charnley hip joint (monolithic femoral and acetabular component, 15-year follow-up). (b) Modular total hip system: head, femoral stem, porous coated proximal wedge, porous coated metal backing for cup, UHMWPE cup, and fixation screws.

Ti alloy or CoCr alloy (316L stainless steel was used earlier) and is fixed into a reamed medullary canal by cementation or press fitting. The femoral head is made of CoCr alloy, alumina, or zirconia. Although Ti alloy heads function well under clean articulating conditions, they have fallen into disuse because of their low wear resistance to third bodies, for example, bone or cement particles. The acetabular component is generally made of ultra-high molecular weight polyethylene (UHMWPE).

The prostheses can be monolithic when they consist of one part, or modular when they consist of two or more parts and require assembly during surgery. Monolithic components are often less expensive, and less prone to corrosion or disassembly. However, modular components allow customizing of the implant intraoperatively, and during future revision surgeries, for example, modifying the length of an extremity by using a different femoral neck length after the stem has been cemented in place, or exchanging a worn polyethylene bearing surface for a new one without removing the well-functioning part of the prosthesis from the bone. In modular implants (Figure 9.8b), the femoral head is fitted to the femoral neck with a Morse taper, which allows changes in head material and size, and neck length. Table 9.5 illustrates the most frequently used combinations of material in total hip replacement.

When the acetabular component is monolithic, it is made of UHMWPE; when it is modular, it consists of a metallic shell and a UHMWPE insert. The metallic shell seeks to decrease the microdeformation of the UHMWPE and to provide a porous surface for fixation of the cup [Skinner, 1992]. The metallic shell allows worn polyethylene liners to be exchanged. In cases of repetitive dislocation of the hip after surgery, the metallic shell allows replacing the old liner with a more constrained one, to provide additional stability. Great effort has been placed on developing an effective retaining system for the insert, as well as on maximizing the congruity between insert and metallic shell (Figure 9.8[b]). Dislodgment of the insert results in dislocation of the hip and damage of the femoral head, since it contacts the metallic shell directly. Micromotion between insert and shell produces additional polyethylene debris which can eventually contribute to bone loss [Friedman, 1994].

The hip joint is a ball-and-socket joint, which derives its stability from congruity of the implants, pelvic muscles, and capsule. The prosthetic hip components are optimized to provide a wide range of motion

TABLE 9.5 Possible Combination of Total Hip Replacements

without impingement of the neck of the prosthesis on the rim of the acetabular cup to prevent dislocation. The design characteristics must enable implants to support loads that may reach more than eight times body weight [Paul, 1976]. Proper femoral neck length and correct restoration of the center of motion and femoral offset decrease the bending stress on the prosthesis–bone interface. High-stress concentration or stress shielding may result in bone resorption around the implant. For example, if the femoral stem is designed with sharp corners (diamond shaped in a cross-section), the bone in contact with the corners of the implant may necrose and resorb.

Load bearing and motion of the prosthesis produces wear debris from the articulating surface, and from the interfaces were there is micromotion, for example, stem–cement interface. Bone chip, cement chip, or broken porous coating are often entrapped in the articulating space and cause severe polyethylene wear (third-body wear). The principal source of wear under normal conditions is the UHMWPE bearing surface in the cup. Approximately 150,000 particles are generated with each step and a large proportion of these particles are smaller than 1 μm. Cells from the immune system of the host, for example, *macrophages*, are able to identify the polyethylene particles as foreign and initiate a complex inflammatory response. This response may lead to rapid focal bone loss (**osteolysis**), bone resorption, loosening, and/or fracture of the bone. Recently, low-wear UHMWPE has been developed using a cross-linking of polyethylene molecular chains. There are several effective methods of cross-linking polyethylene, including irradiation of cross-linking, peroxide cross-linking, and silane cross-linking [Shen et al., 1996]. However, none of the cross-linked polyethylene has been clinically tested yet. Numerous efforts are underway to modify the material properties of articulating materials to harden and improve the surface finish of the femoral head [Friedman, 1994]. There is growing interest in metal–metal and ceramic–ceramic hip prostheses as a potential solution to the problem of osteolysis induced by polyethylene wear debris.

9.1.3.2 Knee Joint Replacements

The prosthesis for total knee joint replacement consists of femoral, tibial, and patellar components. Compared to the hip joint, the knee joint has a more complicated geometry and movement biomechanics, and it is not intrinsically stable. In a normal knee, the center of movement is controlled by the geometry of the ligaments. As the knee moves, the ligaments rotate on their bony attachments and the center of movement also moves. The eccentric movement of the knee helps distribute the load throughout the entire joint surface [Burstein and Wright, 1993].

The prostheses for total knee replacement (Figure 9.9) can be divided according to the extent to which they rely on the ligaments for stability: (1) *Constrained*: these implants have a hinge articulation, with a fixed axis of rotation, and are indicated when all of the ligaments are absent, for example in reconstructive procedures for tumoral surgery. (2) *Semi-constrained*: these implants control posterior displacement

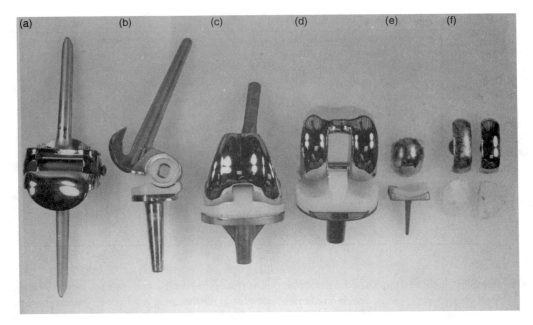

FIGURE 9.9 Various types of knee joints: (a) metal hinged, (b) hinged with plastic liner, (c) intramedullary fixed semiconstrained, (d) surface replacement, (e) uni-compartmental replacement, and (f) bi-compartmental replacement.

of the tibia on the femur and medial-lateral angulation of the knee, but rely on the remaining ligaments and joint capsule to provide the rest of the constraint. Semi-constrained prostheses are often used in patients with severe angular deformities of the extremities, or in those that require revision surgery, when moderate ligamentous instability has developed. (3) *Non-constrained*: these implants provide minimal or no constraint. The prosthesis that provides minimal constraint requires resection of the posterior cruciate ligament during implantation, and the prosthetic constraint reproduces that normally provided by this ligament. The ones that provide no constraint spare the posterior cruciate ligament. These implants are indicated in patients who have joint degeneration with minimal or no ligamentous instability. As the degree of constraint increases with knee replacements, the need for the use of femoral and tibial intramedullary extensions of the prosthesis is greater, since the loads normally shared with the ligaments are then transferred to the prosthesis–bone interface.

Total knee replacements can be implanted with or without cement, the latter relying on porous coating for fixation. The femoral components are typically made of CoCr alloy and the monolithic tibial components are made of UHMWPE. In modular components, the tibial polyethylene component assembles onto a titanium alloy tibial tray. The patellar component is made of UHMWPE, and a titanium alloy back is added to components designed for uncemented use. The relatively small size of the patellar component compared to the forces that travel through the extensor mechanism, and the small area of bone available for anchorage of the prosthesis, make the patella vulnerable.

The wear characteristic of the surface of tibial polyethylene is different from that of acetabular components. The point contact stress and sliding motion of the components result in delamination and fatigue wear of the UHMWPE [Walker, 1993]. Presumably because of the relatively larger particle size of polyethylene debris, osteolysis around a total knee joint is less frequent than in a total hip replacement.

9.1.3.3 Ankle Joint Replacement

Total ankle replacements have not met with as much success as total hip and knee replacements, and typically loosen within a few years of service [Claridge et al., 1991]. This is mainly due to the high load

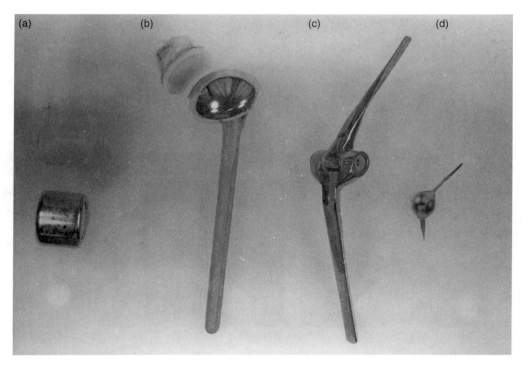

FIGURE 9.10 Miscellaneous examples of prostheses for total joint replacement: (a) ankle, (b) socket-ball shoulder joint, (c) hinged elbow joint, and (d) encapsulated finger joint.

transfer demand over the relatively small ankle surface area, and the need to replace three articulating surfaces (tibial, talar, and fibular). The joint configurations that have been used are cylindrical, reverse cylindrical, and spherical. The materials used to construct ankle joints are usually CoCr alloy and UHMWPE. Degeneration of the ankle joint is currently treated with fusion of the joint, since prostheses for total ankle replacement are still considered to be under initial development. Figure 9.10 shows ankle and other total joint replacements.

9.1.3.4 Shoulder Joint Replacements

The prostheses for total shoulder replacement consist of a humeral and a glenoid component. Like the femoral stem, the humeral component can be divided into head, neck, and shaft. Variations in the length of the neck result in changes in the length of the extremity. Even though the patient's perception of length of the upper extremity is not as accurate as that of the lower extremity, the various lengths of the neck are used to fine-tune the tension of the soft tissues to obtain maximal stability and range of motion.

The shoulder has the largest range of motion in the body, which results from a shallow ball and socket joint, which allows a combination of rotation and sliding motions between the joint surfaces. To compensate for the compromise in congruity, the shoulder has an elaborate capsular and ligamentous structure, which provides the basic stabilization. In addition, the muscle girdle of the shoulder provides additional dynamic stability. A decrease in the radius of curvature of the implant to compensate for soft tissue instability will result in a decrease in the range of motion [Neer, 1990].

9.1.3.5 Elbow Joint Replacements

The elbow joint is a hinge-type joint allowing mostly flexion and extension, but having a polycentric motion [Goel and Blair, 1985]. The elbow joint implants are either hinged, semi-constrained, or unconstrained. These implants, like those of the ankle, have a high failure rate and are not used commonly. The high loosening rate is the result of high rotational moments, limited bone stock for fixation, and minimal

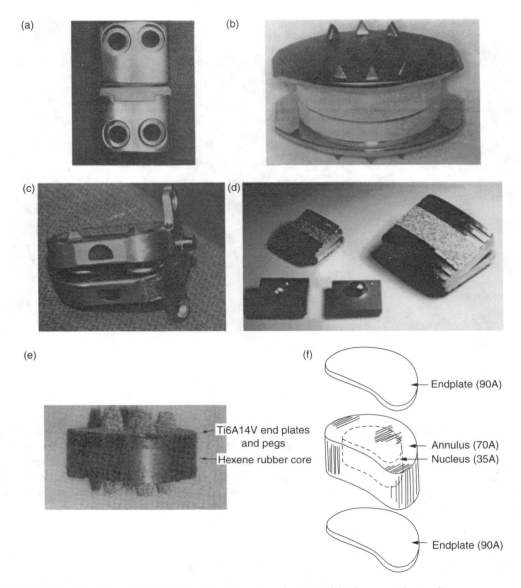

FIGURE 9.11 Experimental artificial discs "used" to restore function of the degenerated spine disc.

ligamentous support [Morrey, 1993]. In contrast to fusions of the ankle which function well, fusions of the elbow result in a moderate degree of incapacitation.

9.1.3.6 Finger Joint Replacements

Finger joint replacements are divided into three types: (1) hinge, (2) polycentric, and (c) space-filler. The most widely used are the space-filler type. These are made of high performance silicone rubber (polydimethylsiloxane) and are stabilized with a passive fixation method. This method depends on the development of a thin, fibrous membrane between implant and bone, which allows pistoning of the prosthesis. This fixation can provide only minimal rigidity of the joint [Swanson, 1973]. Implant wear and cold flow associated with erosive cystic changes of adjacent bone have been reported with silicone implants [Carter et al., 1986; Maistrelli, 1994].

9.1.3.7 Prosthetic Intervertebral Disk

Because of adjacent-level degeneration and other complications, such as non-fusion, alternatives to fusions have been proposed. One of the most recent developments for non-fusion treatment alternatives is replacement of the intervertebral discs [Hedman et al., 1991]. The goal of this treatment alternative is to restore the original mechanical function of the disc. One of the stipulations of artificial disc replacement is that remaining osseous spinal and paraspinal soft tissue components are not compromised by pathologic changes. Several artificial disk prostheses have been developed to achieve these goals (Figure 9.11).

9.1.3.8 Prostheses for Limb Salvage

Prosthetic implant technology has brought new lifestyles to thousands of patients who would lose their limbs due to bone cancer. In the past, the treatment for primary malignant bone cancer of the extremities was amputation. Significant advances in bone tumor treatment have taken place during the last two decades. The major treatment methods for limb reconstruction following bone tumor resection are resection arthrodesis (fusion of two bones), allograft-endoprosthetic composite, and endoprosthetic reconstruction. Endoprosthetic reconstruction is an extension of a total joint replacement(s) component to the resected bone area, and is the most popular option due to an advantage of fast postoperative recovery (Figure 9.12). Therefore, material and fixation methods for limb salvage endoprostheses are exactly the same as for total joint replacements. Femur, tibia, humerus, pelvis, and scapular are often resected and replaced endoprostheses. Similar to total joint replacements, disadvantages of the endoprosthetic

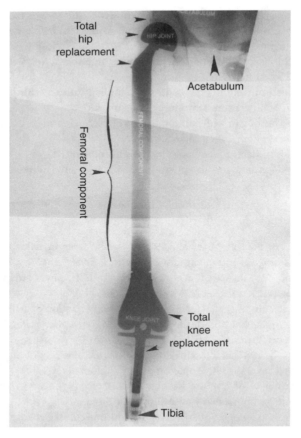

FIGURE 9.12 Radiographic appearance (montage) of a modular endoprosthetic replacement. Entire femur, hip joint, and knee joint of the bone tumor patient were replaced with prostheses for a limb salvage.

reconstruction are prosthesis loosening due to polyethylene wear and cement failure, and mechanical failure of the prostheses.

Most of the endoprostheses for limb salvage are of the expandable type [Ward et al., 1996]. Expandable endoprostheses are required for children who have a potential for skeletal growth. Several expandable prostheses require an open surgical procedure to be lengthened, whereas others have been developed that can be lengthened by servomechanisims within the endoprosthesis. The modular segmental system is a new option for expandable endoprosthesis. They can be revised easily to elongate modular components to gain length over time. The modular segmental system has several advantages over the mechanically expandable one. The use of the modular system allows intraoperative customization of the endoprosthesis during surgery. It minimizes discrepancies between custom implants and actual skeletal defects due to the radiographic magnification and uncertainty of margin of tumor resection. It allows the surgeon to assemble the prosthesis intraoperatively. The cost of the modular system is less than the cost of an expandable custom endoprosthesis. The modular system allows for simpler and less expensive revision when failures occur, obviating the need for an entirely new prosthesis when only one part needs to be replaced. On the other hand, the modular system has a high chance of corrosion failure at the Morse tapers and dislodgment at the Morse taper fittings. It can be lengthened only in certain increments. The modular component system has additional applications: metastatic bone disease, failure of internal fixation, severe acute fractures with poor bone quality, and failure of total joints with insufficient bone stock.

Defining Terms

Bone resorption: A type of bone loss due to the greater osteoclastic activity than the osteogenic activity.
Callus: Unorganized meshwork of woven bone which is formed following fracture of bone to achieve early stability of the fracture.
Fibrous membrane: Thin layer of soft tissue which covers an implant to isolate it from the body.
Necrosis: Cell death caused by enzymes or heat.
Osseointegration: Direct contact of bone tissues to an implant surface without fibrous membrane.
Osteolysis: Dissolution of bone mineral from the bone matrix.
Primary Healing: Bone healing in which union occurs directly without forming callus.
Secondary Healing: Bone union with a callus formation.
Stress shielding: Bone is protected from stress by the stiff implant.
Wolff's Law: Bone develops or adapts its structure to that most suited to resist the forces acting upon it.

References

Agins, H.J., Alcock, N.W., Bansal, M. et al., 1988. Metallic wear in failed Titanium-alloy total hip replacements. A istological and quantitative analysis. *J. Bone. Joint Surg.* 70-A: 347.

Albrektsson, T., Carlsson, L.V., Morberg, P. et al., 1994. Directly bone-anchored implants. In *Bone Implant Interface*, R. Hurley (Ed.), pp. 97–120, Mosby, St. Louis, MO.

Barb, W., Park, J.B., von Recum, A.F. et al., 1982. Intramedullary fixation of artificial hip joints with bone cement precoated implants: I. Interfacial strengths. *J. Biomed. Mater. Res.* 16: 447.

Black, J. 1994. Biological performance of tantalum. *Clin. Mater.* 16: 167.

Brand, R.A. and Rubin, C T. 1987. Fracture healing. In *Scientific Basics of Orthopaedics*, 2nd ed. J. Albert and R. Brand (Eds.), pp. 325–340, Appleton & Lange, Norwalk, CT.

Brighton, C.T. 1984. Principle of fracture healing, In *Instructional Course Lectures*, J. Murray (Ed.), pp. 60–106, The American Academy of Orthopaedic Surgeons.

Buckwalter, J.A., Woo, S. et al., 1993. Soft-tissue aging and musculoskeletal function. *J. Bone Joint Surg.* 75A: 1533.

Burstein, A.H. and Wright, T.H. 1993. Biomechanics. In *Surgery of the Knee*, 2nd ed., Vol. 7, J. Insall, R. Windsor, W. Scott et al. (Eds.), pp. 43–62, Churchill Livingstone, New York.

Cameron, H.U. 1994a. Smooth metal–bone interface. In *Bone Implant Interface*, R. Hurley (Ed.), pp. 121–144, Mosby, St. Louis, MO.

Cameron, H.U. 1994b. The implant–bone interface: porous metals. In *Bone Implant Interface*, R. Hurley (Ed.), pp. 145–168, Mosby, St. Louis, MO.

Capello, W.N. and Bauer, T.W. 1994. Hydroxyapatite in orthopaedic surgery. In *Bone Implant Interface*, R. Hurley (Ed.), pp. 191–202, Mosby, St. Louis, MO.

Carter, P., Benton, L., and Dysert, P. 1986. Silicone rubber carpal implants; A study of the incidence of late osseous complications, *J. Hand Surg.* 11A: 639.

Charnley, J. 1979. *Low Friction Arthroplasty of the Hip*. Springer-Verlag, Berlin.

Claes, L. 1989. The mechanical and morphological properties of bone beneath internal fixation plates of differing rigidity. *J. Orthop. Res.* 7: 170.

Claridge, R.J., Hart, M.B., Jones, R.A. et al., 1991. Replacement arthroplasties of the ankle and foot. In *Disorder of the Foot & Ankle*, 2nd ed. M. Jahss (Ed.), pp. 2647–2664, Saunders, Philadephia, PA.

Cochran, G.V.B. 1982. Biomechanics of orthopaedic structures. In *Primer in Orthopaedic Biomechanics*, pp. 143–215, Churchill Livingstone, New York.

Curtis, M.J., Jinnah, R.H., Wilson, V.D., and Hungerford, D.S. 1992. The initial stability of uncemented acetabular components. *J. Bone Joint Surg.* 74B: 372.

DeCoster, T.A., Heetderks, D.B. et al., 1990. Optimizing bone screw pullout force. *J. Orthop. Trauma* 4: 169.

Dupont, J.A. and Charnley, J. 1972. Low-friction arthroplasty of the hip for the failures of previous operations. *J. Bone Joint Surg.* 54B: 77.

Ebramzadeh, E., Sarmiento, A., McKellop, H.A. et al., 1994. The cement mantle in total hip arthroplasty. Analysis of long-term radiographic results. *J. Bone Joint Surg.* 76A: 77–87.

Friedman, R.F., Black, J., Galante, J.O. et al., 1994. Current concepts in orthopaedic biomaterials and implant fixation. In *Instructional Course Lectures*, J.M. Schafer (Ed.), pp. 233–255. The American Academy of Orthopaedic Surgeons.

Friedman, R.J. 1992. Advanced in biomaterials and factors affecting implant fixation. In *Instructional Course Lectures*, R.E. Eilert (Ed.), pp. 127–135, The American Academy of Orthopaedic Surgeons.

Goel, V.K. and Blair, W. 1985. Biomechanics of the elbow joint. *Automedica* 6: 119.

Goldring, S.R., Schiller, A.L., Roelke, M. et al., 1983. The synovial-like membrane at the bone–cement interface in loose total hip replacements and its proposed role in bone lysis. *J. Bone Joint Surg.* 65A: 575.

Griss, P. 1984. Assessment of clinical status of total joint replacement. In *Functional Behavior of Orthopaedic Biomaterials*, P. Ducheyne and G.W. Hastings (Eds.), pp. 21–48, CRC Press, Boca Raton, FL.

Harris, W.H. and Davies, J.P. 1988. Modern use of modern cement for total hip replacement. *Orthop. Clin. N. Am.* 19: 581.

Hedman, T.P., Kostuik, P.J., Fernie, G.R. et al., 1991. Design of an intervertebral disc prosthesis. *Spine* 16: 256.

Hench, L.L. 1994. Bioactive glasses, ceramics and composites. In *Bone Implant Interface*, R. Hurley (Ed.), pp. 181–190, Mosby, St. Louis, MO.

Hulth, A. 1989. Current concepts of fracture healing. *Clin. Orthop.* 249: 265.

Hutzschenreuter, P. and Brümmer, H. 1980. Screw design and stability. In *Current concepts of Internal Fixation*, H. Uhthoff (Ed.), pp. 244–250, Springer-Verlag, Berlin.

Jeffery, A.K. 1994. Articular cartilage and the orthopaedic surgeon. Part 1: structure and function. *Curr. Orthop.* 8: 38.

Kessler, S.B., Hallfeldt, K.K., Perren, S.M. et al., 1986. The effects of reaming and intramedullary nailing on fracture healing. *Clin. Orthop.* 212: 18.

Ling, R.S. 1992. The use of a collar and precoating on cemented femoral stems is unnecessary and detrimental. *Clin. Orthop.* 285: 73.

Maistrelli, G.L. 1994. Polymer in orthopaedic surgery. In *Bone Implant Interface*, R. Hurley (Ed.), pp. 169–190, Mosby, St. Louis, MO.

McKellop, H.A., Campbell, P., Park, S.H. et al., 1995 The origin of submicron polyethylene wear debris in total hip arthroplasty. *Clin. Orthop.* 311: 3.

Morrey, B.F. 1993. *The Elbow and its Disorders*, 2nd ed, Saunders, Philadelphia, PA.

Morscher, E.W. 1992. Current status of acetabular fixation in primary total hip arthroplasty. *Clin. Orthop.* 274: 172.

Mow, V.C. and Hayes, W.C. 1991. *Basic Orthopaedic Biomechanics*, Raven Press, New York.

Neer, C.S. 1990. *Shoulder Reconstruction*, Saunders, Philadelphia, PA.

Oh, I., Carlson, C.E., Tomford, W.W. et al., 1978. Improved fixation of the femoral component after total hip replacement using a methacrylate intramedullary plug. *J. Bone Joint Surg.* 60A: 608.

O'Slullivan, M.E., Chao, E.Y.S., and Kelly, P.J. 1989. Current concepts review. The effects of fixation on fracture healing. *J. Bone Joint Surg.* 71A: 306.

Park, J.B. and Lakes, R.S. 1992. *Biomaterials: An Introduction*, 2nd ed, Plenum, London.

Park, J.B., Malstrom, C.S., and von Recum, A.F. 1978 Intramedullary fixation of implants precoated with bone cement: A preliminary study. *Biomater. Med. Dev. Artif. Organs* 6: 361.

Paul, J.P. 1976. Loading on normal hip and knee joints and joint replacement., In *Advances in Hip and Knee Joint Technology*, M. Schaldach and D. Hohmann (Eds.), pp. 53–77, Springer-Verlag, Berlin.

Perren, S.M., 1989. The biomechnics and biology of internal fixation using plates and nails. *Orthopaedics* 12: 21.

Perren, S.M. 1991. Basic aspects of internal fixation. In *Manual of Internal Fixation*, 3rd ed., M. Müller, M. Allgöwer, R. Schneider, and H. Willenegger (Eds.), pp. 1–112, Springer-Verlag, Berlin.

Salter, R.B. 1988. The biologic concept of continuous passive motion of synovial joints. The first 18 years of basic research and its clinical application. *Clin. Orthop.* 242: 12.

Sarmiento, A., Ebramzadeh, E., Gogan, W.J. et al., 1990. Cup containment and orientation in cemented total hip arthroplasties. *J. Bone Joint Surg.* 72B: 996.

Schatzker, J., Sanderson, R., and Murnaghan, J.P. 1975. The holding power of orthopaedic screws *in vivo*. *Clin. Orthop.* 108: 115.

Schenk, R.K. 1992. Biology of fracture repair. In *Skeletal Trauma*, B. Browner, J. Jupitor, A. Levine, and P. Trafton (Eds.), pp. 31–75, W.B. Saunders, Philadephia, PA.

Shen, F.W., McKellop, H., and Salovey, R. 1996. Irradiation of chemically crosslinked ultrahigh molecular weight polyethylene. *J. Polym. Sci.: Part B: Polym. Phys.* 34: 1063.

Skinner, H.B. 1992. Current biomaterial problems in implants. In *Instructional Course Lectures*, R.E. Eilert (Ed.), pp. 137–144, The American Academy of Orthopaedic Surgeons.

Swanson, A.B. 1973. Concepts of flexible implant design. In *Flexible Implant Reconstruction Arthroplasty in the Hand and Extremity*, A. Swanson (Ed.), pp. 47–59, Mosby, St. Louis, MO.

Taitsman, J.P. and Saha, S. 1977. Tensile strength of wire-reinforced bone cement and twisted stainless-steel wire. *J. Bone Joint Surg.* 59A: 419.

Tencer, A.F., Johnson, K.D., Kely, R.F. et al., 1993. Biomechanics of fractures and fracture fixation. In *Instructional Course Lectures*, J.D. Heckman (Ed.), pp. 19–55, The American Academy of Orthopaedic Surgeons.

Uhthoff, H.K. and Finnegan, M.A. 1984. The role of rigidity in fracture fixation. *Arch. Orthop. Trauma Surg.* 102: 163.

Walker, P.S. 1993. Design of total knee arthroplasty. In *Surgery of the Knee*, 2nd ed., Vol. 7, J. Insall, R. Windsor, W. Scott et al. (Eds.), pp. 723–738, Churchill Livingstone, New York.

Ward, W.G., Yang, R.S., and Eckart, J.J. 1996. Endoprosthetic bone reconstruction following malignant tumor resection in skeletally immature patients. *Orthop. Clin. N. Am.* 27: 493.

Weinstein, A.M., Spires, W.P. Jr, Klawitter et al., 1979. Orthopaedic implant retrieval and analysis study. In *Corrosion and Degradation of Implant Materials*, B.C. Syrett and A. Acharya (Eds.), pp. 212–228, American Society for Testing and Materials Tech. Pub. No. 684, Philadelphia, PA.

Wolff, J. 1986. *The Law of Bone Remodeling*, R. Maquet and R. Furlong (trans.), Springer-Verlag, Berlin.

9.2 Dental Implants: The Relationship of Materials Characteristic to Biologic Properties

J.C. Keller

As dental implants have become an acceptable treatment modality for partially and fully edentulous patients, it has become increasingly apparent that the interaction of the host tissue with the underlying implant surface is of critical importance for long-term prognosis [Young, 1988; Smith, 1993]. From the anatomical viewpoint, it is generally accepted that dental implants must contact and become integrated with several types of host tissues. Due largely to the work of Branemark and his colleagues [Albrektsson et al., 1983; Branemark, 1983], the importance of developing and maintaining a substantial bone-implant interface for mechanical retention and transmission of occlusal forces was realized. Despite documented long-term success of dental implants, longer-term implant failures are noted due to poor integration of connective and epithelial tissues and subsequent failure to develop a permucosal seal akin to that with natural tooth structures. From a biologic point of view, the characteristics of the implant substrate which permit hard and soft tissue integration and prevent adhesion of bacteria and plaque need to be further understood. It is likely that as a more complete understanding of the basic biologic responses of host tissues becomes known, refinements in the currently employed materials as well as new and improved materials will become available for use in the dental implant field.

It is important to realize that the overall biologic response of host tissue to dental implants can be divided into two distinct but interrelated phases (as given in Table 9.6). Phase I consists of the tissue responses which occur during the clinical healing phase immediately following implantation of dental implants. During this healing phase, the initial biologic processes of protein and molecular deposition on the implant surface are followed by cellular attachment, migration, and differentiation [Stanford and Keller, 1991]. It is therefore important to understand the characteristics of the implant material which affect the initial formation of the host tissue–implant interface. The characteristics include materials selection and the physical and chemical properties of the implant surface. These initial tissue responses lead to the cellular expression and maturation of extracellular matrix and ultimately to the development of bony interfaces with the implant material. After the initial healing phase is complete, usually between 3 and 6 months according to the two-stage Branemark implant design, the bone interface remodels under the occlusal forces placed on the implant during the Phase II functional period [Skalak, 1985; Brunski, 1992]. The overall bioresponses including bone remodeling during the functional phase of implant service life are then strongly influenced by the characteristics of loading and distribution of stress at the interface [Brunski, 1992]. The ability of the maturing interface to "remodel" as stresses are placed on the implant thus depend in large part on the original degree of tissue–implant surface interaction.

TABLE 9.6 Correlation Between the Clinical Phases of Implant Service Life with Biologic Events and Important Implant Materials Characteristics

Clinical phase	Biological events	Influential materials characteristics
I (healing)	Protein deposition	Materials selection
	Cell attachment	Metals
	Cell migration	Ceramics
	Development of extracellular matrix	Chemical and physical characteristics
	Bone deposition	Topography
		Micro
		Macro
		Surface chemistry
		Inert
		Dissolution
II (functional)	Matrix and bone remodeling	

TABLE 9.7 Approximate Room Temperature Mechanical Properties of Selected Implant Materials
Compared to Bone

	Elastic modulus $(MPa \times 10^3)$	Proportional limit (MPa)	Ultimate tensile strength (MPa)	Percent elongation
316L SS				
Annealed	200	240	550	50
Cold worked	200	790	965	20
CoCrMo(ASTM-F75)	240	500	700	10
Ti(ASTM-F67)	100	520	620	18
Ti-6A1-4V(ASTM-F136)	110	840	900	12
Cortical bone	18	130	140	1

Source: Keller, J.C. and Lautenschlager, E.P. 1986. Metals and alloys. In *Handbook of Biomaterials Evaluation*, A. Von Recon (Ed.), pp. 3–23, New York, Macmillan.

This chapter will focus on the factors concerning dental implants which affect biologic properties of currently available dental implant materials. As pertains to each major topic, the influence of the materials properties on biologic responses will be emphasized.

9.2.1 Effects of Materials Selection

9.2.1.1 Metals and Alloys

Previously, dental implants have been fabricated from several metallic systems, including stainless steel and cobalt–chrome alloys as well as from the titanium family of metals. Several studies have reported on the ability of host bone tissues to "integrate" with various metal implant surfaces [Albrektsson et al., 1983; Katsikeris et al., 1987; Johansson et al., 1989]. In current paradigms, the term osseointegration refers to the ability of host tissues to form a functional interface with implant surfaces without an intervening layer of connective tissue akin to a foreign body tissue capsule observable at the light microscopic level [Albrektsson et al., 1983; Branemark, 1983]. By this definition, it becomes apparent that several biomedical materials including Ti and Ti alloy fulfill this general criterion. Ultrastructural investigations using transmission electron microscopy (TEM) approaches have further refined descriptions of the tissue implant interface, and the early work by Albrektsson and colleagues [1983] has become the descriptive standard by which other materials interfaces are compared. When bone was allowed to grow on Ti, a partially calcified amorphous ground substance was deposited in immediate contact with the implant, followed by a collagenous fibril-based extracellular matrix, osteoblast cell processes, and a more highly calcified matrix generally 200 to 300 Å from the implant surface.

However, other metallic materials have fallen from favor for use as dental implants due to widely differing mechanical properties compared to bone (Table 9.7), which can result in a phenomenon termed stress shielding [Slalak, 1985; Brunski, 1992], and the propensity for formation of potentially toxic corrosion products due to insufficient corrosion resistance properties [Van Orden, 1985; Lucas et al., 1987]. Ultrastructurally, the interface between bone and 316L stainless steel was described as consisting of a multiple-cell layer separating the bone from metal. Inflammatory cells were prominent in this layer, and a thick proteoglycan noncollagenous coating was present. This histologic appearance resembled that of a typical foreign body reaction and typifies a nonosseointegration-type response. The poor biologic response to stainless steel alloys has been reconfirmed by recently conducted *in vitro* studies which related the inability of host tissue to attach to the metal surface to the toxicity associated with metal ion release [Vrouwenvelder et al., 1993].

Due in large measure to the introduction and overall clinical success of the Branemark system, the range of metallic materials utilized for dental implants has become limited largely to commercially pure titanium (cpTi > 99.5%) and its major alloy, Ti–6A1–4V [De Porter et al., 1986; Keller et al., 1987]. Controversy remains, largely due to commercial advertising interests, as to which material provides a more suitable

TABLE 9.8 Surface Characterizations of cpTi and Ti Alloy
(means ± standard deviations)

	cpTi	Ti-6A1-4V
Surface roughness (Ra) (μm)		
Sandblasted	0.9 ± 0.2	0.7 ± 0.03
600 grit polish	0.2 ± 0.02	0.1 ± 0.02
Smooth, 1 μm polish	0.04 ± 0.01	0.03 ± 0.01
Atomic ratios to Ti		
C	1.5 ± 0.2	1.2 ± 0.1
O	2.8 ± 0.1	3.1 ± 0.2
N	0.08 ± 0.01	0.05 ± 0.01
Al	—	0.2 ± 0.04
V[a]	—	(0.02)[a]
Oxide thickness (Å)	32 ± 8	83 ± 12
Wetting angles (°)	52 ± 2	56 ± 4

[a] One specimen.

surface for tissue integration. Early work by Johansson and coworkers [1989] reported that sputtercoated Ti alloy surfaces resulted in wide (5000 Å) **amorphous zones** devoid of collagen filaments, compared to the thinner 200 to 400 Å collagen-free amorphous zone surrounding cpTi surfaces. Subsequent studies revealed differences in the oxide characteristics between these sputtercoated cpTi and Ti alloy surfaces used for histologic and ultrastructural interfacial analyses. Significant surface contamination and the presence of V was observed in the Ti alloy surface, which led to an overall woven bone interface compared to the cpTi surface which had a more compact bone interface. Orr and colleagues [1992] demonstrated a similar ultrastructural morphology for interfaces of bone to Ti and Ti alloy, respectively. In each case an afibrillar matrix with calcified globular accretions, similar in appearance to cement lines in haversian systems, were observed in intimate contact with the oxide surface. Any slight differences in morphology were attributed to minor differences in surface topography or microtexture rather than chemical differences between the two surface oxides.

This is an area that is still under investigation, although recent research indicates that the stable oxide of cpTi and Ti alloy provide suitable surfaces for biologic integration [Keller et al., 1994]. Studies involving comprehensive surface analyses of prepared bulk cpTi and Ti alloy indicated that although the oxide on Ti alloy is somewhat thicker following standard surface preparations (polishing, cleaning, and acid passivation), the overall topography, the chemical characteristics, including presence and concentration of contaminants, and surface energetics were virtually identical for both materials as given in Table 9.8. *In vitro* experiments confirmed that the inherently clean condition of these oxides supports significant osteoblast cell attachment and migration and provides a hospitable surface to allow *in vitro* mineralization processes to occur [Orr et al., 1992].

In vitro experiments designed to study the ultrastructural details of bone-implant interfaces made from cpTi and Ti alloy may provide additional clues as to the histologic and ultrastructural differences which have been observed with these materials. Since clinical implants made from both materials appear to be successful [Branemark, 1983; De Porter et al., 1986], it is possible that because of the difference in mechanical properties between unalloyed and Ti alloy material, the longer-term tissue interface results from differences in bone remodeling due to the local biomechanical environment surrounding these materials [Brunski, 1992]. This hypothesis requires continued investigation for more definitive answers.

9.2.1.2 Ceramics and Ceramic Coatings

The use of single crystal sapphire or Al_2O_3 ceramic implants has remained an important component in the dental implant field [Driskell et al., 1973]. Although this material demonstrates excellent biologic compatibility, implants fabricated from Al_2O_3 have not reached a high degree of popularity in the United

States. Morphologic analyses of the soft tissue interface with Al_2O_3 revealed a hemidesmosomal external lamina attachment adjacent to the **junctional epithelium** — implant interface [Steflik et al., 1984]. This ultrastructural description is often used for comparison purposes when determining the extent of soft-tissue interaction with dental implants. Similarly, *in vivo* studies of the bone — Al_2O_3 implant interface reveal high levels of bone-to-implant contact, with areas of intervening fibrous connective tissue. Although fibrous tissue was present at the interface, the implant remained immobile, and the interface was consistent with a dynamic support system. More recent ultrastructural studies have demonstrated a mineralized matrix in immediate apposition to the Al_2O_3 implants similar to that described for Ti implants [Steflik et al., 1993].

An approach to enhancing tissue responses at dental implant interfaces has been the introduction of ceramics like, *calcium-phosphate-containing (CP) materials* as implant devices. The use of calcium-phosphate materials, in bulk or particulate form or as coatings on metal substrates has taken a predominant position in the biomedical implant area and has been the focus of several recent reviews [Koeneman et al., 1990; Kay, 1992]. One of the most important uses of CP materials has been as a coating on metallic (cpTi and Ti alloy) substrates. This approach has taken advantage of thin-film-coating technology to apply thin coatings of hydroxyapatite (HA) and tricalcium phosphate (TCP) materials to the substrate in order to enhance bone responses at implant sites. The most popular method of coating has been the plasma spray process [Herman, 1988]; although this process has some advantages, there are reports of nonuniform coatings, interfacial porosity, and vaporization of elements in the powder [Cook et al., 1991; Kay, 1992].

The use of this class of materials is based on the premise that a more natural hydroxyapatite (HA-like) could act as a scaffold for enhanced bone response — osseointegration — and thereby minimize the long-term healing periods currently required for uncoated metal implants.

Numerous *in vivo* investigations have clearly demonstrated that HA-like coatings can enhance bone responses at implant interfaces [Cook et al., 1991; Jarcho et al., 1997], although the mechanisms responsible for the development of the interface between hard tissue and these ceramic coatings are not well understood [Jansen et al., 1993]. Histologically, the overall bone-coating interface is similar in appearance and chronologic development to that reported for uncoated implant surfaces. Initially, an immature, trabecular, woven bone interface is formed followed by more dense, compact lamellar supporting bone structure. Ultrastructurally, the interface is reported to consist of a globular, afibrillar matrix directly on the HA surface, an electron-dense, proteoglycan rich layer (20 to 60 nm thick) and the presence of a mineralized collagenous matrix [De Bruijn et al., 1993]. Although the morphologic descriptions of the HA and Ti interfaces are similar, numerous studies have shown that the bone response to HA coatings is more rapid than with uncoated Ti surfaces, requiring approximately one-third to one-half the time to establish a firm osseous bed as uncoated Ti. Likewise, the extent of the bone response to HA coatings is superior and, according to some studies, leads to a several-fold increase in interfacial strength compared to uncoated Ti [Cook et al., 1991].

The cellular events which take place and lead to the interfacial ultrastructure with bone tissue and ceramic surfaces are under current investigation. Based upon preliminary findings, the advantageous biologic properties of HA coatings do not appear to be related to recruitment of additional cells during the early attachment phase of healing. Although recent work indicated that bone cells and tissue form normal cellular focal contacts during attachment to HA coatings, the level of initial *in vitro* attachment generally only approximates that observed with Ti [Puleo et al., 1991; Keller et al., 1992]. Rather, the mechanisms for the enhanced *in vitro* cell responses appear to be related, to a certain degree, to the degradation properties and release of Ca^{+2} and PO_4^{-3} ions into the biologic milieu. This surface corrosion is associated with highly degradable amorphous components of the coating and leads to surface irregularities which may enhance the quality of cell adhesion to these roughened materials [Bowers et al., 1992; Chehroudi et al., 1992]. Cellular events which occur following attachment may be influenced by the nature of the ceramic surface. Emerging evidence from a number of laboratories suggests that cellular-mediated events, including proliferation, matrix expression, and bone formation are enhanced following attachment to HA coatings and appear to be related to the gene expression of osteoblasts when cultured on different ceramic materials. These early cellular events lead to histologic and ultrastructural descriptions of bone healing which take place on these surfaces and are very similar to those reported from *in vivo* studies [Orr et al., 1992; Steflik et al., 1993].

As determined from *in vitro* dissolution studies, there is general agreement that the biodegradation properties of the pertinent CP materials can be summarized as α-TCP > β-TCP >>> HA, whereas amorphous HA is more prone to biodegradation than crystalline HA [Koeneman et al., 1990]. Considerable attempts to investigate the effects of coating composition (relative percentages of HA, TCP) on bone integration have been undertaken. Using an orthopedic canine total hip model, Jasty and coworkers [1992] reported that by 3 weeks, a TCP/HA mixed coating resulted in significantly more woven bone apposition to the implants than uncoated implants. As determined by x-ray diffraction, the mixed coating consisted of 60% TCP, 20% crystalline HA, and 20% unknown Ca-PO$_4$ materials. Jansen and colleagues [1993] reported that, using HA-coated implants (90% HA, 10% amorphous CP), bony apposition was extensive in a rabbit tibia model by 12 weeks. However, significant loss of the coating occurred as early as 6 weeks after implantation. Most recently, Maxian and coworkers [1993] reported that poorly crystallized HA (60% crystalline) coatings demonstrated significant degradation and poor bone apposition *in vivo* compared to amorphous coatings. Both these reports suggest that although considerable bioresorption of the coating occurred in the cortical bone, there was significant bone apposition ($81 \pm 2\%$ for amorphous HA at 12 weeks, 77% for crystalline HA, respectively) which was not significantly affected by bioresorption.

From these *in vivo* reports, it is clear that HA coatings with relatively low levels of crystallinity are capable of significant bone apposition. However, as reported in a 1990 workshop report, the FDA is strongly urging commercial implant manufacturers to use techniques to increase the postdeposition crystallinity and to provide adequate adhesion of the coating to the implant substrate [Filiaggi et al., 1993]. Although the biologic responses to HA coatings are encouraging, other factors regarding HA coatings continue to lead to clinical questions regarding their efficacy. Although the overall bone response to HA-coated implants occurs more rapidly than with uncoated devices, with time an equivalent bone contact area is formed for both materials [Jasty et al., 1992]. These results have questioned the true need for HA-coated implants, especially when there are a number of disadvantages associated with the coating concept. Clinical difficulties have arisen due to failures within the coating itself and with continued dissolution of the coating, and to catastrophic failure at the coating–substrate interface [Koeneman et al., 1988].

Recent progress is reported in terms of the improvements in coating technology. Postdeposition heat treatments are often utilized to control the crystallinity (and therefore the dissolution characteristics) of the coatings, although there is still debate as to the relationship between compositional variations associated with differing crystallinity and optimization of biologic responses. Additional coating-related properties are also under investigation in regard to their effects on bone. These include coating thickness, level of acceptable porosity in the coating, and adherence of the coating to the underlying substrate. However, until answers concerning these variables have been more firmly established, HA coatings used for dental implants will remain an area of controversy and interest.

9.2.2 Effects of Surface Properties

9.2.2.1 Surface Topography

The effects of surface topography are different than the overall three-dimensional design or geometry of the implant, which is related to the interaction of the host tissues with the implant on a macroscopic scale as shown in Figure 9.13. This important consideration in overall biologic response to implants is discussed later in this chapter. In this discussion the concept of surface topography refers to the surface texture on a microlevel. It is on this microscopic level that the intimate cell and tissue interactions leading to osseointegration are based as shown in Figure 9.14.

The effects of surface topography on *in vitro* and *in vivo* cell and tissue responses have been a field of intense study in recent years. The overall goal of these studies is to identify surface topographies which

FIGURE 9.13 Examples of current dental implant designs, illustrating the variety of macroscopic topographies which are used to encourage tissue ingrowth. Left to right: Microvent, Corevent, Screw-vent, Swede-vent, Branemark, IMZ implant.

mimic the natural substrata in order to permit tissue integration and improve clinical performance of the implant. In terms of cell attachment, the *in vitro* work by Bowers and colleagues [1992] established that levels of short-term osteoblast cell attachment were higher on rough compared to smooth surfaces and cell morphology was directly related to the nature of the underlying substrate. After initial attachment, in many cases, cells of various origin often take on the morphology of the substrate as shown in Figure 9.15. Increased surface roughness, produced by such techniques as sand or grit blasting or by rough polishing, provided the rugosity necessary for optimum cell behavior.

Work in progress in several laboratories is attempting to relate the nature of the implant surface to cell morphology, intracellular cytoskeletal organization, and extracellular matrix development. Pioneering work by Chehroudi and coworkers [1992] suggests that microtextured surfaces (via micromachining or other techniques) could help orchestrate cellular activity and osteoblast mineralization by several mechanisms including proper orientation of collagen bundles and cell shape and polarity. This concept is related to the theory of **contact guidance** and the belief that cell shape will dictate cell differentiation through gene expression. In Chehroudi's work, both tapered pitted and grooved surfaces (with specific orientation and sequence patterns) supported mineralization with ultrastructural morphology similar in appearance to that observed by Davies and colleagues [1990]. However, mineralization was not observed on smooth surfaces in which osteoblastlike cells did not have a preferred growth orientation. Thus the control of surface microtopography by such procedures as micromachining may prove to be a valuable technology for the control and perhaps optimization of bone formation on implant surfaces.

It is apparent that macroscopic as well as microscopic topography may affect osteoblast differentiation and mineralization. In a recent study by Groessner-Schrieber and Tuan [1992], osteoblast growth, differentiation, and synthesis of matrix and mineralized nodules were observed on rough, textured, or porous coated titanium surfaces. It may be possible therefore, not only to optimize the interactions of host tissues with implant surfaces during the Phase I tissue responses but also to influence the overall bone responses to biomechanical forces during the remodeling phase (Phase II) of tissue responses.

Based on these concepts, current implant designs employ microtopographically roughened surfaces with macroscopic grooves, threads, or porous surfaces to provide sufficient bone ingrowth for mechanical

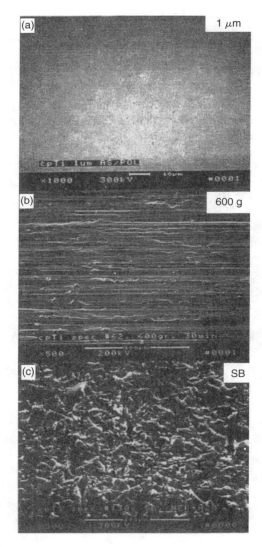

FIGURE 9.14 Laboratory-prepared cpTi surfaces with (a–c, top to bottom) smooth (1 μm polish), grooved (600 grit polish), and rough (sandblasted) surfaces.

stabilization and the prevention of detrimental micromotion as shown in Figure 9.16 and Figure 9.17 [De Porter et al., 1986; Keller et al., 1987; Pilliar et al., 1991; Brunski, 1992].

9.2.3 Surface Chemistry

Considerable attention has focused on the properties of the oxide found on titanium implant surfaces following surface preparation. Sterilization procedures are especially important and are known to affect not only the oxide condition but also the subsequent *in vitro* [Swart et al., 1992; Stanford et al., 1994] and *in vivo* [Hartman et al., 1989] biologic responses. Interfacial surface analyses and determinations of surface energetics strongly suggest that steam autoclaving is especially damaging to titanium oxide surfaces. Depending upon the purity of the autoclave water, contaminants have been observed on the metal oxide and are correlated with poor tissue responses on a cellular [Keller et al., 1990, 1994] and tissue [Meenaghan et al., 1979; Baier et al., 1984; Hartman et al., 1989] level.

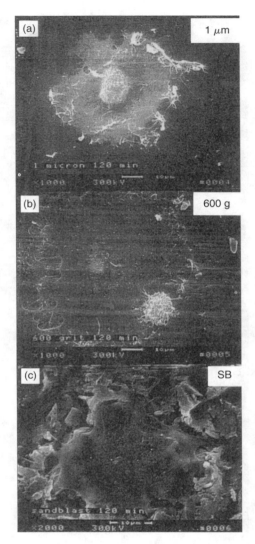

FIGURE 9.15   Osteoblastlike cell morphology after 2 h attachment on (a–c, top to bottom) smooth, grooved, and rough cpTi surfaces.

The role of multiple sterilization regimens on the practice of implant utilization is also under scrutiny. Many implants and especially bone plate systems are designed for repackaging if the kit is not exhausted. However, early evidence indicates that this practice is faulty and, depending on the method of sterilization, may affect the integrity of the metal oxide surface chemistry [Vezeau et al., 1991]. *In vitro* experiments have verified that multiple-steam-autoclaved and ethylene-oxide-treated implant surfaces adversely affected cellular and morphologic integration. However, the effects of these treatments on long-term biological responses including *in vivo* situations remain to be clarified.

Other more recently introduced techniques such as radiofrequency argon plasma cleaning treatments have succeeded in altering metal oxide chemistry and structure [Baier et al., 1984; Swart et al., 1992]. Numerous studies have demonstrated that PC treatments produce a relatively contaminant-free surface with improved surface energy (wettability), but conflicting biologic results have been reported with these surfaces. Recent *in vitro* studies have demonstrated that these highly energetic surfaces do not necessarily improve cellular responses such as attachment and cell expression. This has been confirmed by *in vivo*

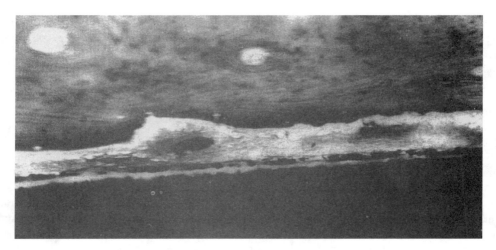

FIGURE 9.16 Light microscopic photomicrograph of a bone–smooth cpTi interface with intervening layer of soft connective tissue. This implant was mobile in the surgical site due to lack of tissue ingrowth. (Original magnification = 50×.)

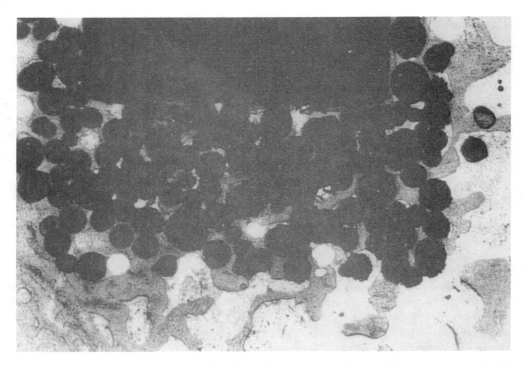

FIGURE 9.17 Light microscopic photomicrograph of a bone–porous Ti alloy implant interface. Note significant bone ingrowth in open porosity at the apical end of the implant. (Original magnification = 10×.)

studies which indicate that the overall histologic and ultrastructural morphology of the bone-implant interface is similar for plasma-cleaned and dry-heat-sterilized implant surfaces [Albrektsson et al., 1983]. Another promising technique for the sterilization of implant materials is the exposure of the implant surface to ultraviolet light [Singh and Schaff, 1989] or gamma irradiation [Keller et al., 1994]. Both these methods of sterilization produce a relatively contaminant-free thin oxide layer which fosters high levels of cell attachment [Keller et al., 1994] and inflammatory-free long-term *in vivo* responses [Hartman et al., 1989]. Currently, gamma irradiation procedures are widely used for the sterilization of metallic dental implant devices.

9.2.3.1 Metallic Corrosion

Throughout the history of the use of metals for biomedical implant applications, electrochemical corrosion with subsequent metal release has been problematic [Galante et al., 1991]. Of the biomedical metal systems available today, Ti and its major medical alloy, Ti–6A1–4V, are thought to be the most corrosion resistant; however, Ti metals are not totally inert *in vivo* [Woodman et al., 1984]. Release of Ti ions from Ti oxides can occur under relatively passive conditions [Ducheyne, 1988]. Whereas other factors such as positioning of the implant and subsequent biomechanical forces may play important roles in the overall tissue response to implants, it is not unreasonable to predict that electrochemical interactions between the implant surface and host tissue may affect the overall response of host bone [Blumenthal and Cosma, 1989]. For example, it has been shown by several groups [De Porter et al., 1986; Keller et al., 1987] that the percentages of intimate bony contact with the implant is inconsistent, at best, and generally averages approximately 50% over a 5-year period. Continued studies involving the effects of dissolution products on both local and systemic host responses are required in order to more fully understand the consequences of biologic interaction with metal implants.

9.2.3.2 Future Considerations for Implant Surfaces

It is clear that future efforts to improve the host tissue responses to implant materials will focus, in large part, on controlling cell and tissue responses at implant interfaces. This goal will require continued acquisition of fundamental knowledge of cell behavior and cell response to specific materials' characteristics. It is likely that a better understanding of the cellular-derived extracellular matrix-implant interface will offer a mechanism by which biologic response modifiers such as growth and attachment factors or hormones may be incorporated. Advancements of this type will likely shift the focus of future research from implant surfaces which as osseoconductive (permissive) to those which are osseoinductive (bioactive).

Defining Terms

Amorphous zone: A region of the tissue-implant interface immediately adjacent to the implant substrate. This zone is of variable thickness (usually <1000 Å), is free of collagen, and is comprised of proteoglycans of unknown composition.

Calcium phosphate: A family of calcium- and phosphate-containing materials of synthetic or natural origin which are utilized for implants and bone augmentation purposes. The most prominent materials are the tricalcium-phosphate- and hydroxyapatite-based materials, although most synthetic implants are a mixture of the various compositions.

Contact guidance: The theory by which cells attach to and migrate on substrates of specific microstructure and topographic orientation. The ability of the cell to attach and migrate on a substrate is related to the cytoskeletal and attachment molecules present on the cell membrane.

Junctional epithelium: The epithelial attachment mechanism which occurs with teeth, and has been observed infrequently with implants by some researchers. Less than 10 cell layers thick, the hemidesmosomal attachments of the basal cells to the implant surface provide a mechanical attachment for epithelium and prevent bacterial penetration into the sulcular area.

Osseointegration: A term developed by P.I. Branemark and his colleagues indicating the ability of host bone tissues to form a functional, mechanically immobile interface with the implant. Originally described for titanium only, several other materials are capable of forming this interface, which presumes a lack of connective tissue (foreign body) layer.

Plasma spray: A high-temperature process by which calcium-phosphate-containing materials are coated onto a suitable implant substrate. Although the target material may be of high purity, the high-temperature softening process can dramatically affect and alter the resultant composition of the coating.

References

Albrektsson, T., Branemark, P.I., Hansson, H.A. et al. 1983. The interface zone of inorganic implants *in vivo*: Titanium implants in bone. *Ann. Biomed. Eng.* 11: 1.

Albrektsson, T., Hansson, H.A., and Ivarsson, B. 1985. Interface analysis of titanium and zirconium bone implants. *Biomaterials* 6: 97.

Baier, R.E., Meyer, A.E., Natiella, J.R. et al. 1984. Surface properties determine bioadhesive outcomes. *J. Biomed. Mater. Res.* 18: 337.

Blumenthal, N.C. and Cosma, V. 1989. Inhibition of appetite formation by titanium and vanadium ions. *J. Biomed. Mater. Res.* 23: 13.

Bowers, K.T., Keller, J.C., Michaels, C.M. et al. 1992. Optimization of surface micromorphology for enhanced osteoblast responses *in vitro*. *Int. J. Oral Maxillofac. Implants* 7: 302.

Branemark, P.I. 1983. Osseointegration and its experimental background. *J. Pros. Dent.* 59: 399.

Brunski, J.B. 1992. Biomechanical factors affecting the bone-dental implant interface. *Clin. Mater.* 10: 153.

Chehroudi, B., Ratkay, J., and Brunette, D.M. 1992. The role of implant surface geometry on mineralization *in vivo* and *in vitro*: a transmission and scanning electron microscopic study. *Cells Mater.* 2: 89–104.

Cook, S.D., Kay, J.F., Thomas, K.A. et al. 1987. Interface mechanics and histology of titanium and hydroxylapatite coated titanium for dental implant applications. *Int. J. Oral Maxillofac. Implants* 2: 15.

Cook, S.D., Thomas, K.A., and Kay, J.F. 1991. Experimental coating defects in hydroxylapatite coated implants. *Clin. Orthop. Rel. Res.* 265: 280.

Davies, J.E., Lowenberg, B., and Shiga, A. 1990. The bone–titanium interface *in vitro*. *J. Biomed. Mater. Res.* 24: 1289–1306.

De Bruijn, J.D., Flach, J.S., deGroot, K. et al. 1993. Analysis of the bony interface with various types of hydroxyapatite *in vitro*. *Cells Mater.* 3: 115.

De Porter, D.A., Watson, P.A., Pilliar, R.M. et al. 1986. A histological assessment of the initial healing response adjacent to porous-surfaced, titanium alloy dental implants in dogs. *J. Dent. Res.* 65: 1064.

Driskell, T.D., Spungenberg, H.D., Tennery, V.J. et al. 1973. Current status of high density alumina ceramic tooth roof structures. *J. Dent. Res.* 52: 123.

Ducheyne, P. 1988. Titanium and calcium phosphate ceramic dental implants, surfaces, coatings and interfaces. *J. Oral Implantol.* 19: 325.

Filiaggi, M.J., Pilliar, R.M., and Coombs, N.A. 1993. Post-plasma spraying heat treatment of the HA coating/Ti–6A1–4V implant system. *J. Biomed. Mater. Res.* 27: 191.

Galante, J.O., Lemons, J., Spector, M. et al. 1991. The biologic effects of implant materials. *J. Orthop. Res.* 9: 760.

Groessner-Schreiber, B. and Tuan, R.S. 1992. Enhanced extracellular matrix production and mineralization by osteoblasts cultured on titanium surfaces *in vitro*. *J. Cell Sci.* 101: 209.

Hartman, L.C., Meenaghan, M.A., Schaaf, N.G. et al. 1989. Effects of pretreatment sterilization and cleaning methods on materials properties and osseoinductivity of a threaded implant. *Int. J. Oral Maxillofac. Implants* 4: 11.

Herman, H. 1988. Plasma spray deposition processes. *Mater. Res. Soc. Bull.* 13: 60.

Jansen, J.A., van der Waerden, J.P.C.M., and Wolke, J.G.C. 1993. Histological and histomorphometrical evaluation of the bone reaction to three different titanium alloy and hydroxyapatite coated implants. *J. Appl. Biomater.* 4: 213.

Jarcho, M., Kay, J.F., Gumaer, K.I. et al. 1977. Tissue, cellular and subcellular events at a bone-ceramic hydroxylapatite interface. *J. Bioeng.* 1: 79.

Jasty, M., Rubash, H.E., Paiemont, G.D. et al. 1992. Porous coated uncemented components in experimental total hip arthroplasty in dogs. *Clin. Orthop. Rel. Res.* 280: 300.

Johansson, C.B., Lausman, J., Ask, M. et al. 1989. Ultrastructural differences of the interface zone between bone and Ti-6A1-4V or commercially pure titanium. *J. Biomed. Eng.* 11: 3.

Katsikeris, N., Listrom, R.D., and Symington, J.M. 1987. Interface between titanium 6A1–4V alloy implants and bone. *Int. J. Oral Maxillofac. Surg.* 16: 473.

Kay, J.F. 1992. Calcium phosphate coatings for dental implants. *Dent. Clinic N. Am.* 36: 1.

Keller, J.C., Draughn, R.A., Wightman, J.P. et al. 1990. Characterization of sterilized cp titanium implant surfaces. *Int. J. Oral Maxillofac. Implants* 5: 360.

Keller, J.C. and Lautenschlager, E.P. 1986. Metals and alloys. In *Handbook of Biomaterials Evaluation*, A. Von Recon (Ed.), pp. 3–23, New York, Macmillan.

Keller, J.C., Niederauer, G.G., Lacefield, W.R. et al. 1992. Interaction of osteoblast-like cells with calcium phosphate ceramic materials. *Trans. Acad. Dent. Mater.* 5: 107.

Keller, J.C., Stanford, C.M., Wightman, J.P. et al. 1994. Characterization of titanium implant surfaces. *J. Biomed. Mater. Res.*

Keller, J.C., Young, F.A., Natiella, J.R. 1987. Quantitative bone remodeling resulting from the use of porous dental implants. *J. Biomed. Mater. Res.* 21: 305.

Koeneman, J., Lemons, J.E., Ducheyne, P. et al. 1990. Workshop of characterization of calcium phosphate materials. *J. Appl. Biomater.* 1: 79.

Lucas, L.C., Lemons, J.E., Lee, J., et al. 1987. *In vivo* corrosion characteristics of Co–Cr–Mu/ Ti–6A1–4V–Ti alloys. In *Quantitative Characterization and Performance of Porous Alloys for Hard Tissue Applications*, J.E. Lemons (Ed.), pp. 124–136, Philadelphia, ASTM.

Maxian, S.H., Zawadsky, J.P., and Durin, M.G. 1993. Mechanical and histological evaluation of amorphous calcium phosphate and poorly crystallized hydroxylapatite coatings on titanium implants. *J. Biomed. Mater. Res.* 27: 717.

Meenaghan, M.A., Natiella, J.R., Moresi, J.C. et al. 1979. Tissue response to surface treated tantalum implants: Preliminary observations in primates. *J. Biomed. Mater. Res.* 13: 631.

Orr, R.D., de Bruijn, J.D., and Davies, J.E. 1992. Scanning electron microscopy of the bone interface with titanium, titanium alloy and hydroxyapatite. *Cells Mater.* 2: 241.

Pilliar, R.M., DePorter, D.A., Watson, P.A. et al. 1991. Dental implant design — effect on bone remodeling. *J. Biomed. Mater. Res.* 25: 647.

Puleo, D.A., Holleran, L.A., Doremus, R.H. et al. 1991. Osteoblast responses to orthopedic implant materials *in vitro*. *J. Biomed. Mater. Res.* 25: 711.

Singh, S. and Schaaf, N.G. 1989. Dynamic sterilization of titanium implants with ultraviolet light. *Int. J. Oral Maxillofac. Implants* 4: 139.

Skalak, R. 1985. Aspects of biomechanical considerations. In *Tissue Integrated Prostheses*, P.I. Branemark, G. Zarb, and T. Albrektsson (Eds.), pp. 117–128, Chicago, Quintessence.

Smith, D.C. 1993. Dental implants: Materials and design considerations. *Int. J. Prosth.* 6: 106.

Stanford, C.M. and Keller, J.C. 1991. Osseointegration and matrix production at the implant surface. *CRC Crit. Rev. Oral Bio. Med.* 2: 83.

Stanford, C.M., Keller, J.C., and Solursh, M. 1994. Bone cell expression on titanium surfaces is altered by sterilization treatments. *J. Dent. Res.*

Steflik, D.E., McKinney, R.V., Koth, D.L. et al. 1984. Biomaterial-tissue interface: A morphological study utilizing conventional and alternative ultrastructural modalities. *Scanning Electron Microsc.* 2: 547.

Steflik, D.E., Sisk, A.L., Parr, G.R. et al. 1993. Osteogenesis at the dental implant interface: High voltage electron microscopic and conventional transmission electric microscopic observations. *J. Biomed. Mater. Res.* 27: 791.

Swart, K.M., Keller, J.C., Wightman, J.P. et al. 1992. Short term plasma cleaning treatments enhance *in vitro* osteoblast attachment to titanium. *J. Oral Implant* 18: 130.

Van Orden, A. 1985. Corrosive response of the interface tissue to 316L stainless steel, Ti-based alloy and cobalt-based alloys. In *The Dental Implant*, R. McKinney and J.E. Lemons (Eds.), pp. 1–25, Littleton, PSG.

Vezeau, P.J., Keller, J.C., and Koorbusch, G.F. 1991. Effects of multiple sterilization regimens on fibroblast attachment to titanium. *J. Dent. Res.* 70: 530.

Vrouwenvelder, W.C.A, Groot, C.G., and Groot, K. 1993. Histological and biochemical evaluation of osteoblasts cultured on bioactive glass, hydroxylapatite, titanium alloy and stainless steel. *J. Biomed. Mater. Res.* 27: 465–475.

Woodman, J.L., Jacobs, J.J., Galante, J.O. et al. Metal ion release from titanium-based prosthetic segmental replacements of long bones in baboons: A long term study. *J. Orthop. Res.* 1: 421–430.

Young, F.A. 1988. Future directions in dental implant materials research. *J. Dent. Ed.* 52: 770.

10

Controlling and Assessing Cell–Biomaterial Interactions at the Micro- and Nanoscale: Applications in Tissue Engineering

Jessica Kaufman
Joyce Y. Wong
Catherine Klapperich
Boston University

10.1   A Need for Understanding Cell–Biomaterial Interactions at the Micro- and Nanoscale

A long-standing goal of tissue engineering has been to create functional tissues, ideally by promoting the regenerative capacity of a patient's autologous cells by controlling cellular response. Regardless of whether this is achieved by using artificial scaffolds or naturally derived extracellular matrices, the engineered tissue must be able to support the necessary physiological loads during the remodeling processes that ultimately lead to generation of the functional tissue. While there has been tremendous progress in engineering a number of tissue systems, namely skin replacements [1–3], an incomplete understanding of the underlying mechanisms that control cell behavior has led to limited clinical success. Specifically, there are many open questions as to how cell–biomaterial interactions impact cellular and ultimately tissue phenotype [4,5].

In order to gain a mechanistic understanding of factors that control cell behavior, molecular and cell biologists have numerous methodologies to systematically unravel specific cell-signaling pathways involved in regulating cell behavior. It has been more difficult to control properties of biomaterials with analogous precision, but this is rapidly changing with recent enabling micro- and nanotechnologies that create scaffolds characterized by features that can now be controlled at length-scales that range from cellular dimensions all the way down to the molecular scale.

A solid base of research in both engineering and the biomedical sciences has demonstrated that tissue regeneration in synthetic environments is possible and often successful, but is difficult to predict and control. Design of these scaffolds has been conducted largely by trial and error, and little is known about how the chemical and mechanical properties of the scaffold affect the biological response of the seeded cells. Due to this lack of understanding, it has proven significantly more difficult to engineer complex tissues like cartilage, bone, nerves, and muscle [6,7] By considering the biological response of cells to scaffolds a material property that can be controlled by altering processing variables, we can begin to build the framework necessary for intelligent de novo design of new scaffold materials. However, in order for biological response to be a useful design variable, it must be quantifiable. Several of the approaches reviewed in this chapter are aimed at quantifying biological responses.

In most tissue engineering applications, a positive result is more likely if the scaffold material can provide the environmental stimuli necessary for healthy tissue regeneration. In most tissues, the molecular nature of these factors is unknown, but is likely some combination of mechanical, chemical, and biological stimuli take place on the micro- and nanoscale. It is largely unknown how mechanical forces and surface chemistry affect the physiology of cells, but this is a very active area of research [8–10]. Recently, many researchers have begun to focus on precisely how cells are responding to chemical and topological features at the micro- and nanoscale. Covered in this chapter are current issues involved in probing cell–biomaterials interactions on the molecular level and their implications for tissue engineering research. Discussed are methods to quantify cell response, engineer cell-surface molecules for targeted cell–biomaterials inter-actions, gain geometric and surface chemical control of cell fate, and directed movement of cells on substrates with mechanical gradients. We also discuss recent enabling technologies in materials pro-cessing that have yielded nanoscale biomaterials as tissue engineering scaffolds for various organ systems. Finally, we discuss the need for determining potential undesirable immunogenic responses to nano- and microsystems.

This review is by no means complete, and we apologize in advance to our colleagues for not including all of the excellent work in this area.

10.2   Genomic and Proteomic Data as a Measure of Cell–Biomaterials Interactions

By combining our ever increasing control over fabricating nanoscale features and the increased stand-ardization of biochemical assays afforded by access to information about the human genome, we can now embark on intelligent tissue engineering scaffold design. Cell adhesion, migration, growth, and

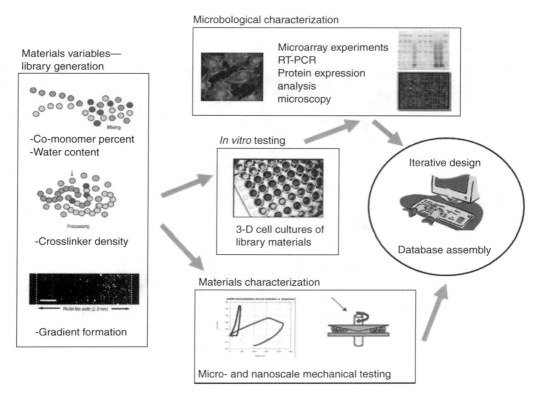

FIGURE 10.1 Schematic of the combinatorial approach using cell responses as variables in synthetic biomaterials design for tissue engineering applications. Libraries of chemically well-defined materials are generated combinatorially. The materials are characterized both chemically and mechanically at the nano- and microscale. *In vitro* cell-based assays are used to assess gene and protein expression. Cell–biomaterial interactions are visualized in 3-D culture using confocal microscopy. Finally, data is stored in an organized database for use in de novo materials design.

extracellular matrix synthesis are four major areas of cell responses that are relevant to tissue engineering applications. These responses can be probed on many levels including mRNA levels (gene expression) and protein levels (proteomics). A schematic of the general approach taken in one of our laboratories is included in Figure 10.1.

We have used the collagen–glycosaminoglycan (collagen–GAG) mesh scaffold to probe molecular level cell–biomaterials interactions [11]. We seeded IMR-90 human fibroblasts onto three-dimensional (3-D) collagen–GAG meshes and control surfaces of tissue culture polystyrene (TCPS). Nucleic acids (mRNA) from cells from each culture were isolated, amplified, and hybridized to human genome microarrays (U133A Gene Chip, Affymetrix, Santa Clara, CA).

Connective tissue growth factor (CTGF) and tissue inhibitor metalloproteinase 3 (TIMP3) were down regulated in 3-D collagen exposed fibroblasts compared to the tissue culture polystyrene grown cells. CTGF, which plays a role in the induction of collagen, has known involvement in matrix accumulation in fibrosis, as well as the development of excess fibrous connective tissue. TIMP3 inactivates metalloproteinases, which degrade components of the extracellular matrix thereby remodeling the tissue. By underexpressing TIMP3, it is suspected that collagen–GAG interaction encourages the reorganization of the fibroblasts extracellular matrix. The 3-D arrangement stimulated the expression of proangiogenic genes including vascular endothelial growth factor (VEGF) and angiopoietin (ANGPTL2). The 3-D mesh environment also yielded high expression levels of the mRNA for proteins involved in matrix remodeling such as type III collagen (COL3A1).

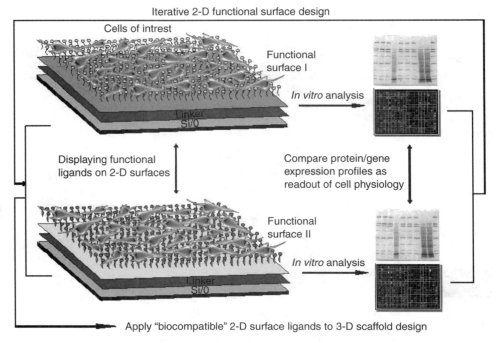

FIGURE 10.2   A functional surface-based strategy to probe cell–ligand interactions — an iterative tissue engineering scaffold design approach. (Reproduced from Song, J. et al., *J. Mater. Chem.*, 2004, **14**: 2643–8. With permission.)

These studies and complementary proteomic work are beginning to more precisely define material biocompatibility for tissue regeneration. We extended this approach to look at how surfaces conjugated with anionic peptides affected gene and protein expression in osteosarcoma cells (Figure 10.2) [12]. Other researchers have taken similar approaches to study the migration and spreading behavior of endothelial cells on small-gauge vascular prostheses [13]. Ku et al. [14] looked at global gene expression of osteoblasts grown on different Ti–6Al–4V surface treatments. They were able to detect differences in inflammatory response of cells on three different surface treatments. Once a more complete understanding of the molecular nature of biocompatibility is achieved, the cell-surface molecule modifications and advanced materials design techniques described later will become much more powerful tools.

10.3   Engineering the Cell Surface to Control Cell-Adhesion Events

Another approach to controlling cell–biomaterials interactions is to modify the cell surface. The surfaces of cells are decorated with polysaccharides, glycolipids, and glycoproteins. These cell-surface molecules are the handles through which cells communicate with their surroundings. Cell-adhesion events can trigger signaling events inside of the cell and lead to changes in gene expression and eventually cell fate. These properties make cell-surface molecules ideal engineering targets. By modifying the molecules presented on a cell surface, it may be possible to control a cell's phenotype and eventual fate.

One way to modify the cell-surface molecules is by introducing nonnatural monosugars into the biosynthetic pathways that build up polysaccharides (Figure 10.3) [15,16]. By introducing monosaccharides with synthetic ketone groups, Bertozzi et al. were able to engineer previously nonadherent Jurkat cells to adhere to lipid bilayers decorated with azides [17]. These researchers have also demonstrated that a modification of a cell-surface molecule, polysialic acid (PSA), in substrata cells supporting the growth of

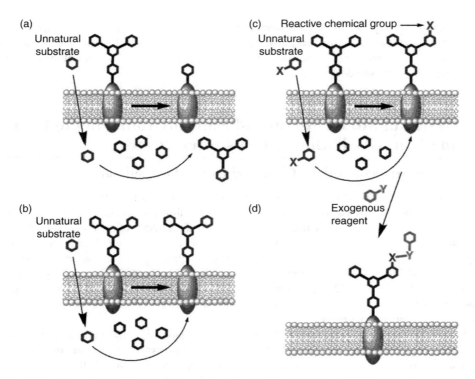

FIGURE 10.3 (Right) Modulating cell-surface glycosylation by metabolic interference. (a) Unnatural substrates fed to cells can divert oligosaccharide biosynthesis away from endogenous scaffolds, reducing the expression of specific carbohydrate structures. (b) Unnatural substrates can be used in biosynthetic pathways and incorporated into cell-surface glycoconjugates. (c) If the unnatural substrates possess unique functional groups, their metabolic products on the cell-surface can be chemically elaborated with exogenous reagents (d). (Reproduced from Bertozzi, C.R. and L.L. Kiessling, *Science*, 2001, **291**: 2357–64. With permission.)

chick dorsal root ganglion sensory neurons were able to differentially affect the outgrowth and plasticity of those cells [18]. In this case a cell-surface molecule that inhibits adhesion and outgrowth of neurons was modified, but it is easy to imagine that engineering cell-surface molecules that encourage such outgrowth would benefit work in nerve tissue regeneration. Since these artificial sugars can be installed through metabolic pathways, the cell culture needs only to be fed the engineered sugar through the media. It has also been demonstrated that feeding these sugars to small animals orally leads to the desired cell-surface modifications in cells of the animal, indicating that these substances may be able to be administered as drugs to patients or injected directly into the site of artificial tissue repair.

Other methods of cell-surface engineering also look promising. Genetic methods have been used to transfect cells with genes that express unique or unnatural cell-surface protein [19,20]. Kato and Mrksich [21] describe a method to modify both the cell and the biomaterial surface in a complimentary manner to encourage specific cell adhesion. They transfected Chinese hamster ovary (CHO) cells to express a chimeric receptor with a carbonic anhydrase IV (CAIV) domain at its terminus. This domain binds selectively to benzenesulfonamide groups that were installed on self-assembled monolayers designed for the study. The study demonstrates the ability to control both sides of the receptor–ligand interaction, which suggests a potentially powerful tool for tissue engineering.

Since cell-adhesion events are almost always the first step in the regeneration of injured or destroyed tissues, the ability to control these events should lead to the generation of better artificial tissues. Complementary modifications of the cell and biomaterial surfaces could allow cells to be directed to a specific location for tissue regeneration. Current problems with cell seeding of artificial scaffolds could be ameliorated in this way. Of course it is essential to remain mindful that at this time only a patient's own cells can provide a preseeded tissue engineering scaffold that will not elicit an immune response. The advent of stem cell use in regenerative medicine and tissue engineering may solve some of these problems, but much research is left to be done [22,23].

10.4   Design of Model Micro- and Nanoenvironments to Probe and Direct Cell–Biomaterials Interactions

10.4.1   Geometric and Surface Chemical Control of Cell Fate

In order to sort out what cues cells are receiving from their three-dimensional extracellular environments, many groups have begun to make model environments with well-defined chemical and topological features. By maintaining control over the feature size and chemical character of the model system, changes in these variables can be made and the resulting cell responses can be measured using microbiological techniques. Most often, these features are made using photolithographic methods developed for the microelectronic industry [24]. Features approaching 1 μm in size are easily fabricated using these methods. In recent work, Dike et al. [25] demonstrated specific control over cell fate in vascular endothelial cells, showing that micropatterned surfaces can determine whether the cells would follow an angiogenic, apoptotic, or differentiation pathway.

Early work by Chen et al. [26] demonstrated that cells grown on subsequently smaller areas of microcontact printed extracellular matrix experienced different fates. In these experiments, changing the substrate extracellular matrix components and cell source, bovine or human, did not override the geometric effects. This work was expanded to three dimensions by examining how capillary endothelial cells grow in solution with fibronectin-coated beads [27]. Cells attached to single 10 μm beads did not improve viability whereas attaching cells to 25 μm beads did. The capillary epithelial cells undergo apoptosis unless the substrate is designed to allow them to adopt a relatively flat conformation by making multiple connections on a relatively planar surface.

Recently, Chen has used microneedle-like posts made of polydimethylsiloxane (PDMS) to measure the mechanical forces that smooth muscle cells exert on each other at cell adhesions (Figure 10.4) [28]. The researchers were also able to control the type of cell-adhesion molecules present on the tip of each elastomeric post. Using this approach of geometric control combined with precisely defined surface chemistries, they were able to demonstrate correlations between mechanical and chemical signals in these cells.

Just as Chen showed that cell fate can be determined by geometry of individual cells, Bhatia et al. [29] showed that the geometry of individual populations within a micropatterned coculture can also affect cell behavior. Cocultures were constructed by culturing hepatocytes on collagen-patterned wafers and then culturing fibroblasts directly on the substrate surrounding the wafers. In the experiment, the contact area between hepatocytes and fibroblasts and the contact area between individual hepatocytes were held constant. As the contact area between fibroblasts was increased, liver-specific functions in hepatocytes also increased. Work by Tang et al. [30] was a first step at using micropatterning techniques to create three-dimensional cocultures in collagen gels (Figure 10.5). Building up three-dimensional cultures with controlled geometries is a critical step toward multicellular tissue engineering constructs. Control of the local or microscale arrangement of cells in a tissue engineering scaffold will be a key to creating functional organs.

Cells do react to changes in their microscale environment, but cell–cell and cell–matrix adhesions are often nanoscale phenomena. Recently, techniques that yield nanoscale topography have emerged. As we are able to probe cells with smaller and smaller changes in surface topography and local surface chemistry, the closer we will come to mimicking the *in vivo* microenvironment. Andersson et al. [31] demonstrated that uroepithelial cells seeded onto substrates with nanoscale pillars expressed smaller amounts of the cytokines IL-6 and IL-8 than cells cultured on flat substrates of the same material. They saw no difference in cytokine

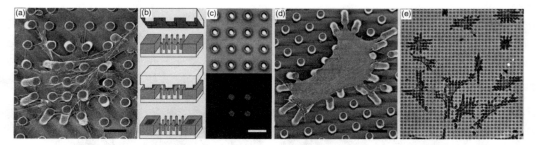

FIGURE 10.4 Cell culture on arrays of posts. (a) Scanning electron micrograph of a representative smooth muscle cell attached to an array of posts that was uniformly coated with fibronectin. Cells attached at multiple points along the posts as well as the base of the substrates. (b) Schematic of microcontact printing of protein (red), precoated on a PDMS stamp, onto the tips of the posts (gray). (c) Differential interference contrast (Upper) and immunofluorescence (Lower) micrographs of the same region of posts where a 2 × 2 array of posts has been printed with fibronectin. (d and e) Scanning electron micrograph (d) and phase-contrast micrograph (e) of representative smooth muscle cells attached to posts where only the tips of the posts have been printed with fibronectin by using a flat PDMS stamp. Cells deflected posts maximally during the 1 to 2-h period after plating, were fully spread after 2 h, and were fixed and critical point dried 4 h after plating. (Scale bars indicate 10 μm.) (Reproduced from Tan, J.L. et al., *Proc. Natl Acad. Sci. USA*, 2003, **100**: 1484–9. With permission.)

production between cells grown on the flat substrate and one with parallel microscale grooves. Corneal epithelial cells were shown to exhibit significantly different phenotypes when seeded onto substrates with nanoscale features [32]. Focal adhesions and cytoskeleton proteins were found to align in the direction of the nanoscale features, a common result when cells are seeded on microscale surface topographies. This result was expected, since the surface of the corneal basal lamina is known to have features in the range of 20 to 200 nm.

10.4.2 Directed Movement of Cells on Mechanical Gradients

In addition to surface chemical micropatterning to control cell positioning, recently one of our laboratories has used micropatterning techniques to create mechanical gradient gels [33,34] (Figure 10.6). This phenomenon was first reported by Wang's and Dembo's groups [35] for 3T3 fibroblasts cultured on a substrate characterized with a step-gradient in mechanical properties. However, the method used to produce these substrates lacked precise control at the microscale, so we developed two methods based on photopolymerization to control substrate mechanical properties at the microscale. In one approach, mask patterns are used to control the degree of polymerization in polyacrylamide gels by controlling the UV light exposure time. Using this system, cell motility of bovine vascular smooth muscle cells were compared on mechanically patterned collagen-coated gels vs. uniformly compliant gels. The cells exhibited a clear migration from soft regions to stiff regions, durotaxis, when compared with cells grown on uniform hydrogels. Furthermore, the vascular smooth muscle cells accumulated in stiff regions of the gel, suggesting that mechanical patterning could be a powerful tool for creating complex cultures with multiple cell types that are stable over long periods of time.

The second approach involves the integration of a microfluidic gradient generator and photopolymerization to create substrate with well-defined compliance gradients [34]. An advantage of this approach is that more complex gradients can be achieved. In addition, variations in steepness of the gradient can be easily tuned in this system in order to quantify the range in which cells respond.

At this point, it is worth noting that there is limited data regarding the actual *local* mechanical properties of native and diseased tissues. Rather, our knowledge of tissue mechanical properties is largely based on *macroscopic* mechanical measurements. The lack of mechanical property characterization at the microscale is in part due to the limited number of techniques (e.g., modified atomic force microscopy [36]) that are

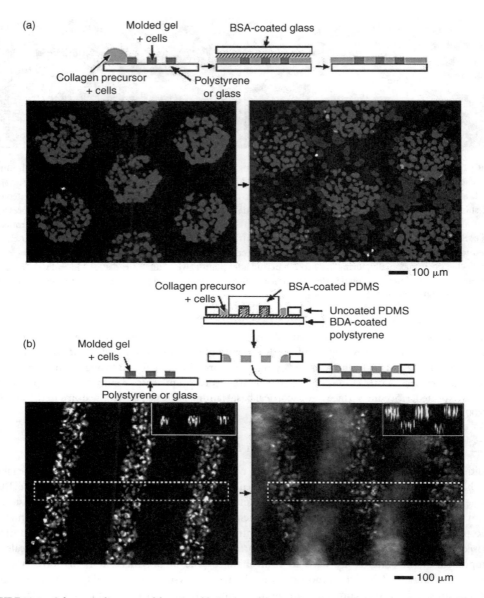

FIGURE 10.5 Schematic diagrams of the microfabrication of composite gels, and fluorescence images of cells in these gels. The shaded regions represent collagen gels in which fibroblasts are embedded. (a) (Left) An array of hexagonal gels (100 m on a side, 100 m thick) and cells on a glass coverslip. (Right) A composite structure that consists of an array of isolated hexagonal gels and a separate gel filling the spaces between the isolated gels on a glass coverslip. Fibroblasts in and between the hexagonal gels were prelabeled with green and red fluorescent dyes, respectively. (b) (Left) Lines (100 m wide, 100 m thick) of gel and cells formed by MIMIC on a glass coverslip. Nuclei of fibroblasts were labeled with the fluorescent dye Hoechst 33342 (1 g/ml; Molecular Probes). (Right) A bilayered structure that is composed of two layers of lines at a relative angle of ∼30 on a glass coverslip. The top layer is in focus, and the layer below is out-of-focus. (Insets) Cross-sectional views (corresponding to the regions outlined by dashed boxes) of deconvolved images. (Reproduced from Tang, M.D., A.P. Golden, and J. Tien, *J. Am. Chem. Soc.*, 2003, **125**: 12988-9. With permission.)

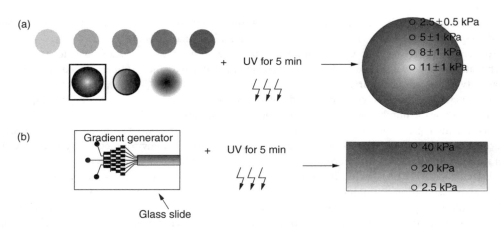

FIGURE 10.6 Methods for generating elastic substrata with gradients in mechanical properties. (a) Acrylamide is photopolymerized under transparency masks with varying degrees of opacity [33]. The radial gradient pattern boxed on the left can be used to create a substrate shown on the right with the 'map' of its mechanical properties (Young's modulus). (b) Acrylamide is photopolymerized using a combination of microfluidics and photopolymerization [34]. The gradient generator is used to create a gradient in the cross-linker (bis-acrylamide). The width of the gel is ~ 3 mm, and the gradient in the Young's modulus (as determined from atomic force microscopy) is approximately 12 Pa/μm. (Reproduced from Wong, J.Y., J.B. Leach, and Z.Q. Brown, *Surf. Sci.*, 2004, **570**: 119–33. With permission.)

able to measure the mechanical properties of hydrated tissues at this length-scale. Recently, we have developed a simple, inexpensive microindentation technique that is able to measure the local mechanical properties of hydrated biomaterials and tissue samples at the microscale [37]. While it is interesting that a number of different cell types have recently been shown to respond to changes in the mechanical properties of the substrate (see Reference 9), the measurement of the *local* mechanical properties of native and diseased tissues will be critical to establish clinical relevance when developing model *in vitro* biomaterial systems with physiologically matched mechanical properties.

10.4.3 Studies of Tissue Engineering Scaffolds with Nanoscale Features

Recently studies have been published that demonstrate enhanced cell growth, adhesion, and differentiation in the presence of nanoparticles or nanostructured coatings. In orthopedics, the fabrication of nanostructured ceramic coatings for total joint replacements has resulted in enhanced cell growth and bone remodeling *in vitro* [38]. Modified plasma spray processes have produced nanoscale hydroxyapatite coatings [39] for the same application. Hydrogel materials that either incorporate [40] or stimulate the nucleation of nanometer sized hydroxyapatite particles [41] have been demonstrated to form reasonable bone mimics. Tan and Saltzman [42] created a hierarchical bone-culture system employing microscale features coated with a nanoscale mineral phase. They were able to control both cell orientation and shape by modulating their surfaces. Further study will show if these materials will be successful as tissue engineering implants.

Several researchers focused on engineering soft tissues have also begun to employ nanostructured materials in the laboratory. Hydrogel materials with specific ligands for cell adhesion have been designed to affect desired cell-signaling events in native or implanted cells. A good review is provided by Boontheekul and Mooney [43]. Nerve tissue regeneration has been driven by the idea that if you can form a sheath of material of the appropriate size, then severed nerve endings will grow toward each other through it. Tubes have been made of silicone, collagen, and other synthetic and natural polymer materials. One common conclusion of much of this research is that not only do the regenerating axons need the mechanical cues provided by the artificial tube, but also most likely require the appropriate chemical signals at the nanoscale to drive complete and functional regeneration. Some researchers have begun to incorporate peptide sequences known to be neurotrophic in their tube-filling materials [44].

Hepatocytes, which are difficult to culture *in vitro* were cultured on gold nanoparticles to create a gold/hepatocyte colloidal solution [45]. The researchers were able to demonstrate increased proliferation and activity of the cells in culture.

Nanostructured poly(lactic-co-glycolic acid) (PGLA) created through chemical etching techniques was shown to increase the number and function of bladder smooth muscle cells in tissue engineering scaffolds intended for the regeneration of the bladder [46,47]. The authors attribute this result to the nanoscale surface topography, which they say more closely mimics the *in vitro* environment of soft tissue cells. Operating under similar assumptions, Zhang and Ma [48] developed a tissue engineering scaffold in poly(L-lactic acid) (PLLA) having micrometer scale pores with walls exhibiting nanofibrous features using a phase separation technique. These scaffolds were compared with scaffolds made from the same material with micropores that had smooth walls. Differences in adsorption of extracellular proteins into the scaffolds were studied using Western blot analysis [49]. The nanofibrous materials adsorbed more proteins than the smooth-walled scaffolds. Specifically, more fibronectin and vitronectin was adsorbed onto the nanostructured materials. The nanofibrous materials also enhanced adhesion of MC3T3-E1 osteoblasts when compared to the smooth-walled scaffolds.

Modified electrospray techniques have been used to produce polymer meshes for tissue scaffolds that have nanometer-scale fibers. This process involves injecting a polymer solution through a charged needle onto a substrate some distance away from the injection point. A mat of material can be made in varying thicknesses, and this mat can be cut or molded into the desired shape. Scaffolds of nanoscale fibers have been formed from poly(D,L-lactide-co-glycolide [50], collagen and elastin [51], and poly(D,L-lactide)-poly(ethylene glycol) (PLA-PEG) [52].

Nanoscale surface roughness can also affect cell differentiation. A recently developed technique for growing bone-like minerals on a hydrolyzed poly(lactide-co-glycolide) substrate by incubating in a salt-rich simulated body fluid was used to create bone mimics by Mooney et al. [53]. The resulting mineral layer displays a plate-like nanostructure similar to natural bone. Mesenchymal stem cells cultured under osteoblast inducing conditions were grown on both smooth substrate and mineralized substrate. Surprisingly, the mineralized substrate inhibited differentiation of the stem cells. In a similar work by Jansen et al. [54], surfaces roughness at the micron scale showed no effect on the differentiation of rat bone marrow cells under osteogenic conditions. Understanding both the chemical and nanoscale topographical signals necessary for differentiation will thus be essential for the development of stem-cell-based tissue-engineered constructs.

10.5   Impact of Nanoscale Processing on Biomaterial Properties

Recently there have been a number of reports of biomaterials that exploit molecular self-assembly processes. For example, repeating peptide sequences and synthetic polymers that form gels at neutral pH and body temperature are attractive materials for injectable scaffold materials [55]. A good review of these materials is found in Zhang et al. [56]. These materials will probably not be able to replace structural tissues in the near term, but are promising as materials for filling defects in these and other soft tissues. While these materials in general do not have nanoscale features, they are products of nanotechnology in that they are designed from the bottom up using solid-phase peptide synthesis. Stupp and coworkers [57] describe the formation of peptide amphiphiles that self-assemble to form nanostructures similar to extracellular matrix. They also demonstrated the ability of these nanostructures to nucleate hydroxyapatite nanocrystals from solution. Deming and coworkers [58] describe hydrogel formation from repeating polypeptide amphiphiles. By using these engineered materials in lieu of naturally occurring proteins, they describe enhanced control over the mechanical properties and processing of these biocompatible gels.

Some of the processing techniques described here require the application of either harsh chemicals or high temperatures. Before these materials can be exposed to cells or implanted into the body, they need to undergo extensive washing and in some cases sterilization. If nanostructured or micro/nanostructured materials prove to be advantageous for tissue engineering applications, more biocompatible manufacturing techniques will need to be developed.

10.6   Issues with Immune Response to Nanoparticles and Tissue Engineering Biomaterials

It has long been known in the orthopedic community that particles formed during the wear of total joint replacements will elicit an immune response in the body that can become chronic and systemic over the lifetime of an implant [59]. These particles are particularly troublesome when they are submicron in size and often lead to osteolysis, or pathological bone resorption [60–62] The cells responsible for digestion of these foreign particles, macrophages, initiate a more aggressive immune response to particles that are submicron in size [63,64] Total joint replacements are made of both polymer and metal components, and wear particles of both classes of materials have been found in tissues and organs throughout the bodies of patients wearing these devices [65].

Studies in the total joint replacement field should be a warning to those researchers attempting to use nanoparticles in imaging, detection, and tissue engineering applications. In fact, a recent study on polymer-based colloidal drug delivery systems found that nanoparticles from these injections are recognized and engulfed by the macrophages of the reticuloendothelial system [66] In addition, Akerman et al. [67] demonstrated that quantum dots (3–5 nm in size) used as fluorescent probes *in vivo* accumulated in the liver and the spleen of mice. This effect was eliminated by coating the quantum dots with polyethylene glycol. Given what we know about how submicron particle stimulate the immune system, it will be very important to consider what side effects nanoparticles of all types might have on all of the systems in the body.

One method of circumventing immune response is to encapsulate tissue-engineered constructs within immunoisolation membranes. By creating pores on the nanoscale, Desai et al. [68] have been able to prevent immune cells, antibodies, and complement from reaching the cells and biomaterials within the capsule, while allowing nutrients and products, such as insulin, to diffuse in and out of the capsule. Although, the cells and scaffold are not subject to an immune response, encapsulation devices, as with all implanted devices, will still have strong immune reactions at the biomaterial surface. To mitigate negative immune reactions, Desai et al. [69] have modified the outer surface of their biocapsule with PEG. *In vivo* studies suggest that the modified biocapsule elicits minimal fibrotic tissue development and no significant host response.

10.7   Summary

Recent advances in biomaterials science have allowed for greater control over materials chemistry and mechanical properties at smaller and smaller size scales. In addition to these advances, it is becoming more straightforward to test the molecular level response of cells to biomaterials. Only through the implementation of highly controlled biomaterials processing and characterization combined with a deeper understanding of cell response to synthetic microenvironments will the regeneration of more complicated tissues be possible.

Acknowledgments

Catherine M. Klapperich and Jessica D. Kaufman gratefully acknowledge the Whitaker Foundation for funding. Joyce Y. Wong gratefully acknowledges funding from the Whitaker Foundation, NSF (CAREER), NIH (NHLBI), and NASA.

References

[1] Jones, I., L. Currie, and R. Martin, A guide to biological skin substitutes. *Br. J. Plast. Surg.*, 2002, **55**: 185–93.

[2] Bello, Y.M., A.F. Falabella, and W.H. Eaglstein, Tissue-engineered skin. Current status in wound healing. *Am. J. Clin. Dermatol.*, 2001, **2**: 305–13.

[3] Boyce, S.T., Design principles for composition and performance of cultured skin substitutes. *Burns*, 2001, **27**: 523–33.

[4] Naughton, G.K., From lab bench to market: critical issues in tissue engineering. *Ann. NY Acad. Sci.*, 2002, **961**: 372–85.

[5] Mason, C., Automated tissue engineering: a major paradigm shift in health care. *Med. Device Technol.*, 2003, **14**: 16–8.

[6] Orban, J.M., K.G. Marra, and J.O. Hollinger, Composition options for tissue-engineered bone. *Tissue Eng.*, 2002, **8**: 529–39.

[7] Cancedda, R. et al., Tissue engineering and cell therapy of cartilage and bone. *Matrix Biol.*, 2003, **22**: 81–91.

[8] Guldberg, R.E., Consideration of mechanical factors. *Ann. NY Acad. Sci.*, 2002, **961**: 312–4.

[9] Wong, J.Y., J.B. Leach, and X.Q. Brown, Balance of chemistry, topography, and mechanics at the cell–biomaterial interface: issues and challenges for assessing the role of substrate mechanics on cell response. *Surf. Sci.*, 2004, **570**: 119–33.

[10] Brown, X.Q., K. Ookawa, and J.Y. Wong, Evaluation of polydimethylsiloxane scaffolds with physiologically-relevant elastic moduli: interplay of substrate mechanics and surface chemistry effects on vascular smooth muscle cell response. *Biomaterials*, 2005, **26**: 3123–9.

[11] Klapperich, C.M. and C.R. Bertozzi, Global gene expression of cells attached to a tissue engineering scaffold. *Biomaterials*, 2004, **25**: 5631–41.

[12] Song, J. et al., Functional glass slides for in vitro evaluation of interactions between osteosarcoma TE85 cells and mineral-binding ligands. *J. Mater. Chem.*, 2004, **14**: 2643–8.

[13] Gerritsen, M.E. et al., Branching out: a molecular fingerprint of endothelial differentiation into tube-like structures generated by affymetrix oligonucleotide arrays. *Microcirculation*, 2003, **10**: 63–81.

[14] Ku, C.H. et al., Large-scale gene expression analysis of osteoblasts cultured on three different Ti–6Al–4V surface treatments. *Biomaterials*, 2002, **23**: 4193–202.

[15] Mahal, L.K. and C.R. Bertozzi, Engineered cell surfaces: fertile ground for molecular landscaping. *Chem. Biol.*, 1997, **4**: 415–22.

[16] Mahal, L.K., K.J. Yarema, and C.R. Bertozzi, Engineering chemical reactivity on cell surfaces through oligosaccharide biosynthesis. *Science*, 1997, **276**: 1125–8.

[17] Yarema, K.J. et al., Metabolic delivery of ketone groups to sialic acid residues. Application to cell surface glycoform engineering. *J. Biol. Chem.*, 1998, **273**: 31168–79.

[18] Charter, N.W. et al., Differential effects of unnatural sialic acids on the polysialylation of the neural cell adhesion molecule and neuronal behavior. *J. Biol. Chem.*, 2002, **277**: 9255–61.

[19] Link, A.J., M.L. Mock, and D.A. Tirrell, Non-canonical amino acids in protein engineering. *Curr. Opin. Biotechnol.*, 2003, **14**: 603–9.

[20] Sampson, N.S., M. Mrksich, and C.R. Bertozzi, Surface molecular recognition. *Proc. Natl Acad. Sci. USA*, 2001, **98**: 12870–1.

[21] Kato, M. and M. Mrksich, Rewiring cell adhesion. *J. Am. Chem. Soc.*, 2004, **126**: 6504–5.

[22] Parenteau, N.L. and J. Hardin-Young, The use of cells in reparative medicine. *Ann. NY Acad. Sci.*, 2002, **961**: 27–39.

[23] Barry, F.P. and J.M. Murphy, Mesenchymal stem cells: clinical applications and biological characterization. *Int. J. Biochem. Cell Biol.*, 2004, **36**: 568–84.

[24] Desai, T.A., Micro- and nanoscale structures for tissue engineering constructs. *Med. Eng. Phys.*, 2000, **22**: 595–606.

[25] Dike, L.E. et al., Geometric control of switching between growth, apoptosis, and differentiation during angiogenesis using micropatterned substrates. *In Vitro Cell Dev. Biol. Anim.*, 1999, **35**: 441–8.

[26] Chen, C.S. et al., Geometric control of cell life and death. *Science*, 1997, **276**: 1425–8.

[27] Chen, C.S., M. Mrksich, S. Huang, G. M. Whitesides, and D. E. Ingber, Micropatterned surfaces for control of cell shape, position, and function. *Biotechnol. Prog.*, 1998. **14**: 356–63.

[28] Tan, J.L. et al., Cells lying on a bed of microneedles: an approach to isolate mechanical force. *Proc. Natl Acad. Sci. USA*, 2003, **100**: 1484–9.

[29] Bhatia, S.N., U.J. Balis, M.L. Yarmush, and M. Toner, Microfabrication of hepatocyte/fibroblast co-cultures: role of homotypic cell interactions. *Biotechnol. Prog.*, 1998, **14**: 378–87.

[30] Tang, M.D., A.P. Golden, and J. Tien, Molding of three-dimensional microstructures of gels. *J. Am. Chem. Soc.*, 2003, **125**: 12988–9.

[31] Andersson, A.S. et al., Nanoscale features influence epithelial cell morphology and cytokine production. *Biomaterials*, 2003, **24**: 3427–36.

[32] Teixeira, A.I. et al., Epithelial contact guidance on well-defined micro- and nanostructured substrates. *J. Cell Sci.*, 2003, **116**(Pt 10): 1881–92.

[33] Wong, J.Y., A. Velasco, P. Rajagopalan, and Q. Pham, Directed movement of vascular smooth muscle cells on gradient-compliant hydrogels. *Langmuir*, 2003, **19**: 1908–13.

[34] Zaari, N. et al., Hydrogels photopolymerized in a microfluidics gradient generator: tuning substrate compliance at the microscale to control cell response. *Adv. Mater.*, 2004, **15**: 2133–7.

[35] Lo, C.M. et al., Cell movement is guided by the rigidity of the substrate. *Biophys. J.*, 2000, **79**: 144–52.

[36] Dimitriadis, E.K. et al., Determination of elastic moduli of thin layers of soft material using the atomic force microscope. *Biophys. J.*, 2002, **82**: 2798–810.

[37] Jacot, J.G., S.W. Dianis, and J.Y. Wong, Probing microscale compliance of soft hydrated materials and tissues using simple microindentation. *Bio Phys. J.*, 2005: in review.

[38] Catledge, S.A. et al., Nanostructured ceramics for biomedical implants. *J. Nanosci. Nanotechnol.*, 2002, **2**: 293–312.

[39] Han, Y. et al., Evaluation of nanostructured carbonated hydroxyapatite coatings formed by a hybrid process of plasma spraying and hydrothermal synthesis. *J. Biomed. Mater. Res.*, 2002, **60**: 511–16.

[40] Liao, S.S. et al., Hierarchically biomimetic bone scaffold materials: nano-HA/collagen/PLA composite. *J. Biomed. Mater. Res.*, 2004, **69B**: 158–65.

[41] Song, J., E. Saiz, and C.R. Bertozzi, A new approach to mineralization of biocompatible hydrogel scaffolds: an efficient process toward 3-dimensional bonelike composites. *J. Am. Chem. Soc.*, 2003, **125**: 1236–43.

[42] Tan, J. and W.M. Saltzman, Biomaterials with hierarchically defined micro- and nanoscale structure. *Biomaterials*, 2004, **25**: 3593–601.

[43] Boontheekul, T. and D.J. Mooney, Protein-based signaling systems in tissue engineering. *Curr. Opin. Biotechnol.*, 2003, **14**: 559–65.

[44] Rekow, D., Informatics challenges in tissue engineering and biomaterials. *Adv. Dent. Res.*, 2003, **17**: 49–54.

[45] Gu, H.Y. et al., The immobilization of hepatocytes on 24 nm-sized gold colloid for enhanced hepatocytes proliferation. *Biomaterials*, 2004, **25**: 3445–51.

[46] Thapa, A. et al., Nano-structured polymers enhance bladder smooth muscle cell function. *Biomaterials*, 2003, **24**: 2915–26.

[47] Thapa, A., T.J. Webster, and K.M. Haberstroh, Polymers with nano-dimensional surface features enhance bladder smooth muscle cell adhesion. *J. Biomed. Mater. Res.*, 2003, **67A**: 1374–83.

[48] Zhang, R. and P.X. Ma, Synthetic nano-fibrillar extracellular matrices with predesigned macroporous architectures. *J. Biomed. Mater. Res.*, 2000, **52**: 430–8.

[49] Woo, K.M., V.J. Chen, and P.X. Ma, Nano-fibrous scaffolding architecture selectively enhances protein adsorption contributing to cell attachment. *J. Biomed. Mater. Res.*, 2003, **67A**: 531–7.

[50] Berkland, C., D.W. Pack, and K.K. Kim, Controlling surface nano-structure using flow-limited field-injection electrostatic spraying (FFESS) of poly(d,l-lactide-co-glycolide). *Biomaterials*, 2004, **25**: 5649–58.

[51] Boland, E.D. et al., Electrospinning collagen and elastin: preliminary vascular tissue engineering. *Front Biosci.*, 2004, **9**: 1422–32.

[52] Luu, Y.K. et al., Development of a nanostructured DNA delivery scaffold via electrospinning of PLGA and PLA–PEG block copolymers. *J. Control Release*, 2003, **89**: 341–53.

[53] Mooney, D.J., W.L. Murphy, S. Hsiong, T.P. Richardson, and C.A. Simmons, Effects of a bone-like mineral film on phenotype of adult human mesenchymal stem cells *in vitro*. *Biomaterials*, 2005, **26**: 303–10.

[54] Jansen, J.A., A.J.E. de Ruijter, P.H.M. Spauwen, and J. van den Dolder, Observations on the effect of BMP-2 on rat bone marrow cells cultured on titanium substrates of different roughness. *Biomaterials*, 2003, **24**: 1853–60.

[55] Zhang, S. et al., Design of nanostructured biological materials through self-assembly of peptides and proteins. *Curr. Opin. Chem. Biol.*, 2002, **6**: 865–71.

[56] Zhang, S., Emerging biological materials through molecular self-assembly. *Biotechnol. Adv.*, 2002, **20**: 321–29.

[57] Hartgerink, J.D., E. Beniash, and S.I. Stupp, Self-assembly and mineralization of peptide-amphiphile nanofibers. *Science*, 2001, **294**: 1684–8.

[58] Nowak, A.P. et al., Rapidly recovering hydrogel scaffolds from self-assembling diblock copolypeptide amphiphiles. *Nature*, 2002, **417**: 424–8.

[59] Campbell, P., F.W. Shen, and H. McKellop, Biologic and tribologic considerations of alternative bearing surfaces. *Clin. Orthop.*, 2004 (418): 98–111.

[60] Ingham, E. and J. Fisher, Biological reactions to wear debris in total joint replacement. *Proc. Inst. Mech. Eng. [H]*, 2000, **214**: 21–37.

[61] Santerre, J.P., R.S. Labow, and E.L. Boynton, The role of the macrophage in periprosthetic bone loss. *Can. J. Surg.*, 2000, **43**: 173–9.

[62] Archibeck, M.J. et al., The basic science of periprosthetic osteolysis. *Instr. Course Lect.*, 2001, **50**: 185–95.

[63] Tomazic-Jezic, V.J., K. Merritt, and T.H. Umbreit, Significance of the type and the size of biomaterial particles on phagocytosis and tissue distribution. *J. Biomed. Mater. Res.*, 2001, **55**: 523–9.

[64] Bainbridge, J.A., P.A. Revell, and N. Al-Saffar, Costimulatory molecule expression following exposure to orthopaedic implants wear debris. *J. Biomed. Mater. Res.*, 2001, **54**: 328–34.

[65] Urban, R.M. et al., Dissemination of wear particles to the liver, spleen, and abdominal lymph nodes of patients with hip or knee replacement. *J. Bone Joint Surg. Am.*, 2000, **82**: 457–76.

[66] Moghimi, S.M. and A.C. Hunter, Capture of stealth nanoparticles by the body's defences. *Crit. Rev. Ther. Drug Carrier Syst.*, 2001, **18**: 527–50.

[67] Akerman, M.E. et al., Nanocrystal targeting *in vivo*. *Proc. Natl Acad. Sci. USA*, 2002, **99**: 12617–21.

[68] Tao, S.L. and T.A. Desai, Microfabricated drug delivery systems: from particles to pores. *Adv. Drug Delivery Rev.*, 2003, **55**: 315–28.

[69] Leoni, L. and T.A. Desai, Micromachined biocapsules for cell-based sensing and delivery. *Adv. Drug Delivery Rev.*, 2004, **56**: 211–29.

Index

Date Due

MAY 0 2 2019			
MAY 0 6 2019			

BRODART, CO.   Cat. No. 23-233   Printed in U.S.A.